GOVERNING THE CLIMATE

Climate change is an issue that transcends and exceeds formal political and geographical boundaries. Social scientists are increasingly studying how effective policies on climate change can be enacted at the global level, 'beyond the state'. Such perspectives take into account governance mechanisms with public, hybrid and private sources of authority. Studies are raising questions about the ways state authority is constituted and practised in the climate arena and about the implications for how we understand the potential and limits for addressing the climate problem. This book focuses on the rationalities and practices by which a carbon-constrained world is represented, categorized and ordered. This book will enable investigations into a range of sites (e.g. the body, home, shopping centre, firm, city, forests, streets, international bureaucracies, financial flows, migrants and refugees) where subjectivities around climate change and carbon are formed and contested. Despite a growing interest in this area of work, the field remains fragmented and diffuse. This edited collection brings together the leading scholarship in the field to cast new light on the question of how, why and with what implications climate governance is taking place. It is the first volume to collect this body of scholarship, and provides a key reference point in the growing debate about climate change across the social sciences.

Governing the Climate is invaluable for three main audiences: social science researchers and advanced students in the field of climate change; the wider research community interested in global environmental politics and global environmental governance; and policy makers and researchers concerned more broadly with environmental politics at international, national and local levels.

JOHANNES STRIPPLE is an associate professor of political science in the Department of Political Science, Lund University, Sweden. His research interests lie at the intersection of international relations theory and global environmental politics. His recent research has covered European and international climate policy, carbon markets, renewable energy, climate adaptation, carbon sinks, and climate policy scenarios and governmentalities around climate change, carbon and the earth system. He has published papers in journals such as *Review of International Studies*, *Global Governance*, *Critical Policy Studies*, *Global Environmental Change*, *International Environmental Agreements*, *Environment and Planning C*, *Environmental Politics*, *Environment and Planning D: Society and Space* and *Climate Policy*.

HARRIET BULKELEY is a professor of geography at Durham University, United Kingdom. Her research interests are in the nature and politics of environmental governance, and focus on policy processes, climate change and urban sustainability. She is coauthor of *Cities and Climate Change* (2003, with Michele Betsill) and of *Governing Climate Change* (2010, with Peter Newell), and coeditor of *Cities and Low Carbon Transitions* (2011, with Vanesa Castan-Broto, Mike Hodson and Simon Marvin). She has published widely on these topics, including articles in *Political Geography*, *Environment and Planning A*, *International Studies Quarterly*, *Global Environmental Politics* and *Environmental Politics*.

GOVERNING THE CLIMATE

New Approaches to Rationality, Power and Politics

Edited by

JOHANNES STRIPPLE

Lund University

HARRIET BULKELEY

Durham University

CAMBRIDGE
UNIVERSITY PRESS

CAMBRIDGE
UNIVERSITY PRESS

32 Avenue of the Americas, New York NY 10013-2473, USA

Cambridge University Press is part of the University of Cambridge.

It furthers the University's mission by disseminating knowledge in the pursuit of education, learning and research at the highest international levels of excellence.

www.cambridge.org
Information on this title: www.cambridge.org/9781107046269

First published 2014

Printed in the United States of America

A catalogue record for this publication is available from the British Library.

Library of Congress Cataloguing in Publication Data
Governing the climate : new approaches to rationality, power and politics / edited by Johannes Stripple, Harriet Bulkeley.
pages cm
Includes bibliographical references and index.
ISBN 978-1-107-04626-9 (hardback)
1. Global environmental change – Government policy. 2. Global environmental change – International cooperation. 3. Climatic changes – Government policy. 4. Climatic changes – International cooperation. 5. Environmental policy. 6. Environmental policy – International cooperation. I. Stripple, Johannes, author, editor of compilation. II. Bulkeley, Harriet, 1972– author, editor of compilation.
GE170.G688 2013
363.738′74561–dc23 2013030421

ISBN 978-1-107-04626-9 Hardback

Contents

Figures and Tables

Figures

Tables

Contributors

Moritz Albrecht is a postdoctorate researcher in the Department of Geographical and Historical Studies at the University of Eastern Finland. His research focuses on relational approaches to environmental governance and sustainable spaces with a current focus on forestry and bioenergy.

Anders Blok is an assistant professor in sociology at the University of Copenhagen, Denmark. His research focuses on the knowledge politics of global environmental change, and he is the author (with Torben E. Jensen) of *Bruno Latour: Hybrid Thoughts in a Hybrid World* (Routledge, 2011).

Harriet Bulkeley is a professor of geography at Durham University. Her research is concerned with the practice and politics of environmental governance, and focuses particularly on issues of climate change, energy and urban sustainability.

Olaf Corry is a lecturer in international relations at The Open University, United Kingdom. He has published articles on risk and security, climate change politics, environmental politics and social movements as well as the book *Constructing a Global Polity: Theory, Discourse and Governance* (Palgrave Macmillan, 2013).

Carl Death is a senior lecturer in international political economy at the University of Manchester. Prior to moving there in August 2013 he was at Aberystwyth University. His work focuses on environmental politics in Africa and Foucauldian perspectives on international politics.

Sally Eden is a reader in geography at the University of Hull, United Kingdom. She researches how diverse actors understand and shape the environment and has focused particularly on sustainable consumption, eco-certification and public engagement in environmental management.

Juliet J. Fall is a professor in geography at the University of Geneva, Switzerland. She has written about transboundary governance, protected areas, biodiversity policy and invasive species, as well as the history of geographical thought.

Tom Hargreaves is a lecturer in environmental science and policy at the University of East Anglia, United Kingdom. His recent research has focused on sustainability transitions, pro-environmental behaviour change and sustainable lifestyles, and the impact of energy feedback on everyday life and social practices.

Rhys Jones is a professor of historical geography at Aberystwyth University. His work looks at various aspects of national identity and emerging forms of citizenship.

Jarmo Kortelainen is a professor of human geography at the Department of Geographical and Historical Studies, University of Eastern Finland. His research broadly covers issues related to the spatiality of natural resource governance, especially forest questions.

Eva Lövbrand is an associate professor at the Centre for Climate Science and Policy Research at Linköping University in Sweden. Her research is located at the interface of political science, science and technology studies and environmental studies. Her research interests revolve around the role of science and expertise in global environmental governance in general, and climate governance in particular.

Heather Lovell is a reader in geography at the University of Edinburgh, United Kingdom. Her research analyses contemporary socio-technical transitions catalysed by climate change (low-energy housing, carbon markets/accounting), paying close attention to the policies, practices and politics of innovation.

Chris Methmann is a Fellow at the Institute of Political Science at the University of Hamburg, Germany. His recent research has focused on poststructuralist approaches to global climate governance, the climate security nexus and Green Economy discourses. His most recent publication is *Deconstructing the Greenhouse: Interpretive Approaches to Global Climate Governance* (Routlege, 2013, with B. Stephan and D. Rothe).

Angela Oels is a temporary full professor in international politics at the University of Hagen, Germany. Her recent research has focused on climate governmentality studies, critical security studies and the climate-induced migrant.

Jessica Pykett is a lecturer at the University of Birmingham. Her work focuses on citizenship and the practices of governing, the geographies of education and the formation of neurological and psychological citizen subjectivities.

Samuel Randalls is a lecturer in geography at University College London. His research explores diverse approaches to climate change, including science-policy dynamics, security and ethics.

Delf Rothe is a researcher at the Helmut-Schmidt-University Hamburg. He has recently worked on security discourses of climate change, risk and security from a poststructuralist perspective and on methodology.

Benjamin Stephan is a research Fellow at the Centre for Globalisation and Governance, part of the KlimaCampus at the University of Hamburg. His research is concerned with REDD+ and the role of deforestation in global climate governance as well as carbon markets. He recently coedited a special issue of *Environmental Politics* 21(4) with Matthew Paterson on *The Politics of Carbon Markets*.

Johannes Stripple is an associate professor in the Department of Political Science, Lund University, Sweden. His recent research has covered European and international climate policy, carbon markets, renewable energy, sinks, scenarios and governmentalities around climate change, carbon and the earth system.

Mark Whitehead is a professor in environmental geography at Aberystwyth University. His work focuses on various aspects of environmental governance, and incorporates research on sustainable cities, pro-environmental behaviour and eco-governmentality.

Foreword: Order! Order in the House!

Governmentality is about order, and the chapters in this book report on a particular type of climatic order, one that threatens disaster if we do not obey but whose power to deliver cuts directly against normal human behaviours and social practices. But why? Human beings, it would seem, desire order in their lives. This is as much a neurological claim as a social one: we are, after all, constituted as human beings through our social relations with others, through our individual locations within social orders, through the shared practices that render us human. Humans have never been solitary animals, notwithstanding parables of individualistic 'states of nature'. In the trees and on the plains of Africa, survival depended on relations with others. Our situation today is very different, yet some verities have not changed: we face a world of risk and uncertainty – or, so we are reminded daily – one whose order is subject to disruption and destruction and which depends, therefore, on self-comportment and common behaviour in ways that seek to stave off the chaos of an unpredictable universe and an unknowable future. As individuals in (neo)liberal societies, however, we are reluctant to recognize our complex social constitution. We are constantly implored to seek our own interests, to 'innovate', to act in ways that contribute to the very disorder we fear so much. As Marx and Engels warned, 'all that is solid melts into air'. How, then, can order be maintained?

In the 'olden days', religious authority filled this role. If the world showered misfortune on some and good fortune on others, surely this was due to the will of whatever deity governed the cosmos (recall Job). When god(s) proved insufficient for this purpose, as in seventeenth-century England, responsibility for household order was taken on by Hobbes's 'mortal god', who governed and disciplined in the (capitalist) interest of his subjects. This entity, in turn, became the nation-state, whose nationalist structures and demands provided a governmental grid within which citizens and subjects could exist in properly oriented and orderly fashion. By the middle of the twentieth century, the great nationalist explosions and implosions were, increasingly, rendering fragile the authority of the state and its capacity to order the polity. Somewhat paradoxically, the post–World War II 'golden era of capitalism and socialism' loosed onto the world the current wave of high individualism, in which authority has been delegated to the sovereign consumer, as it were, and order, both at home and across the world, has come to depend on her or his 'right conduct'. It hardly seems a coincidence that capitalist globalization, climate concerns and communist collapse all took place more or less concurrently, or that the rise of what we call religious and market fundamentalisms also emerged over roughly the same time.

Climate governance and governmentality are not simply analytical concepts or a set of disciplining and ordering practices; they are also near-*functional* epistemes that (re)construct social power and social relations through certain types of ontological stories, myths, fabulosities of necessary authority and order. It is here that social science might play its most important role: explaining how such narratives emerge and why. Consider *apocalypse* (Ἀποκάλυψις), a Greek term found in the New Testament that means 'an unveiling or unfolding of things not previously known and which could not be known apart from the unveiling'.[1] In more commonsense parlance, 'apocalypse' has come today to be associated with the world's end, whether Divine, biogeophysical, celestial, stellar, biological or nuclear: that end will come, inevitably, although we can never know until it happens. For those fatalistic or religious, good human behaviour now might either put off or hasten what must come, sooner or later. Those less resigned to such a fate seek to predict, anticipate and prepare for apocalypse, thereby preventing it and allowing the business of life to proceed, in confidence and an expectation of order (much as Hobbes's Leviathan). Like the apocalyptics of religion, however, those arising from science require 'good' human behaviour as well. Otherwise, we shall almost surely experience a future of mortal threat.

But people tend to be a disorderly lot in the absence of discipline and *habitus*; they do not obey well and, notwithstanding the tale of Adam Smith's Invisible Hand, they rarely generate order in the pursuit of individual interests and sovereignty. It is to this end, we may surmise, that apocalypse, whatever its cause or occurrence, fulfils a necessary social need for an all-encompassing ontology, an account of *why* humans are here and *what* their purpose is in the natural order of things. Apocalypse warns, in effect, that if we do not get our shit in order, the shit will hit the fan. The stories told about climate change offer such apocalyptics and the morals to those tales instruct listeners on what is to be done. Yet, while science can tell us what we might (or should) do, it cannot compel us to do what it commands. It is the social sciences that may shed some light on this conundrum.

I do not think it is necessary to review the many apocalyptic narratives of climate change with which most readers of this book will be familiar. But it is probably important to point out that we live in a world of disenchantment, one mostly devoid of supernatural deities and in whose hands our futures lie. Notwithstanding the resurgence of religiosity in many parts of the world, moreover, we humans have become

[1] Goswiller, R., (1987). *Revelation*, Pacific Study Series. Melbourne.

imbued with the notion that we possess the power of the gods to transform and even destroy the world through our material actions and activities. Hence, it is also incumbent upon us to save it. This is a considerable task and would seem to require a good deal more than environmentally conscious consumers or prescient scientists. Less evident is *why* the world needs saving, because it will continue to exist whether we are here or not. Nor is it evident *what* might mobilize humanity in this effort – if the nuclear era ontology 'One world or none' with its threat of rapid annihilation could not bring us together, why should another apocalyptic, however real, do any better?

Dear reader, you can probably see where this is going: whether true or not, narratives of climate catastrophe are not merely about saving the world, they also provide a near-theological *meaning* to human existence and, in so doing, propose particular forms of authority and order as the prerequisites for fulfilment of human purposes on Earth. While this conclusion might seem far from and irrelevant to a discussion of how climate governmentality operates or what practices it mandates or why it seems not to work at the international level, I would propose that the ontologization of climate change helps to clarify why those who dissent from it are called 'sceptics' and 'nonbelievers' and why we have such faith in the authority of the IPCC's 'scientific consensus' even if we seem reluctant to act on it.

So far as I am aware, Michel Foucault did not make connections between the concept and organization of governmentality and its theological antecedents and contemporary elements, although the role of religious authority over the individual body is only too evident in works such as *Discipline and Punish*. Foucault did make clear the extent to which liberal governmentality relies on self-comportment under the strictures of the social power that constitutes each of us as individuals in societies. At the same time, self-discipline is a thin reed on which to base a household and world order (and the production) necessary for societal reproduction and world salvation. That may be why emergent forms of climate governmentality – carbon calculators, neuroliberalism and the 'nanny state', so-called smart meters, climate conflicts, even tradable emission permits – have become a few of the many 'technologies of surveillance' that, in concert with political economy and security, are being deployed to ensure that prescribed practices are followed faithfully. People are not to be trusted with their own fate (or faith). Whether this latter proposition is an accurate one or not – I hesitate to use the word 'true' – I cannot say. But an old and troubling paradox is embedded in this claim about human 'nature': free will is too dangerous to be permitted, especially if it leads to the diminution or overthrow of authority and further disruption of order – or, in this case, climate disaster.

Here, I want to turn briefly to one more contradictory feature of climate governmentality as it appears in this volume: the enterprise of modelling and its foundation in old theological debates about free will and determinism. Models, we are frequently reminded, are not 'reality', even though they often come to be treated as such. We know that the real, material world is chaotic and disorderly, notwithstanding 'laws of nature'; even human will and determination are not enough to ensure that intended outcomes will be accomplished purposefully and reliably (I take this point to be the essence of Foucault's method of genealogy). Nevertheless, people's ability and capacity to imagine the future and make choices about it are, aside from the constraints of social power, relatively open, and the well-meaning pursuit of self-interest by many can be disastrous for all. Free will could lead to apocalypse.

Modelling seems to offer an escape from this fate. What if we were really able to calculate the consequences of our activities; could we then not counsel, command or coerce individuals to act properly (as would Leviathan and as predicted by Robert Heilbroner)? Could we not turn a world of uncertainty and danger into one of predictability and safety? Could we not then do away with the risks of free will and replace them with the comfort of a known and determinable future? If uncertainty about the future could be eliminated, we could get on with the business of life (Hobbes again).

A known future may well be preferable to the unknown one, but even the effort to get there rests on the identification of critical 'variables' and political and social efforts to ensure these remain within necessary bounds (physical bounds and limits might stand in the way of many futures, but that does not mean we will not try to exceed them). Where human behaviours are concerned, consequently, modes of social discipline may be required, ones that place limits on free will in the name of desired, safe futures. Of course, as social scientists, we know only too well that it is difficult to make things foolproof, inasmuch as fools are such clever people. Often, moreover, the best-laid plans can lead to outright disasters, whether through ignorance, oversight or just plain *fortuna*. Control and determinism are limited. None of this is to say that it would be good public policy to leave well enough alone, to adopt a laissez faire approach to climate change and let happen what may. But hubris in human endeavour is always a good idea; pursue simple and straightforward approaches rather than complex, arcane ones. That is why, in a society such as ours, carbon taxes appear preferable to carbon markets, although both rest on notions of altering human practices through governmentality.

At the end of the day, governmentality has to be regarded as a feature of modern life and society, a perspective well illustrated by the chapters in this volume. Governmentality is not a particular form of discipline and punishment characteristic of specific types of government and governance, as is often thought. It is not a specific disposition of power that, somehow, renders dissent, opposition and resistance impotent. It is not so much like a steel web as a spider's. We cannot wriggle free, to be sure, but neither are we so trapped as to be immobilized. At the same time, some manifestations of governmentality seek to be more like steel webs than spiders' from fear that too much freedom may lead to the collapse of the social web. Navigating between the Scylla and Charybdis of those two possibilities will require all the political and social wisdom and action of which we humans are capable.

If libertarian economics are the governmental set of hegemonic ideologies and practices that have ruled the world from the 1970s to the present, climate governmentality seems poised as a potential replacement, although the victory of this 'new' hegemony over the older one is by no means certain (the new is struggling to be born from the old, but it is not nearly that new). Where the ideologies of the twentieth century promised one sort of paradise or another – none of which were delivered – climate governmentality offers a restoration of an imagined golden age when climate was benign and human impacts were limited. We should be under no illusions that, in the pursuit of climate governmentality, enlightenment and reason may, somehow, sweep across the planet and put an end to the irrationalities of the past millennia. There has never been a real golden age, whether of climate or government. And Leviathan is a myth, too.

Ronnie D. Lipschutz
Santa Cruz, California

Preface

This book arose from an increasing sense of frustration with how the social sciences have engaged with the climate change issue. Starting with assumptions about the sovereign state and the sovereign individual, the social science of climate change has come to be understood broadly either as a matter of international politics or as concerning personal preferences. Enquiries have thus been polarized according to alternative scales of action (international or personal) and slipped into particular disciplinary areas, which has defined and guarded what the politics within those scales of action is, and could be, about. In bringing this volume together, our intention has been to explore what an engagement with critical social and political theory could instead bring to the study of climate change. Our intention was to seek a new set of insights into the ways climate change is creating new forms of social order, and the ways these are structured through the workings of rationality, power and politics. Governing the climate, as both a scholarly and practical effort, we feel, has to be about interrogating the many diverse sites in which a carbon-constrained and/or rapidly warming world is represented, categorized and ordered. Households, markets, forests, migratory species, renewable energy and displaced persons are all examples of sites where subjectivities around climate change and carbon are formed, forged and contested. Climate governance, when approached in the general, tends to become an empty concept, unable to grapple with the way the climate issue becomes interwoven in our lives and how we organize our societies. Our response is instead to populate this terrain and to explore the specificities through which climate governance takes place.

The impetus for organizing a workshop on 'rethinking climate governance' came from Philipp Pattberg, VU University Amsterdam, who chaired the EU COST Action on 'Transformation of Global Environmental Governance' on which Harriet served as a founding member. We approached and encouraged a diverse set of critical social and political scholars, many of whom did not previously know about each other or who did not see their work as part of a larger stream of thought, to come to a workshop at Lund University. In June 2011 we hosted 'Governing the Global Climate

Polity: Rationality, Practice and Power,' which drew together around forty scholars from across many disciplines and with a diverse set of backgrounds. The meeting was made possible through generous funding by the EU COST Action IS0802 and BECC, the Swedish government's strategic focus on climate change hosted by the Faculty of Science at Lund University. We would like to extend our heartfelt thanks to all scholars for two vibrant days of excellent presentations and inspiring conversations. It was clear to us that while a growing body of work is turning to critical political and social theories – including those derived from Foucault, those termed 'Actor-Network Theory', and various forms of discourse analysis – the field remains fragmented and diffuse. The idea for this volume emerged as one means to try to bring these perspectives into conversation, and we are indebted to the chapter authors who enthusiastically answered our call for bringing our efforts together. Through several iterations of the chapters and through the possibility to share, learn and comment on each other's work, the authors have with considerable energy and passion made the book possible. Ronnie D. Lipschutz, who has inspired each of us in opening up this debate, was gracious enough to provide the thought-provoking foreword. As has been clear while working on this book, the climate issue may challenge existing theoretical conventions, approaches and methods and have thus much to offer to mainstream social science and the way we understand our social and material present, but this is an enterprise that requires much puzzling through and we are grateful to our colleagues for sharing this journey with us.

Matt Lloyd at Cambridge University Press responded to our book proposal with great enthusiasm and has, together with Sarika Narula at Cambridge, been very supportive throughout the book production process. We are also very thankful to the reviewers of this book who, while being encouraging and supportive, also challenged us to articulate our arguments and specify our contribution. While putting the volume together, we have benefitted from the excellent editorial assistance of Adis Dzebo and Angie Sohlberg, both in the Department of Political Science, Lund University. Our work with this book has been made possible through financial assistance from the Low-Carbon Energy and Transport Systems (LETS) project, financed by the Swedish Environmental Protection Agency (among others); the Fair and Feasible Climate Change Adaptation (Fair-Ad) Project, financed by the Swedish Research Council and the Swedish Research Council FORMAS; and the Leverhulme Prize Fellowship held by Harriet over the years 2008–12 from The Leverhulme Trust.

Finally, but as always, our families (Gunilla and Pete) have not only supported and encouraged our work, but also provided meaning and context. Our kids – Siri, Alma, Walter, Sigge, Elodie and Théa – who are on track to grow up in a world four degrees warmer than today (if the trajectories of greenhouse gas emissions are not dramatically altered) remind us of why we feel it is important that social science renew and widen its engagement with the climate issue. They also have the burden and the curiosity of having 'book writers' in the house, a world that can often seem strange, isolated and distracted from daily life. One day when Johannes was walking to daycare and school with the kids, one of them asked, 'why do you write a book if you are a researcher?' While her idea of doing research revolved around 'measuring' and 'discovering' things, he tried to answer that it is also important to tell others about your findings and insights in order to inspire them in their research and their activities. She sceptically responded, 'but isn't that boring to write a whole book?' He responded that this book has not been boring at all to write, not least because of the great help from our friends in writing it. In that spirit, we hope that we have offered inspiration in the growing debate about climate change across the social sciences, and that our children may have different stories to tell.

Johannes Stripple and Harriet Bulkeley
Lund University
Durham University

Abbreviations

AGW	Anthropogenic global warming
ANT	Actor-network theory
BEG	Bioenergy governance
CCP	Cities for climate protection
CDM	Clean development mechanism
CO_2	Carbon dioxide
COP	Conference of the Parties
CPS	Secretariat of the Swiss Commission for Wild Plant Conservation
CRAG	Carbon rationing action group
DECC	Department of energy and climate change (UK)
DEFRA	Department for environment, food and rural affairs (UK)
ETS	Emission trading scheme
EU	European Union
GCM	Global circulation model
GEO	Green energy options
GHG	Greenhouse gas
GISP	Global invasive species programme
GMO	Genetically modified organism
GOFC-GOLD	Global observation of forest and land cover dynamics
GTOS	Global terrestrial observing system
ICLEI	International Council for Local Environmental Initiatives
IMO	International Meteorological Organization
IPCC	Intergovernmental Panel on Climate Change
IR	International relations
LULUCF	Land Use, Land Use Change and Forestry
MRV	Measurement, reporting and verification
NGO	Nongovernmental organization
NREAP	National renewable energy action plans

ODE	Ordonnance fédérale sur la dissemination des organisms dans l'environnement
PES	Payments for ecosystem services
RED	Directive on renewable energy
REDD	Reducing emissions from deforestation and forest degradation
RTD	Real-time display
SBSTA	Subsidiary body on scientific and technical advice
SOA	Spheres of authority
STS	Science and technology studies
UN	United Nations
UNEP	United Nations Environment Programme
UNFCCC	United Nations Framework Convention on Climate Change
WMO	World Meteorological Institute
WWF	World Wide Fund for Nature (previously named the World Wild-life Fund)

Introduction

On Governmentality and Climate Change

Johannes Stripple and Harriet Bulkeley

Introduction

> *Climate change is everywhere. ... The idea that humans are altering the physical climate of the planet through their collective actions, an idea captured in the simple linguistic compound 'climate change', is an idea as ubiquitous and powerful in today's social discourses as are the ideas of democracy, terrorism, or nationalism.*
>
> Mike Hulme 2009: 322

What are we to make of the apparent ubiquity of climate change? From faith-based movements to celebrities, the halls of the United Nations (UN) to advertising hoardings, climate change appears to be written through the contemporary social world. What started as a matter of scientific enquiry in the nineteenth century, and became a battleground for intergovernmental negotiations in the late 1980s, has now become a part of people's everyday lives and livelihoods. As Hulme (2009) argues, the idea of climate change has become as powerful as the issue itself. Gerson (2012), for example, writing in *The Washington Post* and commenting on the 2012 U.S. presidential campaign, noted that a theory about the role of carbon dioxide in shaping climate patterns has now joined abortion and gay marriage as a culture war controversy. Against this seemingly unending encounter with climate change, however, there is also a sense that its meaning has been lost. With familiarity comes a sense of the taken for granted, and what climate change might mean – socially, politically, culturally, economically – fades into the background. In this book, we seek to recover and renew our understanding of climate change as a matter for the social sciences. Drawing together twenty scholars from across five disciplines, we seek to interrogate the social meaning and implications of climate change through a sustained engagement with critical political and social theory.

Over the past two decades, the growing social scientific interest in climate change has been dominated by forms of enquiry in which concepts and constructs of the social and political world were largely taken as given. At stake in this body of work was charting the social, political and economic effects of climate change as a complement to the dominant scientific understanding of the issue. More recently, a burgeoning area of scholarship has sought to articulate key social scientific problematics – surrounding, for example, the nature of power, agency, the state, market, individual responsibility and so on – with the climate change phenomenon. Such approaches are 'critical' in their disposition in that they seek to probe and deconstruct the assumed building blocks of the social world, not only to bring new forms of meaning to the climate change issue, but also as a means to advance social science theorization itself. As we set out later in this chapter, one particular body of work that has attracted attention in this regard is the scholarship on Foucault's notion of governmentality, which is now increasingly widely applied in the climate change domain. In this chapter, by way of an introduction to this volume, we chart the emergence of these approaches and consider why and how they offer new illumination for understanding climate change. At the same time, we argue that engaging with climate change provides new challenges for critical social science theory that at once raise questions of scale, of nature and of materiality.

Climate Change, Politics and the Social Sciences

Since climate change emerged as a public concern, the social science community has largely focussed its analytical attention on two of the policy 'puzzles' posed by this complex issue. The first, and arguably dominant, set of concerns has been with the emergence, workings and challenges of international climate policy. In this field, traditionally dominated by the discipline of international relations, the focus of attention has been on the interaction between nation-states and the international institutions created to address climate change (Sprinz and Luterbacher 2001; Aldy and Stavins 2007). In contrast, a second set of concerns with which social science disciplines has engaged has included the attitudes and behaviours of individuals. Here, the puzzle is why, given apparent public concern, exhortations to individuals to reduce energy consumption and 'save the planet' have largely failed (Gardner and Stern 1996; Nilsson, Von Borgstede and Biel 2004). Across the disciplines of psychology, economics and sociology, the research community has sought to engage with how and why individual behaviour can change and the barriers that are encountered. In this body of work, marked by diversity in its interest in markets, education, regulations, voting patterns and so on as the means through which preferences are expressed, the predominant focus of study is on the individual agent and the ways climate change has come to enter into decision-making (Stern 2000; Senecah 2004; Shove 2010).

It has been through these two predominant approaches that the social science of climate change has come to be understood broadly either as a matter of international politics or as concerning individual preferences. These particular concerns have served to polarize social scientific enquiry according to alternative scales of action (international or personal), and into disciplinary areas. Here, we take our point of departure from the standpoint of climate change as an international concern and consider what this has meant for our understanding of the issue, before examining the ways this approach has been challenged and extended. The emergence of perspectives drawn from critical political and social science has, we suggest, begun to disrupt this division, leading to a rethinking of the nature and practices of climate governance on one hand, and to questions of agency on the other.

Climate Politics and the International

The discipline of International Relations started to consider the climate issue precisely at the moment when the climate issue entered the realm of international politics. There is a consensus in the literature that this happened in 1988, when the 'Toronto Conference', the first international conference that included government officials, industry representatives, scientists and environmentalists, took place (Mintzer and Leonard 1994; Paterson 1996; Elliott 1998). The Intergovernmental Panel on Climate Change (IPCC) was established in the same year to provide assessments of the results of climate change science to policy makers. Miller argues that IPCC's endorsement of a systemic view of the climate – the earth's climate imagined as a global system – was conducive to engaging international organizations and states (2004: 51). Emerging in this period of climate politics, as national governments came together and scientific advice was expected to provide 'truth' for 'power', climate change became regarded as a transboundary scientific problem that required an international solution.

At the root of this conceptualization is a notion of the international as comprised through the interaction of territorially demarcated entities. As Walker (1993) has persuasively argued, the territorial organization of world politics creates two types of spaces – within boundaries (the domestic) and between boundaries (the international). The political within these spaces is understood to be very different. The domestic connotes a realm of politics ordered by the authoritative allocation of value within a society, democracy and legitimate authority structures. The international, on the other hand, is marked by the absence of overarching authority, of competition and rivalry between states. Framed in these terms through the discipline of International Relations, the question of 'governing the climate' thus became a question of how to overcome the fundamental structures of world politics. For example, Barrett writes about the clash between a 'world of one atmosphere and a world of two hundred or so countries' (2005: 286), while Schneider, Rosencranz and Niles suggest that the 'planetary condition necessitates a planetary response' (2002: XIV) and Sprintz and Luterbacher argue in the introductory chapter to their edited volume 'International Relations and Global Climate Change' that the critical issues of climate politics concern states' cooperative problems, free-riding and the ineffectiveness of retaliation threats for making states comply with an emission reduction treaty (2001: 13).

Initially, most social science research efforts within this field focussed on how to address this challenge of the international nature of climate change, and particularly on the emergence of the international climate regime and the questions of science policy, political economy and justice that it raised. For example, Breitmeier (1996) drew attention to the role of international organizations (UNEP, WMO, UN General Assembly) during the creation of the climate regime; while on the other hand, Paterson (1996) showed how the deliberations of the UN's climate convention can display the conceptual weaknesses of both neorealism and neoliberalism. Debates on ethics and justice, particularly distributive justice, for example how to allocate the costs of preventing climate change and coping with the consequences of climate change, as well as what a fair bargaining process and a just allocation of emissions of GHG would be, were also critical during this period (Toth 1999; Ikeme 2003; Rosa and Munasinghe 2003; Tonn 2003). This focus on the nation-state as the unit of political analysis has also permeated beyond these particular concerns with international negotiations, so that within the context of national economies and sovereign citizenship, research has primarily focussed on the extent and nature of national policy responses. The work of authors such as O'Riordan and Jäger (1996), Helm (2005) and Compston and Bailey (2008) has, over the past two decades, provided case studies on how governments in various countries have developed climate change policies and the particular merits of different policy instruments. These include voluntary agreements to taxes, emission trading systems and legal standards in meeting national and international commitments.

More recently, while the scholarship on international climate politics has continued to focus on the development of international co-operation, with arguments on 'Architectures for Agreement' (Aldy and Stavins 2007) and the extent to which such architectures are 'fragmenting' (Biermann et al. 2009), there has been a growing interest in the relationship between climate change, security and violence. Thomas Shelling, in his epilogue at the end of 'Architectures for Agreement', considers NATO as a possible organizational form for governing the climate. NATO, as Shelling puts it, has demonstrated 'outstanding success on a huge scale with "commitments" unenforceable, related to actions rather than consequences' (2007: 349). Within academia, a growing body of recent work has analysed how climate change can affect traditional concerns such as national security and armed conflicts. The result has been a variety of predictions. While some have claimed that warming increases the risk of civil war in Africa, other scholars point out the lack of evidence for this claim (Burke et al. 2009; Buhaug 2010). Some studies suggest that political instability will follow when subsistence farmers are forced to leave their livelihoods because of drought, flooding and famine (Busby 2008). Others foresee a series of 'regional hotspots' around the globe, such as possible water shortages in densely populated areas of Central Asia and drought in the Sahel that might, for example, aggravate the conflict in Darfur (German Advisory Council on Global Change 2008). Melting of the ice caps at the poles opens up new shipping routes and increases ease of access to prospective oil, coal and gas deposits, potentially exacerbating interstate conflicts (Lee 2009). Other scenarios foresee that 'climate wars' might even include a nuclear war between India and Pakistan over water shortages after the Himalayan glaciers have disappeared (Dyer 2008).

The engagement of International Relations with climate change has therefore enabled both cooperative and violent imaginaries when it comes to the politics of a warming world. In both imaginaries, however, the state remains the central unit of analysis and conceptualized as a coherent, autonomous and purposeful entity wielding certain basic forms of power (Bulkeley 2005). As Weber reminds us, disciplines and the theories they construct are themselves a site of cultural practice. International relations, as discipline and practice, is a 'place where stories that make sense of our world are spun, where signifying practices about international politics take place, where meanings about international life are produced, reproduced, and exchanged' (2005: 182). The 'imagining of the state' (Conca 2003) as at the heart of the political response to climate change has served to advance our understanding of the international dynamics of climate change, and also to reproduce the powerful metaphors that have guided our understanding of environmental politics as beset by the tragedy of the commons, north-south divides and the rightful sites of political authority.

Reimagining Global Climate Politics

Alongside this continuing interest in the international dimensions of climate politics, the past decade has witnessed the development of a new imagination of the spaces of international politics through the broad concept of global governance. Conca argues that many scholars of international relations have 'equated the idea of governance with an authoritative inscription of norms that impose constraints on state behaviour' (2005: 195) and have therefore not paid attention to how institutions and governance emerge at the intersection of several influences such as law, advocacy, expert networks, market relations and stakeholder bargaining processes. By drawing attention to the rise of hybrid, nonhierarchical and network-like modes of governing on the global stage, the concept of global governance has enabled an engagement with the multiple ways authority for climate governance is being created and sustained. In so doing, this work has challenged both the centrality of the state and the distinction and separation between domestic and international realms that has underpinned the traditional conceptualization of (climate) politics (Bulkeley 2005; Barnett and Sikkink 2008; Jönsson 2010).

Central to the emergence and development of work in the field of global climate governance has been a concern with the fragmentation of authority and the multiplication of the range of actors involved in agenda-setting, policy development and implementation (see Lövbrand and Stripple, this volume). Early contributions in this field focussed on the ways nonstate actors were influencing the state through the development of expertise, financial power and moral authority (Rowlands 1995; Arts 1998; Corell and Betsill 2001). As the climate change issue itself began to metamorphose, scholarship began to look beyond the regime and the specifics of national responses to illuminate the manifold ways climate change is becoming an issue and agenda across multiple levels of government and in different political sites (conferences, camps) and for a range of nonstate actors, from churches and celebrities, to nongovernmental organizations (NGOs) and corporations (Stripple and Pattberg 2010). After a decade of climate governance research, the academic community has been able to demonstrate the critical roles played by a range of actors in shaping the landscape of international politics, and has begun to demonstrate the ways this politics plays out over a multilevel governance landscape (Carpenter 2001; Kolk and Levy 2001; Lund 2012). Central to this perspective is the recognition of new forms of agency in the international arena and, for some at least, the argument that this heralds a shift in authority from the state to the nonstate, or private, realm (Green 2008; Pattberg and Stripple 2008).

This move towards a global governance perspective on the politics of climate change has involved not only a reengagement with questions of agency, but also a reimagination of its spatial orderings that moves beyond the international/national dichotomy. In its place has emerged a more nuanced sense of spatial hierarchy, where multiple sites of climate politics nest within one another and where it is possible for actors, issues and authority to 'jump' scales (so that urban climate politics bypasses the nation-state, for example) (Bulkeley 2005). For the most part, however, the bounded notions of political space and its territoriality remain taken for granted. For example, there have been extensive discussions of the interplay between different levels of political authority, while many interpretations of multilevel governance have focussed on a set of vertical interactions between more or less territorially bound entities, such as local authorities, regional governments, nation-states and international institutions. There are, however, alternatives recognized within this emerging field. Notable has been the growth of interest in forms of transnational governance, where the borders of the political are not territorially constituted but rather are constituted through new forms of institutional architecture and networks of actors that cut across national boundaries and the divisions between public and private authority. There is also a growing interest in 'governance experiments' (Hoffmann 2011) that seek to fill the policy void around climate change and operate across political boundaries at multiple levels, such as intermunicipal co-operation, social networks and public-private partnerships. For example, Ruggie articulates the 'new global public domain' as existing 'in transnational non-territorial spatial formations and … anchored in norms and expectations as well as institutional networks and circuits within, across and beyond states' (2004: 519). Here, 'socially constructed practices of political authority … may be exercised nonterritorially or in scattered pockets connected by flows across space-spanning networks' (Agnew 2005: 441).

This body of work has undoubtedly had a profound effect in opening up how we understand the politics of climate change and in moving beyond an international orientation to examine the multiple scales through which this is being constituted. In so doing, this work has successfully drawn attention to the ways power, authority and governance are emerging in new sites and has begun to interrogate the consequences and implications. However, for the most part the core elements of the social and political world, such as agency, authority, the state and so on, have remained taken for granted. In this manner, while global governance may have added more complexity to our notion of the political, it has not challenged its fundamental orthodoxies. As a result, the research community has begun to question the extent to which the global governance approach can provide an adequate conceptual toolkit with which to explore the shifting ground of climate politics. For Sending and Neumann it is not so much that global governance literatures are inattentive to these concerns, but that 'their ontology and concomitant analytical tools are not equipped' (2006: 653) to deal with their conceptual implications such that they 'inadvertently perpetuate the very state-centric framework that they seek to transcend' (655). Building on this critique, Okereke, Bulkeley and Schroeder (2009: 72) suggest that current theorization of climate politics in global governance terms has reached a 'conceptual impasse'. To address this, they argue that conceptual development is particularly pressing in four areas – the understanding of power; the relationship between public and private authority; the dynamics between structure and agency and, finally, the rationalities and actual processes of governance (Okereke et al. 2009: 72). These issues, they suggest, require a more critical engagement with what constitutes the political realm, and the development of theoretical perspectives that offer alternative conceptualizations of power and agency. While some mainstream social scientists have started to address this challenge, notably in the recent works of Giddens (2009), this work has remained curiously tied to the dominant framing of climate change as an international problem, albeit cascaded down to other levels of social organization such as the state, community and individual. Giddens's starting assumption is that we 'do not have a developed analysis of the political innovations that have to be made' (2009: 4) to limit climate change, but his book unfortunately has little to say about what those innovations might look like. Rather, the analysis centres on the most economically developed capitalist states and a familiar set of policy measures, such as taxes, planning and international treaty-making under the auspices of the UN. While Giddens locates his analysis to rather few sites of action (e.g. intergovernmental conferences) and few sets of actors (national governments), Urry (2011) uses a broader sociological imagination to characterize how political power currently resides in a 'military-industrial carbon complex', but also how post-carbon systems innovate in pockets and alongside our existing society through politicization and transformation of certain practices, for example mobility and high-carbon living. Climate change is making inroads into mainstream social science and in so doing is itself opening up to a wider range of concerns and questions. Yet such contributions remain relatively few and far between, so that for the most part social science continues as if climate change was of limited interest or concern.

New Engagements with Critical Perspectives: Governmentality and Beyond

Against this backdrop, a new set of engagements between climate change and critical perspectives from the social sciences are now emerging that forms the focus of our attention in this volume. This move beyond the traditional concerns of international politics and global governance, on one hand, and the preferences of individuals, on the other, can be regarded as stemming from two main sources: first, seeking to address the limitations of the theoretical perspectives that have dominated the debate to date, climate change scholars have begun to engage a range of different social scientific theories; second, social scientists in disciplines beyond international relations and political science have begun to engage with the climate change issue, bringing alternative theorizations and disciplinary perspectives with them. These new engagements seek to deploy different ways of theorizing the social and the political that are fundamentally critical in their disposition – that is, that they start from a perspective that what constitutes the political or the social is not regarded as given, but as open to question. Through these new engagements, the field of climate change politics now encompasses perspectives and approaches drawn from international political sociology, actor-network theory (ANT), science and technology studies (STS), Foucauldian governmentality studies, critical theory, posthumanism, critical geopolitics, poststructural security studies, neo-Gramscian political economy and discursive approaches derived from Laclau-Mouffe, Lacan or Habermas. These strands of social scientific theorizing are increasingly invoked to answer the deceptively simple question of how to make political sense of climate change. What is starting to emerge across these diffuse perspectives is a problematization of the nature of the political with regard to climate change, through rethinking the nature of power, a renewed attention to the ways institutions are produced and maintained and a thorough engagement with the processes and practices around climate change and carbon – as a materiality and a social and cultural artefact.

Of particular importance to this emerging body of work has been the concept of governmentality, originally advanced by Foucault in the late 1970s but since adopted and adapted by a range of scholars. Conceived both as a particular modality of power with particular contemporary significance and as a perspective or analytical tool through which to view power in general, governmentality has offered scholars in the climate change community a means through which not only to develop a more critical understanding of the processes at work in governing climate change but also a means through which to connect the spaces of climate politics to more fundamental (and one could argue mainstream) concerns with the workings of the contemporary state and political order. Here, we review the different ways governmentality perspectives are put to work in relation to climate change, and the connections and critiques that are emerging across the multiple strands of critical social theory. We pay particular attention to the ways governmentality, in tandem with other perspectives, can provide new insights into questions of rationality, power and politics, as a means through which we can start to reconstitute the notion of climate politics. We also consider the limitations of such perspectives and how an engagement with climate change may be able to advance conceptual understanding more broadly.

Making Climate Change Governmental

Paul R. Brass once wrote in the *Annual Review of Political Science* that 'Foucault Steals Political Science' (2000). By 'stealing' Brass had in mind the way Foucault has engaged with themes and concepts central for the discipline of political science (e.g. power, government and governance) while the discipline at large has 'departed markedly from serious engagement with those topics' (2000: 305). Brass had hoped to write an article on 'The Foucauldian Turn in Political Science', but because such a turn had yet to appear at the time of his writing, he instead settled on demonstrating the 'importance of Foucault's insights into the nature of power and governance for a discipline that calls itself political science' (2000: 305). Brass convincingly shows that a Foucauldian approach to power and government is not marginal, dealing with esoteric phenomena unimportant for the discipline at large, but that Foucault's work 'ought by now to have become a focal point for the resurrection of these topics and their restoration to centrality in the discipline' (2000: 305). While this volume is not about a resurrection and restoration of political science per se, the different chapters nevertheless provide ample illustrations of how new perspectives on the nature of power and government can be advanced in relation to a contemporary empirical phenomena. However, if we can identify and define Foucault-inspired analysis of the climate issue as a 'stream of thought', we may think of it as an overarching effort to redirect the analysis in new empirical directions and to ask a new set of questions.

Foucault first outlined his governmentality approach in a series of lectures at the Collège de France in the late 1970s, and a range of social scientists has since developed these ideas in multiple directions. As Joseph notes, Foucault did not apply the concept of governmentality in a systematic manner; instead he thought through his ideas during the lectures, offering very general definitions and applying the concept in different ways and in different contexts (2009: 53). As indicated by the semantic linking of the words *governing* and *mentality*, broadly speaking *governmentality* deals with how particular mentalities – ways of thinking and acting – are invested in the process of governing. While recovering an older usage of *government*, governmentalities were understood 'in the broad sense of techniques and procedures for directing human behaviour. Government of children, government of souls and consciences, government of a household, of a state, or of oneself' (Foucault 1997: 82). Despite Foucault's rather fragmented treatment of governmentality, this work, coupled with other parts of Foucault's thinking, has led to the emergence of a vibrant field across disciplines such as sociology, geography, anthropology, history, gender studies and science and technology studies concerned with understanding the governing of human conduct. Walters, reflecting on the past twenty years of governmentality studies, conceives of governmentality as a 'cluster of concepts that can be used to enhance the think-ability and criticize-ability of past and present forms of governance' (2012: 2). It is a field of enquiry that problematizes the collective and often taken for granted systems of thought that make governing strategies appear natural and given at certain times in history (Dean 1999: 16; MacKinnon 2000; Lemke 2002). Governance, according to Walters, should hence not be understood as a set of institutions of ideologies, but a practical activity that can be 'historicized and specified at the level of the rationalities, programmes, techniques and subjectivities which underpin it and give it form and effect' (2012: 3). Hence, it is doubtful whether it is meaningful to speak of 'climate governance' in general but rather we should contextualize and examine particular articulations, rationalities and programmes. As a conceptual approach, governmentality is highly capable of registering even subtle shifts in these programmes and techniques and, according to Walters, it is this sensitivity to the 'intricacies of contemporary governance and the nuances of political change' (2012: 3) that explains the very wide take-up of governmentality across social science and the humanities.

Viewed in this manner, governmentality neither offers us a substantive theory about the forces that shape climate policy and politics, nor does it assume a particular ontology of what constitutes climate policy. Instead, governmentality provides an analytical toolbox that can advance new perspectives on the climate as a political space, and enables us to grasp and highlight the existence of changing discursive productions of a warming world and their effects in mitigating or adapting to that world. How did, for example, tropical rainforests, carbon markets and climate refugees become domains amenable to governmental intervention? What are the rationalities and subjectivities by which a carbon-constrained world is ordered, categorized and represented? In addressing these and other questions, the research community has just begun to deploy a Foucauldian analytics of government to climate change. This body of work does not yet constitute or consolidate a distinct analytical field, but it has nevertheless provided a significant contribution to the debate on how we might understand and reconsider climate politics. In the following, we identify three main interrelated 'studies of governmentality' that, respectively, focus on (1) the climate imagined as a historical and political object that is possible to govern; (2) advanced liberal climate government; (3) subjectivity and the personal conduct of carbon.

Imagining Climate Change

In the first category, we find a literature that has worked on the historical production of the atmosphere, the climate, of flows and stocks of carbon as social and political objects. From the conceptual viewpoint of governmentality, the climate must be represented, depicted and ordered before it can be governed in various ways. Lövbrand and Stripple (2013, in press), in an overview of 'environmental governmentalities', underline how statistical and graphical inscriptions such as geological surveys, inventories, maps and accounting schemes all represent powerful and productive 'regimes of practices' that have constituted particular ways of seeing, knowing and acting upon nature at particular times in history (c.f. Scott 1998; Agrawal 2005; Whitehead 2009). In the same vein, the climate, the atmosphere, carbon stocks, and emission trajectories are possible to understand as political spaces, artefacts that are brought into being and that gain meaning through representational practices and technologies (Baldwin 2003). Edwards considers contemporary climate science a 'vast machine' or a socio-technical system that collects data, models physical processes, tests theories, and thereby generates a widely shared understanding of the global carbon cycle (2010: 8). This 'knowledge infrastructure' not only offers specifics about the past and future of Earth's climate, but also enables the imagining of the planetary climate as something observable, measurable and governable (Edwards 2010: 8). Gabrys (2009), by drawing on waste and material studies, traces the metabolism of carbon and how it flows through the air, ocean, land and bodies, which in turn gives rise to a new natural – cultural relations and imaginings. Lövbrand, Stripple and Wiman (2009) trace the scientific practices that have produced the earth system, including the climate system, as a thinkable analytical category back to the International Geophysical Year in 1957, identifying 'the Anthropocene' as a central, but ambiguous, system of thought. Scholarly work on the constitution of the climate as a governable domain has explored spaces such as tropical forests and carbon markets. Lövbrand and Stripple (2006), Boyd (2010) and Stephan (2012) show how deforestation became an object of climate governance through advances in carbon cycle science, particularly remote sensing technologies, that enables new ways of seeing and imaging climate governance. Lövbrand and Stripple (2011) develop an 'analytics of carbon accounting' that draws attention to the calculative practices that turn stocks and flows of carbon into objects of governance. Carbon accounting as a rationality of government is primarily concerned with the ways carbon can be measured, quantified, demarcated and statistically aggregated. Their article outlines three different regimes of carbon accounting – 'the national carbon sink', 'the carbon credit' and 'the personal carbon budget' – to illustrate how stocks and flows of carbon are constructed as administrative domains amenable to certain forms of political and economic rationality, such as government regulation, market exchanges and self-governance by responsible individual subjects. While most work has examined climate change mitigation, researchers have also more recently considered the governmentalities of adaptation, risk and security. Oels (2012) describes how climate change is rendered governable by modes of risk management and contingency. She argues that 'the governmental rationale since 2007 has been to prepare for and manage the "inevitable" primary and secondary impacts of unmitigated climate change' (2012: 1). Methmann and Rothe (2012) show how the technologies of climate risk management rely on apocalyptic imaginaries that increasingly proliferate in discourses around mitigation, adaptation and the security implications of climate change.

Advanced Liberal Climate Government

The second category engages a particular set of constitutive processes of governing the climate, namely those loosely characterized as technologies of advanced liberal government (Dean 1999: 150–9, 164–71; Rose 1999), for example markets, participatory forms of steering, public-private partnerships, product standards, city networks and stakeholder dialogues. Lövbrand and Stripple (Chapter 1) describe how climate governmentality started to emerge in dialogue with the literature on global climate governance. The increased participation of nonstate actors, public policy networks, supranational jurisdiction, partnerships and cooperation across the public-private divide that characterize how climate change is governed in the early twenty-first century was regarded as mirroring the Foucauldian insistence on heterogeneous and dispersed governing practices – a micro-physics of power that unfolded in seemingly disparate places. It also spoke to the anglo-Foucauldian school of governmentality that emerged in the 1980s and 1990s and the concern with rethinking liberal forms of government and statehood (e.g. Miller and Rose 1990; Barry, Osborne and Rose 1996). Scholars working in this vein have been able to problematize the public and the private, the state versus nonstate dichotomy that has preoccupied studies of global governance in recent years, and instead asked how the state is articulated into the practices of climate government and how the private emerges as a terrain for intervention (Bulkeley and Shroeder 2012; Lövbrand and Stripple 2011, 2012). The government of carbon markets reflects many of the features governmentality scholars identify as key to 'advanced liberal government' where governing is not primarily about governing in a totalizing fashion, but about self-regulation and governing at a distance through technologies such as standards and accounting practices. Oels (2005) and Methmann (2013) both identify an 'advanced liberal' government of the climate. There is an emerging literature on the carbon economy which finds inspiration in poststructural approaches, including Actor-Network Theory, that have started to diagnose how carbon markets become established as governable domains. Mackenzie (2009), with the elegant phrase, 'making things the same', focuses on the commensuration process, and Paterson and Stripple (2012) draw attention to carbon as a 'virtuous commodity', the close affinity between virtuality and virtue – the technological and the ethical – in the government of carbon markets.

The Conduct of Carbon Conduct

A third stream of literature, in a deeper sense, explores subjectifications around climate change and carbon (Rutland and Aylett 2008; De Goede and Randalls 2009; Paterson and Stripple 2010; Rice 2010). Central to Foucault's thinking on governmentality was the notion of 'self-government', of the ways particular forms of conduct could be accomplished through shaping the self and subject positions. The explosion of projects designed to enable individuals to 'do their bit' in line with government and industry objectives to limit climate change can be regarded as one means through which such subject positions are enacted. Paterson and Stripple (2010) outline and compare five different contemporary practices that aim at calculating, measuring and accounting for emissions of greenhouse gases at the level of the individual (footprinting, offsetting, dieting, rationing and trading with Carbon Rationing Action Groups or through a system of Personal Carbon Allowances). What can be seen in these practices is an emergent government of carbon that entails the 'conduct of carbon conduct'. This 'carbon governmentality' moulds and mobilizes individual subjects to govern their own emissions in various ways – as counters, displacers, dieters, communitarians or citizens. Paterson and Stripple (2010) call this governmental rationality 'My Space' – individuals as agents managing their own carbon practice in relation to an articulated global public goal of minimizing climate change. The research community has just begun to explore the processes of subjectification around carbon, or 'how the status of hydrocarbons or standing biomass as resources for the carbon economy hinges on people orienting themselves towards them in particular ways' (Bridge 2010: 8). Skoglund (2011) launched the figure of 'Homo Clima' to describe the emergence of a new ideal citizen, one that invests in its own vulnerability, adaptability and changeability to amend as well as protect a climate-authorized, aestheticized life.

Most work has been on the subjectivities of end-users, those who consume fossil energy and produce emissions. Rice (2010) points to a process of subjectification in her study of greenhouse gas mitigation practices in Seattle. In this urban setting, local governments engage individuals in their climate policies by appealing to 'the good carbon citizen'. Rice illustrates how carbon accounting practices enable 'carbon territories', which she defines as the 'active creation and quantification of bounded and ordered spaces of carbon-producing activities and simultaneous reproduction of local government jurisdictional capacities' (2010: 930). Rutland and Aylett argue that these territorial practices articulate the responsible, carbon-calculating individual. When individuals internalize the carbon metrics of the state, and 'base their actions on these metrics, they become part of a network of self-regulating actors' (Rutland and Aylett 2008: 631). De Goede and Randalls run through different examples of how subjects are constituted through contemporary climate change policies; from the 'neurotic citizen', who is promised the impossible, absolute security from climate change, to 'conscious consumers' knowledgeable of the carbon content of each purchase (2009: 871). Finally, Dowling (2010) cautions about reading too much of a neoliberal (read calculating, self-focussed, rational) subject into these developments, instead suggesting that research should avoid thinking in terms of a close unitary alignment between governmental processes and the identifications of individuals. People are argumentative subjects and research must thus accommodate ambiguity and alternative subject positions.

Limitations and Frontiers

Foucault's concepts and the broader field of governmentality to which this work has given rise provide us with powerful ways to think about the rationalities, politics and practice through which the governing of the climate is taking place. In each sense, the conceptual lexicon of governmentality could act as a means to both broaden and deepen our understanding of the ways climate change has come to be a contemporary social, cultural and political phenomenon. Moving beyond notions of social construction and the workings of discourse as a means through which actors give voice to their essential interests or values, the notion of rationalities enables us to engage with the ways particular logics are produced and serve to structure and orchestrate the social order and the continual process of positioning that takes place across the climate change field. Politics, as the working of power, shifts from a matter of the clash of interests or ideas between ontologically separate entities arising from uneven capacities or resources to an immanent process emerging through social relations. Rather than a held and finite capacity, this suggests, power is an effect mediated through social relations (Allen 2003), in turn implying that new spaces of governance may generate new forms of power, rather than operating in some zero-sum fashion (Hajer 2009; Allen 2010). Critically, power, and the processes of governing through which it is manifest, are not simply held by agents, but realized through practice, encompassing the routine and mundane, calculative techniques, and the material artefacts that serve to translate and mobilize these practices through different forms of socio-technical relations. Establishing this rich tapestry of the interwoven social and material elements, the concept of governmentality provides a means through which to engage with how governing climate change operates through both the strategic and the prosaic, and to offer an explanation for how and why particular forms of authority, knowledge and practice have come to dominate the climate change field.

Such approaches are, of course, not without their limitations. The first and perhaps most readily levelled critique is voiced as a concern about the all-encompassing, seemingly uninterruptable, nature of governmentality as a political order. There is, as Death (Chapter 4) puts it, following Dean (2007: 83), a danger in adopting the 'programmer's view' such that governmentality is a neatly ordered and universalizing phenomenon, in which 'the capacities of subjects to challenge, contest and modify their contexts of governance' (Raco 2003: 91) are circumscribed and in which forms of 'government from below' (Herbert-Cheshire 2003: 458) are neglected. While 'neo-Foucauldian theory provides a framework for examining how governmental programmes and technologies are received and experienced by sub-national institutions' (MacKinnon 2000: 311), the concepts may all too readily be used to negate the geographical variation of regimes of practice. Together, these points mean that the governmentality perspective can appear largely insensitive to what might be regarded as the actual experience of power – in the form of class divisions, overt power struggles and resistances that can characterize sociopolitical relations (Raco and Imrie 2000). Furthermore, the notion of governmentality was developed primarily with the purview of the nation-state in mind, and the field of governmentality studies has been dominated by accounts that seek to understand how different forms of governmentality have been forged and worked through the national arena with little consideration for the potential multiscalar nature of governmentalities (Uitermark 2005). A second strand of critique emerges with regard to the extent to which the conceptual toolkit of governmentality is sufficiently equipped to address the core processes and practices with which it is concerned. As several of the chapters in this book attest, governmentality is rarely used as the sole framework through which to approach matters of climate change governance. This may in part reflect a certain kind of theoretical eclecticism that pervades the social science of climate change, arising in part from its position at the intersection of diverse disciplines and concerns. It may also reflect a sense that governmentality, in its original formulation, was only briefly sketched out and as such lends itself to further development and forms of 'in-filling'. From a more critical view, however, it may suggest that the tools of governmentality are unable to live up to their promise of engaging with both the strategic and everyday working of power.

This collection is positioned at the intersection of these debates about the promise and limitations of governmentality and the wider critical social sciences to contribute to our understanding of climate change. Through conceptual exploration and in-depth engagement with a diverse array of cases, the contributing authors develop and question the ways such theoretical perspectives may extend or challenge current approaches, and offer new ways of engaging with the climate change problematic. In so doing, we suggest that they also offer a valuable contribution to these ongoing theoretical debates in two particular ways. First, as suggested earlier, to date there has been limited engagement within the field of governmentality with matters of governing that operate beyond particular territorially defined political units. Work on climate governmentality has provided a means of thinking through how the conduct of conduct may take place transnationally, or operate between different forms of political space, in ways that could contribute to wider debates on the nature of governmentality. Second, while 'nature' has been a subject for the engagement of a generation of governmentality scholarship, this has tended to operate at the level of understanding how governmentality serves to fix particular entities as nature and govern them accordingly. Working across and between governmentality and approaches derived from ANT, as several of the chapters in this volume do, provides an alternative account, one in which the materiality of nature and its unruly essence come to play a role in the configuring, disrupting and reconfiguring of how governing takes place.

Conclusion: A New Contribution

It is perhaps fitting that as climate change politics appears to be at something of a policy crossroads (Hoffmann 2011), the social science research community should too reflect on the road ahead. As this chapter makes clear, to date, there has been a significant research effort across the social sciences to develop our understanding of the social, political, cultural and economic response to climate change, and to seek to make some sense of what this might mean for contemporary and future societies. In bringing together this collection, we seek to contribute to this ambition, but also to engage the potential of a more critical approach, in which taken for granted categories can be examined and alternatives advanced. This is not only a matter of theoretical interest and academic rigour, but, at least from our perspective, remains a normative project – a means through which to understand and intervene such that the challenge of climate change might be addressed in a more just and effective manner.

The four chapters in Part I of this volume develop the sketch of the theoretical debates outlined here in more detail, each demonstrating the ways governmentality may (and may not) be used in relation to other forms of critical social and political theory. The first chapter, by Eva Lövbrand and Johannes Stripple, takes up the debate where this leaves off, discussing the analytical purchase and added value of applying a Foucauldian analytics of government to the study of global climate governance, offering an overview of the emerging literature in this field and analysing how this body of work may recast the contemporary rethinking of world politics and statehood. In Chapter 2, Anders Blok explores the potentials of ANT in reimagining the analytics of climate governance. From this ANT outset, climate governance becomes a dynamic, heterogeneous, variably scalar and internally fragmented set of assemblages, hovering between the proliferation of climatic problems (carbon markets; adaptation in developing countries; technological innovation; etc.) and a publicly overflowing issue of concern (global warming). Using three cases, he demonstrates the value of ANT in rendering governmentality assumptions more open-ended, situated and politically experimental. Chapter 3, by Chris Methmann, Delf Rothe and Benjamin Stephan, turns attention to the connections between the concepts of hegemony and governmentality. Through the development of an integrated (poststructural) theoretical approach and its application to the case of Reducing Emissions from Deforestation and Degradation (REDD+), the chapter identifies the consequences the current mode of governing deforestation entails, showing the depoliticizing moves of governmentality. In Chapter 4, Carl Death revisits the use of governmentality approaches, suggesting that while early literature on environmental or eco-governmentality tended to critique the statist and domineering nature of environmental governance, more recent writings have emphasized the productive and often politically contradictory character of global climate governance. This chapter identifies a number of key issues and potential limits of these more recent usages of governmentality. Specifically, it argues that in three areas – its global scope, the subjects of government and the ethics and politics of action – climate governmentality faces certain challenges. Yet it also suggests that the idea of climate governmentality is highly adaptive, and that through engagement with these apparent limitations greater theoretical precision and insight can be developed.

In Part II, the contributors seek to engage these debates through more substantive analysis of a range of cases and sites of climate politics. Mark Whitehead, Rhys Jones and Jessica Pykett (Chapter 5) explore the ways attempts to govern the global climate polity have increasingly involved the development of systems of neurological governance. Charting the rise of the so-called environmental behaviour change agenda in the United Kingdom, the chapter outlines the increasingly influential role of behavioural economists and cognitive psychologists in the design of climate change mitigation policy. Developing on the idea of neuroliberalism, this chapter argues that behavioural approaches to climate change mitigation are ultimately about the psychological correction of irrational human conduct, which seeks to reposition the subject as a more capable market actor. The chapter concludes by outlining the ethical and pragmatic flaws associated with this policy regime, and considers alternative, and seemingly more progressive, approaches to neurological governance. In Chapter 6, Sally Eden examines the nature of carbon calculators. Using ideas from ANT, the chapter examines how carbon calculations are coproduced as complex assemblages and as imperfect carbon metrologies. The chapter argues that the analysis of climate governance needs to go beyond a reading of control because, for carbon calculators to work as technologies of government, they must first be co-constructed as socio-technological assemblages, making the subjects of government active users and co-creators of environmental calculation. Tom Hargreaves (Chapter 7) similarly interrogates the forms of climate governance emerging at the household level through the use of real-time display monitors (RTDs) that enable householders to 'see' their energy use (and its associated carbon emissions). The chapter shows how carbon dioxide appears to be a relatively weak rationality of government, seemingly unable to challenge entrenched household practices; how RTDs introduce new forms of surveillance and discipline to household practices that are mediated by preexisting householder relations; and how households resist the monitors through appeals to longer-standing household ethics and aesthetics, and through a partial rejection of the territorialization of climate change at the household level. In Chapter 8, Jarmo Kortelainen and Moritz Albrecht shift scales to consider transnational bioenergy governance. In their chapter, they argue for an approach to governance from a governmentality perspective as a system of translation loops – successive rounds of negotiations and operations transporting and adapting bioenergy governance to various contexts. Bioenergy governance circulates from the loop of European Union (EU) policy design through the national and regional loops to the loops of materialization. A problematic 'rationality jump' occurs when rationalities of climate change policies shift to rationalities of economic calculations, creating tensions and challenges that remain unresolved through the process of bioenergy governance. In Chapter 9, Juliet Fall engages the notion of 'invasive species' as a means of scrutinizing the ways environmental governance is conducted. Fall shows how the calculative practices of listing and categorizing redefine 'nature' as 'biodiversity' and render species as more or less desirable in particular places. This becomes controversial when considering the issue of invasive species on a global scale, within a context of changing climate. The identification of invasive species as a problem requiring global governance relies on the maintenance of the category of global 'invasive species' as internally coherent and global in reach. Yet this coherence is continually challenged and resisted by the very vitality, diversity and scale-jumping characteristics and biographies of the individual species collected under this label – friend or foe in different places.

Heather Lovell (Chapter 10) likewise considers how the governing of climate change is taking place across boundaries, examining the race amongst international science organizations to accurately measure forest carbon in order to constitute the role of forests in relation to climate change policy. Attending to the profession of forest scientists and the practices through which forest carbon is measured and translated, the chapter works through the ways climate politics takes place in practice. In Chapter 11, Angela Oels turns to consider the ways governmentality, as a particular political order, has been mobilized in relation to security. The chapter uses Foucault's work to develop a typology of ways of securing in global politics, each linked to a technology of power: sovereign power, discipline, liberal biopower and advanced liberal government. The chapter illustrates the ways these forms of securitization are shaping the climate agenda. Oels specifically traces an overall change in international climate policy; from a risk-based configuration that she calls 'tolerable levels of climate change' (characteristic of the 1990s and early 2000s) to a recent configuration of 'climate apocalypse' which insists on the fundamental unpredictability of climate change and proposes resilience in the face of sudden disruptions. The articulation of climate change as an uncontrollable phenomenon, Oels argues, makes it difficult to politically approach current levels of greenhouse gas emissions in industrialized countries.

In Part III, the contributors develop reflections on the potential directions that such engagements with climate change may provoke. In Chapter 12, Olaf Corry reflects on if and how the governmental techniques covered in the various chapters 'hang together'. Corry considers 'the bigger picture' in terms of the notion of a 'global climate polity'. If the climate is emerging as a governance-object through apparatuses, discourses and technologies rendering the climate governable, then a 'global climate polity' can also be said to be congealing. As a dynamic assemblage of the technical (remote sensing satellites, computer modelling, measurement stations, carbon calculators) and the social (identities and practices of various actors, e.g. international organizations, states, regions, cities, NGOs, companies, individuals), the chapter considers how the 'global climate polity' emerged and how it might change if it becomes reconfigured as just one element within a broader 'Earth System Polity'. In contrast, in Chapter 13, Sam Randalls makes the case for 'climate change multiple', reflecting on what we can learn from the diverse climate governmentalities discussed in the book for reopening the question of the 'good life' envisaged by proposals to 'solve the climate problem'. To do this, Randalls asks a simple but challenging question: Are people talking about the same thing, climate change, and the same good outcome? Randalls finds three diverse enactments of climate change within practices of security, individualization and economics. Considering the extent to which climate change discourses represent different views on the subject or whether they are enacting climate change multiple, Randalls argues that the latter implies a commitment to assessing these governance projects as ontological interventions as well as politico-philosophical ones. While it may be impossible to definitively commit one way or the other, the chapter concludes by evaluating the extent to which these questions are important and are reflected (directly or indirectly) in the diverse approaches to exploring climate governmentality. Chapter 14 concludes the volume by reflecting on the main themes emerging through these chapters, and considers the implications for the future development of the field of climate governance and politics.

References

Agnew, J. (2005). Sovereignty Regimes: Territoriality and State Authority in Contemporary World Politics. *Annals of the Association of American Geographers* **95**(2), 437–61.

Agrawal, A. (2005). *Environmentality: Technologies of Government and the Making of Subjects*. Durham, NC: Duke University Press.

Aldy, J. E. and R. N. Stavins (2007). *Architectures for Agreement: Addressing Global Climate Change in the Post-Kyoto World*. Cambridge; New York: Cambridge University Press.

Allen, J. (2003). *Lost Geographies of Power*. RGS-IBG book series. Malden, MA: Blackwell Publishing Ltd.

Allen, J. (2010). Powerful City Networks: More than Connections, Less than Domination and Control. *Urban Studies* **47**(13), 2895–911.

Andonova, B. L., M. M. Betsill and H. Bulkeley (2009). Transnational Climate Governance. *Global Environmental Politics*, **9**(2), 52–73.

Arts, B. (1998). *The Political Influence of Global NGOs: Case Studies on the Climate and Biodiversity Conventions*. Utrecht: International Books.

Baldwin, A. (2003). The Nature of the Boreal Forest: Governmentality and Forest-Nature. *Space and Culture* **6**(4), 415–28.

Barnett, M. N. and K. Sikkink. (2008). From International Relations to Global Society. In *The Oxford Handbook of International Relations*, ed. C. Reus-Smit and D. Snidal. Oxford: Oxford University Press, 62–83.

Barrett, S. (2005). *Environment and Statecraft: The Strategy of Environmental Treaty-Making*. Oxford, UK: Oxford University Press, paperback edition, original edition in 2003.

Barry, A., T. Osborne and N. S. Rose (1996). *Foucault and Political Reason: Liberalism, Neo-Liberalism, and Rationalities of Government*. Chicago, IL: University of Chicago Press.

Biermann, F., P. Pattberg, H. van Asselt and F. Zelli. (2009). The Fragmentation of Global Governance Architectures: A Framework for Analysis. *Global Environmental Politics* **9**(4), 14–40. doi: 10.1162/glep.2009.9.4.14.

Boyd, W. (2010). Ways of Seeing in Environmental Law: How Deforestation Became an Object of Climate Governance. *Ecology Law Quarterly* **37**, 843–914.

Brass, P. R. (2000). Foucault Steals Political Science. *Annual Review of Political Science* **3**(1), 305–30.

Breitmeier, H. (1996). Stockholm, Rio and Beyond: Lessons from Two Decades of International Environmental Politics. *Climatic Change* **32**(1), 1–6.

Bridge, G. (2010). Resource Geographies 1. *Progress in Human Geography* **35**(6), 820–34. doi: 10.1177/0309132510385524.

Buhaug, H. (2010). Climate Not to Blame for African Civil Wars. *Proceedings of the National Academy of Sciences, USA* **107**(38), 16477–82.

Bulkeley, H. (2005). Reconfiguring Environmental Governance: Towards a Politics of Scales and Networks. *Political Geography* **24**(8), 875–902.

Bulkeley, H. and H. Schroeder (2012). Beyond State/Non-State Divides: Global Cities and the Governing of Climate Change. *European Journal of International Relations* **18**(4), 743–66.

Burke, M. B., E. Miguel, S. Satyanath, J. A. Dykema and D. B. Lobell (2009). *Warming Increases the Risk of Civil War in Africa. Proceedings of the National Academy of Sciences* **106**, 20670–4 doi: 10.1073/pnas.0907998106.

Busby, J. W. (2008). Who Cares about the Weather?: Climate Change and U.S. National Security. *Security Studies* **17**(3), 468–504.

Carpenter, C. (2001). Businesses, Green Groups and The Media: The Role of Non-Governmental Organizations in the Climate Change Debate. *International Affairs* **77**(2), 313–28. doi: 10.1111/1468–2346.00194.

Compston, H. and I. Bailey (2008). *Turning Down the Heat: The Politics of Climate Policy in Affluent Democracies*. Basingstroke, UK: Palgrave Macmillan.

Conca, K. (2003). Imagining the State. In *Encountering Global Environmental Politics: Teaching, Learning and Empowering Knowledge*, ed. M. Maniates. Lanham: Rowman and Littlefield Publishers Inc, 71–85.

Conca, K. (2005). Old States in New Bottles? The Hybridization of Authority in Global Environmental Governance. In *The State and the Global Ecological Crisis*, ed. J. Barry and R. Eckersley, Cambridge, MA: MIT Press, 181–205.

Corell, E. and M. M. Betsill (2001). A Comparative Look at NGO Influence in International Environmental Negotiations: Desertification and Climate Change. *Global Environmental Politics* **1**(4), 86–107.

De Goede, M. and S. Randalls (2009). Precaution, Preemption: Arts and Technologies of the Actionable Future. *Environment and Planning D: Society and Space* **27**(5), 859–78.

Dean, M. (1999). *Governmentality: Power and Rule in Modern Society*. London; Thousand Oaks, CA: Sage Publications.

Dean, M. (2007). *Governing Societies: Political Perspectives on Domestic and International Rule*. Maidenhead: Open University Press.

Dowling, R. (2010). Geographies of Identity: Climate Change, Governmentality and Activism. *Progress in Human Geography* **34**(4), 488–95.

Dyer, G. (2008). *Climate Wars*. Toronto: Random House.

Edwards, P. N. (2010). *A Vast Machine: Computer Models, Climate Data, and the Politics of Global Warming*. Cambridge, MA: MIT Press.

Elliott, L. (1998). *The Global Politics of the Environment*. London: Macmillan.

Foucault M. (1997). *Ethics: Subjectivity and Truth. Essential Works of Michel Foucault, 1954–1984*. Vol. 1. New York: New Press.

Gabrys, J. (2009). *Sink: The Dirt of Systems. Environment and Planning D: Society and Space* **27**(4), 666–81.

Gardner, G. T., and P. C. Stern (1996). *Environmental Problems and Human Behavior*. Boston, MA: Allyn and Bacon.

German Advisory Council on Global Change (2008). *Climate Change as a Security Risk*. London: Earthscan.

Gerson, M. (2012). Climate and the Culture War. *Washington Post*, 17 January 2012.

Giddens, A. (2009). *The Politics of Climate Change*. Cambridge; Malden, MA: Polity.

Green, J. (2008). Delegation and Accountability in the Clean Development Mechanism: The New Authority of Non-State Actors. *Journal of International Law and International Relations* **4**(2), 21–55.

Hajer, M. A. (2009). *Authoritative Governance: Policy-Making in the Age of Mediatization*. Oxford; New York: Oxford University Press.

Helm, D. (2005). *Climate-Change Policy*. Oxford: Oxford University Press.

Herbert-Cheshire, L. (2003). Translating Policy: Power and Action in Australia's Country Towns. *Sociologia Ruralis* **43**(4), 454–73.

Hoffmann, M. J. (2011). *Climate Governance at the Crossroads: Experimenting With a Global Response after Kyoto*. Oxford; New York: Oxford University Press.

Hulme, M. (2009). *Why We Disagree about Climate Change: Understanding Controversy, Inaction and Opportunity*. Cambridge, UK; New York: Cambridge University Press.

Ikeme, J. (2003). Equity, Environmental Justice and Sustainability: Incomplete Approaches in Climate Change Politics. *Global Environmental Change-Human and Policy Dimensions* **13**(3), 195–206.

Jönsson, C. (2010). Theoretical Approaches to International Organization. In *The International Studies Encyclopedia*, Vol. XI, ed. R. Denemark, 7028–45. Oxford: Wiley-Blackwell.

Joseph, J. (2009). Governmentality of What? Populations, States and International Organisations. *Global Society* **23**(4), 413–19. Reprinted in *Foucault and International Relations: New Critical Engagements*, ed. N. Kiersey and D. Stokes. Abingdon: Routledge.

Kolk, A. and D. Levy (2001). Winds of Change: Corporate Strategy, Climate change and Oil Multinationals. *European Management Journal* **19**(5), 501–9. doi: http://dx.doi.org/10.1016/S0263–2373(01)00064–0.

Lee, J. (2009). *Climate Change and Armed Conflict: Hot and Cold Wars*. New York: Routledge.

Lemke, T. (2002). Foucault, Governmentality, and Critique. *Rethinking Marxism* **14**(3), 49–64. doi: 10.1080/089356902101242288.

Lund, E. (2012). Environmental Diplomacy: Comparing the Influence of Business and Environmental NGOs in Negotiations on Reform of the Clean Development Mechanism. *Environmental Politics*: 1–21. doi: 10.1080/09644016.2012.737253.

Lövbrand, E. and J. Stripple (2006). The Climate as Political Space: On the Territorialisation of the Global Carbon Cycle. *Review of International Studies* **32**(2), 217–35.

Lövbrand, E. and J. Stripple, (2011). Making Climate Change Governable: Accounting for Carbon as Sinks, Credits and Personal Budgets. *Critical Policy Studies* **5**(2), 187–200.

Lövbrand, E., and J. Stripple. (2012). Disrupting the Public-Private Distinction: Excavating the Government of Carbon Markets Post-Copenhagen. *Environment and Planning C: Government and Policy* **30**(4), 658–74.

Lövbrand, E. and J. Stripple (2013). Governmentality. In *Critical Environmental Politics*, ed. C. Death, Routledge, in press.

Lövbrand, E, J. Stripple and Bo Wiman. (2009). Earth System Governmentality: Reflections on Science in the Anthropocene. *Global Environmental Change* **19**(1), 7–13.

MacKenzie, D. (2009). Making Things the Same: Gases, Emission Rights and the Politics of Carbon Markets. *Accounting, Organizations and Society* **34**(3–4), 440–55.

MacKinnon, D. (2000). Managerialism, Governmentality and the State: a Neo-Foucauldian Approach to Local Economic Governance. *Political Geography* **19**(3), 293–314.

Methmann, C. and D. Rothe (2012). Politics for the Day after Tomorrow: The Logic of Apocalypse in Global Climate Politics. *Security Dialogue* **43**(4), 323–44. doi: 10.1177/0967010612450746.

Methmann, C. P. (2013). The Sky is the Limit: Global Warming as Global Governmentality. *European Journal of International Relations* **19**(1), 69–91. doi: 10.1177/1354066111415300.

Miller, C. A. (2004). Climate Science and the Making of a Global Political Order. In *States of Knowledge: The Co-Production of Science and Social Order*, ed. S. Jasanoff. New York: Routledge.

Miller, P. and N. Rose (1990). Governing Economic Life. *Economy and Society* **19**(1), 1–13.

Mintzer, I. and A. Leonard (1994). eds. *Negotiating Climate Change*. Cambridge: Cambridge University Press.

Nilsson, A., C. von Borgstede and A. Biel (2004). Willingness to Accept Climate Change Strategies: The Effect of Values and Norms. *Environmental Psychology* **24**(3), 267–77.

O'Riordan, T., and J Jäger (1996). *Politics of Climate Change: A European Perspective*. London; New York: Routledge.

Oels, A. (2005). Rendering Climate Change Governable: From Biopower to Advanced Liberal Government? *Journal of Environmental Policy and Planning* **7**(3), 185–207.

Oels, A. (2012). Rendering Climate Change Governable by Risk: From Probability to Contingency. *Geoforum*, in press.

Okereke, C., H. Bulkeley and H. Schroeder (2009). Conceptualizing Climate Governance beyond the International Regime. *Global Environmental Politics* **9**(1): 58–78.

Paterson, M. (1996). *Global Warming and Global Politics*. London: Routledge.

Paterson M. and J. Stripple (2010). My Space: Governing Individuals' Carbon Emissions. *Environment and Planning D: Society and Space* **28**(2), 341–62.

Paterson, M. and J. Stripple (2012). Virtuous Carbon. *Environmental Politics* **21**(4), 563–82.

Pattberg, P., and J. Stripple (2008). Beyond the Public and Private Divide: Remapping Transnational Climate Governance in the 21st Century. *International Environmental Agreements-Politics Law and Economics* **8**(4), 367–88.

Raco, M. (2003). Governmentality, Subject-Building and the Discourse and Practices of Devolution in the UK. *Transactions of the Institute of British Geographers* NS **28**(1), 75–95.

Raco, M. and R. Imrie (2000). Governmentality and Rights and Responsibilities in Urban Policy. *Environment and Planning A* **32**(12), 2187–204.

Rice, J. L. (2010). Climate, Carbon, and Territory: Greenhouse Gas mitigation in Seattle, Washington. *Annals of the Association of American Geographers* **100**(4), 929–37.

Rosa, L. P. and M. Munasinghe (2003). *Ethics, Equity, and International Negotiations on Climate Change*. Cheltenham, UK; Northampton, MA: Edward Elgar Pub.

Rose, N. (1999). *Powers of Freedom: Reframing Political Thought*. Cambridge: Cambridge University Press.

Rowlands, I. H. (1995). *The Politics of Global Atmospheric Change*. Manchester; New York: Manchester University Press; distributed exclusively in the United States and Canada by St. Martin's Press.

Ruggie, J. G. (2004). Reconstituting the Global Public Domain – Issues, Actors, and Practices. *European Journal of International Relations* **10**(4), 499–531.

Rutland, T. and A. Aylett (2008). The Work of Policy: Actor Networks, Governmentality, and Local Action on Climate Change in Portland, Oregon. *Environment and Planning D: Society and Space* **26**(4), 627–46.

Schneider, S., A. Rosencranz and J. O. Niles (2002). *Climate Change Policy: A Survey*. Washington DC: Island Press.

Scott, J. (1998). *Seeing Like a State: How Certain Schemes to Improve the Human Condition have Failed*. New Haven, CT: Yale University Press.

Sending, O. J. and Neumann I. B. (2006). Governance to Governmentality: Analyzing NGOs, States, and Power. *International Studies Quarterly* **50**(3), 651–72.

Senecah, S. (2004). *The Environmental Communication Yearbook*. Mahwah, NJ; London: Lawrence Erlbaum Associates.

Shelling, T. (2007). Epilogue. In *Architectures for Agreement: Addressing Global Climate Change in the Post-Kyoto World*, ed. J. Aldy and R. Stavins, Cambridge: Cambridge University Press, 343–9.

Shove, E. (2010). Beyond the ABC: Climate Change Policy and Theories of Social Change. *Environment and Planning A* **42**(6), 1273–85.

Skoglund, A. (2011). *Homo Clima: Climate Man and Productive Power – Government through Climate Change as Bioaesthetic Frame*. Doctoral thesis at the KTH Royal Institute of Technology, Stockholm. Available at: http://urn.kb.se/resolve?urn=urn:nbn:se:kth:diva-34285.

Sprinz, D. and U. Luterbacher (2001). ed. *International Relations and Global Climate Change*. Cambridge, MA: MIT Press.

Stephan, B. (2012). Bringing Discourse to the Market: The Commodification of Avoided Deforestation. *Environmental Politics* **21**(4), 621–39. doi: 10.1080/09644016.2012.688357.

Stern, P. C. (2000). New Environmental Theories: Toward a Coherent Theory of Environmentally Significant Behavior. *Journal of Social Issues* **56**(3), 407–24. doi: 10.1111/0022–4537.00175.

Stripple, J. and P. Pattberg (2010). Agency in Global Climate Governance: Setting the Stage. In *Global Climate Governance Post 2012: Architectures, Agency and Adaptation*, ed. F. Biermann, P. Pattberg and F. Zelli, Cambridge University Press, Cambridge, 137–45.

Tonn, B. (2003). An Equity First, Risk-Based Framework for Managing Global Climate Change. *Global Environmental Change* **13**(4), 295–306.

Toth, F. L. (1999). *Fair Weather? Equity Concerns in Climate Change*. London: Earthscan.

Uitermark, J. (2005). The Genesis and Evolution of Urban Policy: A Confrontation of Regulationist and Governmentality Approaches. *Political Geography* **24**, 137–63.

Urry, J. (2011). *Climate Change and Society*. Cambridge, UK; Malden, MA: Polity Press.

Walker, R. B. J. (1993). *Inside/Outside: International Relations as Political Theory*. Cambridge [England]; New York: Cambridge University Press.

Walters, W. (2012). *Governmentality: Critical Encounters*. New York: Routledge.

Weber, C. (2005). *International Relations Theory: A Critical Introduction*. 2nd ed. New York: Routledge.

Whitehead, M. (2009). *State, Science and the Skies: Governmentalities of the British Atmosphere*, RGS-IBG book series. Chichester, U.K.; Malden, MA: Wiley-Blackwell.

Part I

Governmentality, Critical Theory and Climate Change

1

Bringing Governmentality to the Study of Global Climate Governance

Eva Lövbrand and Johannes Stripple

Introduction

Since the 1990s when global governance emerged as a new and powerful research agenda, scholars have sought to understand the changing role of the state in a time of globalization (Rosenau and Czempiel 1992; Cerny 2010). By drawing attention to the rise of hybrid, nonhierarchical and network-like modes of governing on the global stage, global governance studies have taught us that the state is no longer the sole, or even the principal, source of authority in the international system. As nonstate actors have taken on an increasing number of governance functions in world politics, the sources and institutional locus of authority have changed (Cutler, Haufler and Porter 1999; Hall and Biersteker 2002). The field of global climate governance is no exception. In a time when United Nations (UN) negotiations on a future climate deal have lost momentum, students of international relations have turned their attention to the multiple ways transnational actors and networks such as environmental nongovernmental organizations (NGOs) (Wapner 1996; Betsill and Corell 2008), corporations (Levy and Egan 2003; Newell and Paterson 2010) and city networks (Bulkeley and Betsill 2003) contribute to public rule-setting and steering. Rather than approaching the state as the only actor with purpose and power, the growing field of climate governance studies has sought to establish a broader conception of politics that captures the richness and complexity of climate governance 'beyond the international regime' (Okereke, Bulkeley and Schroeder 2009; Bernstein et al. 2010; Hoffmann 2011).

In this chapter we contrast the conceptual rethinking of world politics offered by global governance studies with that of Foucauldian analytics of government. Since the late 1970s, when French philosopher Michel Foucault launched the awkward concept of 'governmentality', a growing scholarship has asked how thought is linked to the government of human conduct. Although heterogeneous and far from a unified theory, Foucault-inspired governmentality studies share a common concern for critically examining the role that knowledge production plays in the formation of modern governmental practices (Rose, O'Malley and Valverde 2006; Rose-Redwood

2006: 469). For Foucault, knowledge was central to the activities of government and the very function of its objects. Every government programme either articulates or presupposes knowledge of the field of reality upon which it is to intervene (Gordon 1980: 248). In this chapter, we discuss the analytical purchase and added value of applying a Foucauldian analytics of government to the study of global climate governance. We offer an overview of the emerging literature in this field and discuss how this body of work may recast contemporary rethinking of world politics and statehood.

Our argument is organized as follows. We begin by offering an overview of contemporary global governance scholarship and its efforts to map the locus and sources of authority of global climate governance. Secondly, we introduce the analytical legacy of Foucauldian governmentality studies and examine how this poststructuralist account of government differs from and challenges dominant liberal understandings of the contemporary climate governance order. By turning from 'who' to 'how' questions, we find that governmentality studies help us to draw attention to the knowledge practices by which the climate is made governable. In the third section, we discuss how this analytical shift is reflected in the emerging climate governmentality literature and what it does to the understanding of contemporary climate governance. We end by reflecting on the prospects and limitations of this emerging field of analysis.

The Rise of Global Governance Studies

Theoretical and conceptual developments in the discipline of International Relations (IR) have often been related to unfolding events in world politics. The interwar period, the Cold War and the 11 September 2001 attack on the World Trade Center have all given rise to distinct ideas about the character of world politics. When the Berlin Wall fell and the Cold War order quickly waned in 1989, the demand was high for new concepts that could capture a rapidly changing world. Like few other concepts, global governance seemed, at this point in time, to make immediate sense both as an analytical category for comprehending the 'turbulence in world politics' (Rosenau 1990)[1] and as a normative aspiration for the new kind of multilateralism that emerged in the wake of collapsed superpower rivalry. *Global governance* is a suggestive term that quickly became established as a key concept, particularly in UN circles (Jönsson 2010). Weiss reads the conceptual history of the term as a 'shotgun wedding between academic theory and practical policy in the 1990s' (2011: 9) and notes that the seminal publication of *Governance without Government* (Rosenau and Czempiel 1992) coincides with Sweden's launch of the policy-oriented *Commission on Global Governance* (2005).

[1] During the 1990s, Rosenau wrote a trilogy that explored the shift from international to 'postinternational' politics. *Turbulence in World Politics* (1990) is the first, followed by *Along the Domestic-International Frontier* (1997) and *Distant Proximities* (2003). He first introduced the concept of governance in the introductory chapter 'Governance, Order and Change in World Politics' to his edited book (1992) *Governance without Government*.

While IR scholars in the 1990s made the concept of governance bear on the international and the global, the concept as such has a much longer historical lineage in other areas of social science. Introduced in the 1940s, the term *governance* has been used systematically since the 1960s in areas such as business organization, economics and neo-corporatism (Pierre 2000). The conceptual history of governance is, surprisingly, not particularly well covered. Căjvăneanu shows how the concept historically coincides with the representation of society as highly complex and functionally differentiated. The governance concept emerged as a reflection on areas of 'private government' (corporations, labour unions and, later on, the university) as 'autonomous, self-governing units within society that share the governing of individuals with the public government' (2011: 70). The notion that governance could refer to systems of rule that organize the functioning of entities in both the public and the private spheres (Merriam 1944; Eells 1960) is important for understanding the way 'global governance' became articulated in the mid 1990s. Căjvăneanu (2011) shows how developments within, for example, cybernetics and systems theory, enabled the concept of governance to denote control in the context of high complexity.[2] Furthermore, developments within transaction cost/neo-institutional economics enabled governance to be imbued with meanings concerning markets and networks as alternative governance structures that can enhance economic efficiency. By the late 1980s, the concept of governance had 'gained currency as a means of understanding the changing nature of the state and the proliferation of actors and mechanisms involved in the governing of societies' (Bulkeley and Newell 2010: 11).

Building on these bodies of thought and understandings, Rosenau articulated the concept of governance with regard to world politics in an encompassing fashion including the worlds of states, but also with regard to the dynamics that characterized the emergent new postinternational order – the technologies that shrink the world and make the cultural innovations and movements so much faster; that give the transnational and transboundary issues of the worlds greater interdependence; that strengthen subgroups and nonsovereign polities; that promote the greater skills and participation of the world's adults. Rosenau had already in 1969 tried to make sense of a shrinking world by 'recognizing the underlying order that national boundaries obscure' (1969: 5). In the edited book *Linkage Politics* a new theory on national-international linkages was launched that aspired to break through national boundaries and restructure the study of political processes (Rosenau 1969: 5). Twenty years later, the time was ripe for a much broader grip on the processes of politics. The concept of governance proved useful to that end. In 1992 Rosenau argued that governance 'embraces governmental institutions, but it also subsumes informal, non-governmental mechanisms … governance is thus a system of rule that is dependent on intersubjective meanings as on formally sanctioned constitutions and charters' (1992: 4–5).

[2] Actually, the term *governance* derives from the Greek *kybernetes*, which is etymologically linked to navigation and helmanship (Stokke 1997: 28).

Amidst all the paradigmatic changes that abounded in the early 1990s, global governance studies emerged as a new research agenda that tried to account for the changing political order. In a trademark formulation, Rosenau states that 'it is possible to conceive of governance without government – of regulatory mechanisms in a sphere of activity which function effectively even though they are not endowed with formal authority' (1992: 4–5). The ontology of the emergent world order was, in Rosenau's terms, defined by 'Spheres of Authority (SOA) '. The emergence of such new authority structures led governance scholars to identify two (separate) political worlds, one 'state-centric' consisting of 'sovereignty-bound states' and the other 'multi-centric' consisting of 'sovereignty-free' actors. SOAs are authorities that govern within their respective and often overlapping domains. It is important to recognize that the concept of 'global governance' is thus used in both a summative and a descriptive sense – to convey the sum total of the world's many orders (Whitman 2009: 5). Thus, governance can and does exist within, across and beyond the jurisdictions of sovereign states, and 'global governance' refers to the sum of polity domains in the world and not to a form of world government (Ferguson and Mansbach 1996). But there is also the sense that global governance is about governing 'something' – the global governance of what – as Whitman poses the question. Hence, he defines global governance as an activity concerning the 'efforts to bring more orderly and reliable responses to social and political issues that go beyond the capacities of states to address individually' (2009: 8).

Global Climate Governance Studies

When global governance emerged as a new and salient research agenda in the 1990s, IR scholars interested in climate politics were still preoccupied with the nascent international climate regime. In 1992, the UN Framework Convention on Climate Change (UNFCCC) was signed in Rio de Janeiro and an era of interstate negotiations under the UN auspice was set in motion. For liberal scholars, the principles, norms, rules and decision-making procedures developed to guide international behaviour (Krasner 1983) under the UNFCCC and later the Kyoto Protocol represented a fruitful area of research for many years to come (Aldy and Stavins 2007). However, in parallel to the academic interest in UN climate multilateralism, a growing number of scholars also explored the climate as an area governed by a diverse set of actors and a 'vast array of rule systems that exercise authority in the pursuit of goals that function outside normal national jurisdictions' (Rosenau 2000: 172). Inspired by Rosenau's general writings, Jagers and Stripple, for instance, built an early case around the activities of the insurance industry's attempt to govern climate change 'beyond the state'. They argued that global climate governance comprises 'all purposeful mechanisms and measures aimed at steering social systems towards preventing, mitigating, or adapting to the risks posed by climate change' (2003: 385).

Since then IR scholars have suggested that there are many ways climate change is being governed 'beyond the state', through different kinds of emerging institutionalized arenas. Following the global governance agenda outlined by Rosenau and colleagues, a growing scholarship has set out to categorize new sites of climate governance according to the types of actors and sources of authority involved. Most studies have identified a continuum of governance arrangements from those that 1) only involve public actors (e.g. RGGI, Cities for Climate Protection, C40), to 2) hybrid arrangements established in collaboration between public and private actors (e.g. the CDM, the EU ETS, the WSSD partnerships), and finally 3) purely private governance arrangements (e.g. the Carbon Disclosure Project, the Gold Standard, the VCS) (Pattberg and Stripple 2008, Andonova, Betsill and Bulkeley 2009). Beyond these typologies, we have in recent years seen a number of case studies on particular 'climate governance experiments' (Hoffmann 2011) such as public-private partnerships (Bäckstrand 2008), municipal networks (Betsill and Bulkeley 2006) and carbon market arrangements (Lovell 2010; Newell and Paterson 2010). This institutionalization of climate governance 'beyond the state' resembles what Ruggie has called the reconstitution of a global public domain. As a domain, it does not replace states but does 'embed systems of governance in broader global frameworks of social capacity and agency that did not previously exist' (2004: 519).

While global governance studies today represent a well-established research agenda that has prompted students of climate politics to extend their analysis beyond the international climate regime (Okereke et al. 2009; Bernstein et al. 2010), this field of research does not challenge the regime analysis per se or the long-standing interest in UN climate multilateralism. By allowing scholars interested in international organizations to conceptualize 'regimes as governance systems' (Stokke 1997), global governance studies have helped them to rethink international rules and decision-making procedures as 'enmeshed in broader systems of governance instead of issue areas' (Jagers and Stripple 2003: 386). In the concluding chapter of the influential 1997 volume *Global Governance: Drawing Insights from the Environmental Experience*, Oran Young, however, marks the boundary of the study of international regimes and governance. In response to Paul Wapner's suggestion to regard global civil society as a mechanism for the performance of governance, Young argues that we still need 'a clear distinction between regimes, on the one hand, and both international society and global civil society on the other' (1997: 283). This effort to restrict the analytical terrain of global governance studies tells us something about the saliency and recent dominance of this research agenda in the study of climate politics.

Governmentality Studies

If the global governance concept was invented to make sense of a changing world order in the early 1990s, the governmentality concept arose from French philosopher Michel Foucault's investigations into political power and the changing status of liberal government in the late 1970s (1991, 2007, 2008). In a famous lecture series held at Collège de France in Paris, Foucault outlined a new way of analysing government that draws attention to political reason itself, to the underlying mentalities of politics that have shaped our present and worked upon those who have been the subjects of government (Barry, Osborne and Rose 1996: 2). Tracing the development of modern European rule, Foucault detached the understanding of government from the activities of a central, united form of state sovereignty. More loosely defined as 'the conduct of conduct' (Gordon 1991: 2), government instead emerged as a more or less rationally reflected 'way of doing things' or an 'art' for acting upon the actions of individuals so as to shape, guide and correct the ways they conduct themselves (Burchell 1996: 19). Foucault himself launched the governmentality concept to denote a particular 'art of government', which he termed *biopolitics*, that took form in late eighteenth-century Europe and concerned the administration of life, particularly as it appears at the level of populations (1998). However, in the years to come the concept was more broadly adopted as a new mode of analysis that, in the words of Rose and colleagues, 'seeks to identify these different styles of thought, their conditions of formation, the principles and knowledges they borrow from and generate, the practices they consist of, how they are carried out, their contestations and alliances with other arts of governing' (2006: 84).

Since Foucault held his 'Governmentality' lecture in 1978 (1991), governmentality studies have emerged as a diverse field of research that cuts across disciplines such as sociology, geography, history, gender studies and science and technology studies. It is an analytical tradition that has spurred investigations into the 'arts of government' as they are articulated in a range of social domains such as the psy-disciplines, education, poverty and welfare, economics and accounting (for an overview, see Rose and colleagues 2006). Rather than resting on a unified theory, Burchell (1996: 34) describes Foucauldian analytics of government as a way of asking questions about government that highlight the historically situated and reciprocal relationships between truth, subjectivity, techniques of domination and techniques of the self. In contrast to global governance studies' interest in the 'who' and the loci of governing, students of governmentality typically ask 'how' questions of government (Dean 1999). The focus on 'how' questions arises both from a rejection of any a priori understanding of the distribution of power or location of rule, and instead an interest in the very practices, rationalities and identities by which governing operates. As Dean explained, power is from this point of view not a zero-sum game played within an a priori structural distribution. 'It is rather the (mobile and open) resultant of the loose and changing assemblage of governmental techniques, practices and rationalities' (1999: 29).

Miller and Rose (2008) have introduced 'rationalities of government' and 'technologies of government' as useful analytical categories that help us to understand the gist of Foucauldian analytics of government. 'Rationalities of government' refers to the collective and taken for granted body of knowledge and styles of thinking that render aspects of reality thinkable and governable. Following Foucault's awkward neologism, the general assumption is that all government depends on knowledge or modes of 'representation', that is 'the elaboration of a language for depicting a domain in question that claims both to grasp the nature of that reality represented, and literally to *re*present it in a form amenable to political deliberation, argument and scheming' (Miller and Rose 2008: 31, italics in original). Hence, to study rationalities of government involves an analysis of the discursive fields within which the exercise of power is conceptualized, the forms of knowledge authorities draw upon to justify particular ways of exercising power, the division of labour among authorities which they presuppose and enact (Rose and Miller 1992: 177). 'Technologies of government' in turn refers to the vast assemblage of techniques, devices, tools, instruments, materials and apparatuses that render rationalities operable and hereby enable 'the conduct of conduct'. These regimes of practices function as a medium of thought that translates ideas about people and things into reality and thereby makes government possible. For Miller and Rose (2008) rationalities and technologies of government, thought and intervention, represent two inseparable dimensions through which one might characterize and analyse governmentalities.

Against this analytical backdrop, governmentality studies have offered a distinct take on contemporary liberal forms of rule. Instead of approaching liberalism as a theory, an ideology or a judicial philosophy of individual freedom, Foucault himself understood liberalism as a particular way the activity of government has been made both thinkable and practicable as an art (Burchell 1996: 21; Foucault 2008). As 'rationality of government', neoliberalism is sceptical about the excess, inefficiencies and injustices of the welfare state, and instead articulates ideas about a 'free market' and a 'civil society' in which a plurality of groups, organizations and individuals interact in liberty. 'Only "beyond the State", it appears, can life worthy of free human individuals begin' (Rose and Miller 1992: 173). The 'technologies of government' that give shape to liberal arts of government seek to deploy the 'free subject' as a technical instrument in the achievement of governmental purposes and objectives (Dean 1999: 155). As explained by Burchell (1996: 24), the very principle of advanced liberal government requires of the governed that they freely conduct themselves in a certain rational way. However, to act freely, the individual must first be shaped, guided and moulded into one capable of responsibly exercising that freedom. To that end, neoliberal rule rests upon what Dean (1999) calls 'technologies of agency' and 'technologies of performance'. Through practices of self-esteem, empowerment, community development, schooling, and health campaigns, 'technologies of agency' seek to engage us as active and free citizens, as informed and responsible consumers, as agents capable of taking care of our own risks. 'Technologies of performance', in turn, seek to shape and optimize that agency 'at a distance' through establishing indicators and benchmarks for societal expertise and service provision (Dean 1999: 168).

This approach to advanced liberal rule has allowed Foucauldian governmentality scholars to challenge the political vocabulary organized around an opposition between the state and civil society, the public and private, government and market. When conceptualizing the free economic-rational individual both as the object and subject of neoliberal government, personal autonomy is no longer the antithesis of political power but rather a key term in its exercise (Rose and Miller 1992: 174). Translating this assumption to the world of international relations, Sending and Neumann (2006) have challenged the rethinking of world politics offered by global governance studies. Instead of interpreting the recent rise and influence of nonstate actors and civil society organizations in world politics as a relocation of power from the sovereign state to private 'spheres of authority', they conceptualize private self-association and will formation as an expression of neoliberal forms of government. Through 'technologies of agency', civil society organizations and private agencies are today constituted as self-associating units 'whose political significance resides both in their capacity to convey and mobilize the preferences and concerns of individuals and communities, and in their capacity to carry out regulatory functions' (Sending and Neumann 2006: 658). From this vantage point, the question is no longer one of accounting for government in terms of 'the power of the State' but of analysing how and to what extent the state is articulated into the activity of government (Rose and Miller 1992: 177). Within Foucauldian analytics of government the state has no essential necessity or function. Rather, the state can be seen as a particular way the problem of government is discursively codified, a way of delineating a 'public sphere' from one of 'private authority and autonomy' (Rose and Miller 1992: 177).

Towards an Analytics of Climate Government

In contrast to global climate governance studies, climate governmentality studies remain a young and nascent field of research without any distinct conceptual or analytical 'core'. While informed by general debates in Foucauldian governmentality studies, climate governmentality scholars do not primarily speak back to that community. If we can define Foucault-inspired analyses of climate government at all as a school of thought, we may think of it as an effort to redirect the analysis and understanding of power and government in the contemporary study of climate politics. Whereas global governance studies have helped us to ask questions about the *who* and *locus* of climate governance, students of climate governmentality typically draw attention to the *how* of climate governance; how the climate is conceptualized and construed as a domain of government, how climate governance is accomplished in practical and technical terms; and how agent categories and subjectivities are constituted through the practices of climate government. This analytical interest in the *processes* that 'render climate change governable' (Oels 2005) stems from a productive understanding of power that challenges zero-sum representations of the distribution of power in world politics (Okereke et al. 2009).

From a productive vantage point, the proliferation of different climate governance arrangements 'beyond the state' does not imply a relocation of power and authority from public to private spheres of authority, from the state to civil society, as often implied in studies of global governance (Sending and Neumann 2006). Rather than starting from an a priori assumption about the distribution of power in world politics, a certain location of government (the state), or a certain understanding of the locus of the political (local, national, international) with regard to climate change, students of climate governmentality have sought to elucidate 'how different locales are constituted as authoritative and powerful, how different agents are assembled with specific powers, and how different domains are constituted as governable and administrable' (Dean 1999: 29). To that end the analysis typically starts from the constitutive *processes* of governing themselves (e.g. forms of calculation, ways of categorizing, standard-setting practices) rather than with the *entities* (e.g. public or private authorities) that set out to govern the climate (Stripple and Lövbrand 2010). This analytical shift implies that the understanding of climate government and statehood is broadened. When approaching government as the sum of all processes and activities that seek to shape, guide and affect the conduct of persons (Gordon 1991: 2), climate government is no longer confined to the apparatus of the state per se. Neither is the state understood as a universal, fixed or unchanging phenomenon. In the words of Bob Jessop, climate governance instead emerges as 'a set of practices and strategies, governmental projects and modes of calculation that operate on something called the state' (2007: 37).

Interestingly, this Foucauldian analysis of climate government 'in the plural' (Okereke et al. 2009: 67) has tended to mirror the empirical terrain of climate governance studies. To date, students of climate governmentality have primarily examined the role of nonstate actors in shaping and carrying out climate governance functions (Okereke et al. 2009; Bulkeley and Shroeder 2012), the growth of standards and codes of conduct in distinct social spheres such as the carbon marketplace and urban life (Lovell and Liverman 2010; Paterson and Stripple 2010; Rice 2010; Stripple and Lövbrand 2010; Letell, Sundqvist and Elam 2011; Lövbrand and Stripple 2011), as well as the role of ideas (such as liberalism) in shaping the character of international relations (Oels 2005; Death 2010; Methmann 2011; Lövbrand and Stripple 2012). While all these phenomena have been subject to extensive analysis by students of climate governance, governmentality scholars understand and read them in new ways. Let us offer some examples.

In a recent article Bulkeley and Schroeder (2012), for instance, examine the constitution of actor categories and different forms of authority in the urban governance of climate change in the two global cities of London and Los Angeles. Rather than taking the boundaries between the state and the nonstate for granted, their analysis shows how these categories are forged through the very processes of governing themselves. Bulkeley and Schroeder (2012) draw upon the practices of urban climate governance to enrich the scholarly understanding of how the contours and nature of 'the state' and 'non'-state are accomplished. In a similar vein, Rice (2010) has examined the constitution of carbon subjectivities in urban Seattle. By turning attention to the city's practical efforts to map and monitor greenhouse gas emissions by spatially referencing them to transportation, energy production and consumption, Rice shows how Seattle is involved in the creation of a new form of carbon governmentality that shapes and moulds responsible carbon citizens ready to reduce their carbon footprints in accordance with the goals of the state. Moreover, Rice suggests that this 'carbonization' of the urban environment is a central way contemporary climate governance is accomplished. The territorial ordering of climate via carbon makes climate legible and hereby constitutes the city as a central domain of climate government (2010).

Stripple and Lövbrand (2010) and Lövbrand and Stripple (2011, 2012) have, in their analyses of the carbon economy, developed a similar argument. In contrast to the global governance literature's efforts to conceptualize the contemporary trade in carbon credits and allowances according to the actors and sources of authority involved (e.g. Bernstein et al. 2010; Stripple and Pattberg 2010), Lövbrand and Stripple have turned attention to the everyday practices and ways of thinking that make up the carbon economy as a thinkable and governable domain that turns carbon emissions (avoided, sequestered or reduced) into a commodity like any other. Inspired by British governmentality scholars such as Peter Miller and Nicolas Rose (2008), they propose an 'analytics of carbon accounting' that 'investigates how the every-day work of UN officials, scientific experts, auditors, verifiers, consultants and brokers give carbon emissions, produced by different people in different geographical sites, a certain degree of stability and comparability, and, at the same time, translate stocks and flows of carbon into artefacts (e.g. sinks, tradable credits, footprints, benchmarks) amendable to political intervention' (Lövbrand and Stripple 2011: 188). This interest in the practical dimensions of carbon market governance resonates with the work of market sociologists and actor-network theorists. Drawing upon what Callon has called 'the performativity of economics' (1998: 2), a growing scholarship has in recent years examined how economic theory and the everyday practices and products of financial actors *perform* or constitute the carbon economy (MacKenzie and Millo 2003; Knox-Hayes 2009; Lovell and Liverman 2010; Bumpus 2011; Descheneau and Paterson 2011).

By examining the 'technologies of government' at play in the making of different domains of climate governance and the constitution of different actor categories and carbon subjectivities, climate governmentality studies have begun to carve out an analytical space that helps us to shift the analysis of climate politics from the power struggles over 'who gets what, when, how' (Lasswell 1950) within and beyond the international climate regime to the dispersed and heterogeneous ways power operates in the climate domain (Okereke et al. 2009; Lövbrand and Stripple 2011). By advancing a productive account of power, this literature allows us to transcend the state versus nonstate dichotomy that has preoccupied studies of global governance in recent years and instead ask how the state is articulated into the practices of climate government. If we accept that the state has no essence, no pre-given function or autonomous source of power, climate governmentality studies emerges as a field of analysis that may help us to reflect on how contemporary statehood is discursively codified and enacted. While recent analyses of the technologies of climate government have taken important steps in that direction, climate governmentality studies have only begun to link these practices to the particular forms of knowledge and rationalities they give rise to and depend on. Recent studies have given us a sense of the political imaginaries underpinning climate science and modelling (see for instance, Yusoff 2010; Lövbrand, Stripple and Wiman 2009), interstate diplomacy (Death 2011) and climate risk and adaptation governance (Oels 2012). However, we have yet to explore, in a more general sense, what these discursive representations of the climate as governable domain tell us about how climate governance is made practicable and thinkable as an art.

Conclusions

Since the late 1980s when climate change was depicted as a transboundary issue, the idea of 'the international' (a discrete policy domain distinctively different from 'the domestic') has shaped how climate politics has been theorized. Influential theories on regimes, common pool resources and interstate diplomacy all share a particular spatial and statist understanding of 'the international' as constituted by territorially formed authority structures. Governing climate change thus becomes a question of how to achieve cooperation and coordination on the global scale under conditions of fragmentation and conflict (Hurrell and Kingsbury 1992: 1). As we have seen in this chapter, a related, but less state-centric literature locates the climate issue within a globalized society – understood as a political space that explicitly recognizes non-state authority in the governing of collective affairs (e.g. Rosenau 1990). Biermann (2006) thus characterizes global environmental governance as a phenomenon departing from international politics in terms of, for example, increased participation of nonstate actors, public policy networks, supranational jurisdiction, partnerships and cooperation across the public-private divide. Biermann's approach echoes Rosenau's (1990) postinternational ontology of two separate political worlds, one 'state-centric' consisting of 'sovereignty-bound states' and one 'multi-centric' consisting of 'sovereignty-free' actors. From this vantage point, climate governance becomes either a question of what states do together, or a question of what goes on 'beyond the state'.

In contrast, we have in this chapter sought to outline how a Foucauldian analytics of climate government may help us to recast this dominant understanding of world politics in the post–Cold War era. Drawing upon an antifoundational understanding of power and government, this growing body of work rejects any a priori assumption about the distribution of power in world politics, a particular functionality of the state, or a certain understanding of the locus of the political (e.g. local, national, international) with regard to climate change. Instead it sets out to investigate the different ways the activity or 'art' called *climate governance* is made thinkable and practicable. While this analytical exercise to date has produced fragmented insights (there is no unified reading of climate governmentality), a fragmentation or destabilization of our carbon constrained present may be the very ethos of this approach (Barry et al. 1996). By beginning the study of politics and government from below, in the heterogeneous and dispersed micro-physics of power (Jessop 2007: 36), Foucauldian analytics of climate government challenges the received fixedness and inevitability of what governing climate change is all about. In a time when climate change has left the domain of science and become an object with cultural significance and meaning, we conclude that political research needs to take this analytical challenge seriously to maintain a critical attentiveness to new practices of power and carbon control unfolding in dispersed political spaces such as the financial service centres of London, downtown Seattle or in the world's tropical rainforests.

References

Aldy, J. E. and R. N. Stavins (2007). *Architectures for Agreement: Addressing Global Climate Change in the Post-Kyoto World.* Cambridge; New York: Cambridge University Press.

Andonova, B. L., M. M. Betsill, and H. Bulkeley (2009). Transnational Climate Governance. *Global Environmental Politics* **9**(2), 52–73.

Barry, A, T. Osborne and N. S. Rose (1996). *Foucault and Political Reason: Liberalism, Neo-Liberalism, and Rationalities of Government.* Chicago, IL: University of Chicago Press.

Bernstein, S., M. Betsill, M. Hoffmann, and M. Paterson (2010). A Tale of Two Copenhagens: Carbon Markets and Climate Governance. *Millennium – Journal of International Studies* **39**(1), 161–73. doi: 10.1177/0305829810372480.

Betsill, M. M. and H. Bulkeley (2006). Cities and the Multilevel Governance of Global Climate Change. *Global Governance* **12**(2), 141–59.

Betsill M. M. and E. Corell (2008) eds. *NGO Diplomacy: The Influence of Nongovernmental Organizations in International Environmental Negotiations.* Cambridge, MA: MIT Press.

Biermann, F. (2006). Global Governance and the Environment. In *Palgrave Advances in International Environmental Politics*, ed. M. M. Betsill, K. Hochstetler and D. Stevis. New York: Palgrave Macmillan, 237–61.

Bulkeley, H. and M. M. Betsill (2003). *Cities and Climate Change: Urban Sustainability and Global Environmental Governance*, Routledge Studies in Physical Geography and Environment; 4. New York: Routledge.

Bulkeley, H. and P. Newell (2010). *Governing Climate Change.* London: Routledge.

Bulkeley, H. and H. Schroeder (2012). Beyond State/Non-State Divides: Global Cities and the Governing of Climate Change. *European Journal of International Relations* **18**(4), 743–66. doi: 10.1177/1354066111413308.

Bumpus A. G. (2011). The Matter of Carbon: Understanding the Materiality of tCO2e in Carbon Offsets. *Antipode* **43**(3), 612–38.

Burchell, G. (1996). Liberal Government and Techniques of the self. In *Foucault and Political Reason: Liberalism, Neo-Liberalism and Rationalities of Government*, ed. A Barry, T. Osborne and N. Rose. Chicago, IL: University of Chicago Press, 19–36.

Burchell, G., C. Gordon and P. Miller (1991) eds. *The Foucault Effect: Studies in Governmentality*. London: Harvester Wheatsheaf.

Bäckstrand, K. (2008). Accountability of Networked Climate Governance: The Rise of Transnational Climate Partnerships. *Global Environmental Politics* **8**(3), 74–102.

Callon, M., (1998). *The Laws of the Markets*. Oxford & Malden, MA: Blackwell Publishers/ Sociological Review.

Căjvăneanu, D. (2011). *A Genealogy of Government. On Governance, Transparency and Partnership in the European Union*. School of International Studies: University of Trento.

Cerny, P. G. (2010). *Rethinking World Politics: A Theory of Transnational Neopluralism*. New York: Oxford University Press.

Cutler, A. C., V. Haufler and T. Porter (1999). *Private Authority and International Affairs, SUNY Series in Global Politics* Albany: State University of New York Press.

Commission on Global Governance (1995). *Our Global Neighbourhood: Report of the Commission on Global Governance*. Oxford: Oxford University Press.

Dean, M. (1999). *Governmentality: Power and Rule in Modern Society*. London; Thousand Oaks, CA: Sage Publications.

Death, C. (2010). *Governing Sustainable Development: Partnerships, Protests and Power at the World Summit, Interventions*. Milton Park, Abingdon, Oxon, England; New York: Routledge.

Death, C. (2011). Summit Theatre: Exemplary Governmentality and Environmental Diplomacy in Johannesburg and Copenhagen. *Environmental Politics* **20**(1), 1–19. doi: 10.1080/09644016.2011.538161.

Descheneau P. and M. Paterson (2011). Between Desire and Routine: Assembling Environment and Finance in Carbon Markets. *Antipode* **43**(3), 662–81.

Eells, R. S. F. (1960). *The Meaning of Modern Business: An Introduction to the Philosophy of Large Corporate Enterprise*. New York: Columbia University Press.

Ferguson, Y. H. and R. W. Mansbach (1996). *Polities: Authority, Identities and Change*. Columbia: University of South Carolina Press.

Foucault, M. (1991). Governmentality. In *The Foucault Effect*, ed. G. Burchell, C. Gordon and P. Miller. Chicago, IL: University of Chicago Press, 87–104.

Foucault, M. (1998). *The History of Sexuality: Volume 1*. London: Penguin Books.

Foucault, M. (2007). *Security, Territory, Population: Lectures at the Collège de France 1977–1978*. Ed. Michel Senellart, translated by Graham Burchell. New York: Picador.

Foucault, M. (2008). *The Birth of Biopolitics: Lectures at the Collège de France 1978–1979*. Ed. Michel Senellart, translated by Graham Burchell. Basingstoke: Palgrave Macmillan.

Gordon, C. (1980). Afterword. In *Power/Knowledge: Selected Interviews and Other Writings, 1972–1977 by Michel Foucault*, ed. C. Gordon. New York: Pantheon Books, 229–59.

Gordon, C. (1991). Governmental Rationality: An Introduction. In *The Foucault Effect*, ed. G. Burchell, C. Gordon and P. Miller. Chicago, IL: University of Chicago Press, 1–51.

Hall, R. B. and T. J. Biersteker (2002). *The Emergence of Private Authority in Global Governance*. Cambridge: Cambridge University Press.

Hoffmann, M. J. (2011). *Climate Governance at the Crossroads: Experimenting with a Global Response after Kyoto*. Oxford University Press.

Hurrell, A. and B. Kingsbury (1992). The International Politics of the Environment: An Introduction. In *The International Politics of the Environment: Actors, Interests, and Institutions*, ed. A. Hurrel and B. Kingsbury. New York: Oxford University Press, 1–47.

Jagers, S. and J. Stripple (2003). Climate Governance beyond the State. *Global Governance*, **9**(3), 385–99.
Jessop, B. (2007). From Micro-Powers to Governmentality: Foucault's Work on Statehood, State Formation, Statecraft and State Power. *Political Geography* **26**(1), 34–40.
Jönsson, Ch. (2010). Theoretical Approaches to International Organization. In *The International Studies Encyclopedia*, Volume XI, ed. R. Denemark. Chichester: Wiley-Blackwell, 7028–45.
Knox-Hayes J. (2009). The Developing Carbon Financial Services Industry: Expertise, Adaptation and Complementarity in London and New York. *Journal of Economic Geography* **9**(6), 749–77.
Krasner, S. (1983). *International Regimes*. Ithaca, NY: Cornell University Press.
Lasswell, H. D. 1950. *Politics: Who Gets What, When, How*. New York: P. Smith.
Letell, M., G. Sundqvist and M. Elam (2011). Steering through the Neighbourhood: Towards an Advanced Liberal Risk Society? *Environment and Planning A* **43**(1), 106–25.
Levy, D. L. and D. Egan (2003). A Neo-Gramscian Approach to Corporate Political Strategy: Conflict and Accommodation in the Climate Change Negotiations. *Journal of Management Studies* **40**(4), 803–30.
Lovell, H. (2010). Governing the Carbon Offset Market. *Wiley Interdisciplinary Reviews: Climate Change* **1**(3) 353–62.
Lovell, H. and D. M. Liverman (2011). Understanding Carbon Offset Technologies. *New Political Economy* **15**(2), 255–73.
Lövbrand, E., J. Stripple and B. Wiman (2009). Earth System Governmentality: Reflections on Science in the Anthropocene. *Global Environmental Change* **19**(1), 7–13.
Lövbrand, E. and J. Stripple (2011). Making Climate Change Governable: Accounting for Carbon as Sinks, Credits and Personal Budgets. *Critical Policy Studies* **5**(2), 187–200.
Lövbrand, E. and J. Stripple (2012). Disrupting the Public-Private Distinction: Excavating the Government of Carbon Markets Post-Copenhagen. *Environment and Planning C – Government and Policy*, **30**(4), 658–74.
MacKenzie D and Y. Millo (2003). Constructing a Market, Performing Theory: The Historical Sociology of a Financial Derivatives Exchange. *American Journal of Sociology* **109**(1), 107–45.
Methmann, C. P. (2011). The Sky is the Limit: Global Warming as Global Governmentality. *European Journal of International Relations*. first published online 27 October 2011 as doi: 10.1177/1354066111415300.
Merriam, C. E. (1944). *Public and Private Government*. New Haven, CT: Yale University Press.
Miller, P. and N. Rose (2008). *Governing the Present*. Cambridge: Polity Press.
Newell, P. and M. Paterson (2010). *Climate Capitalism: Global Warming and the Transformation of the Global Economy*. Cambridge; New York: Cambridge University Press.
Oels, A. (2005). Rendering Climate Change Governable: From Biopower to Advanced Liberal Government. *Journal of Environmental Policy and Planning* **7**(3), 185–207.
Oels, A. (2012): From the 'Securitization' of Climate Change to the 'Climatization' of the Security Field: Comparing Three Theoretical Perspectives. In *Climate Change, Human Security and Violent Conflict: Challenges for Societal Stability*, ed. J. Scheffran, H. G. Brauch, M. Brzoska, P. M. Link and J. Schilling. Berlin, Germany, Springer Verlag.
Okereke, C., H. Bulkeley, and H. Schroeder (2009). Conceptualizing Climate Governance beyond the International Regime. *Global Environmental Politics* **9**(1), 58–78.
Paterson, M. and J. Stripple (2010). My Space: Governing Individuals' Carbon Emissions. *Environment and Planning D: Society and Space* **28**(2), 341–62.
Pattberg, P. and J. Stripple (2008). Beyond the Public and Private Divide: Remapping Transnational Climate Governance in the 21st Century. *International Environmental Agreements* **8**(4), 367–88.

Pierre, J. (2000). Introduction: Understanding Governance. In *Debating Governance: Authority, Steering, and Democracy*, ed. J. Pierre. Oxford: Oxford University Press, 1–12.

Rice, J. L. (2010). Climate, Carbon, and Territory: Greenhouse Gas mitigation in Seattle, Washington. *Annals of the Association of American Geographers* **100**(4), 929–37.

Rose, N. and P. Miller (1992). Political Power beyond the State: Problematics of Government. *British Journal of Sociology* **43**(2), 173–205.

Rose, N., P. O'Malley and M. Valverde (2006). Governmentality. *Annual Review of Law and Social Science* **2**(1), 83–104.

Rose-Redwood, R. S. (2006). Governmentality, Geography and the Geo-Coded World. *Progress in Human Geography* **30**(4), 469–86.

Rosenau, J. N., Woodrow Wilson School of Public and International Affairs. Center of International Studies (1969). *Linkage Politics: Essays on the Convergence of National and International Systems*. New York: Free Press.

Rosenau, J. N. (1990). *Turbulence in World Politics*. Princeton, NJ: Princeton University Press.

Rosenau, J. N. (1992). Governance, Order, and Change in World Politics. In *Governance without Government: Order and Change in World Politics*, ed. J. N. Rosenau and E. O. Czempiel. Cambridge: Cambridge University Press, 1–29.

Rosenau, J. N. (1997). *Along the Domestic-Foreign Frontier: Exploring Governance in a Turbulent World*. Cambridge: Cambridge University Press.

Rosenau, J. N. (2000). Change, Complexity and Governance in a Globalizing Space. In *Debating Governance*, ed. J. Pierre. Oxford: Oxford University Press, 167–200.

Rosenau, J. N. (2003). *Distant Proximities: Dynamics beyond Globalization*. Princeton, NJ: Princeton University Press.

Rosenau, J. N. and E. O. Czempiel, eds. (1992). *Governance without Government: Order and Change in World Politics*. Cambridge: Cambridge University Press.

Ruggie, J. G. 2004. Reconstituting the Global Public Domain: Issues, Actors, and Practices. *European Journal of International Relations* **10**(4), 499–531.

Sending, O. J. and I. B. Neumann (2006). Governance to Governmentality: Analyzing NGOs, States, and Power. *International Studies Quarterly* **50**(3), 651–72.

Stokke, O. S. (1997). Regimes as Governance Systems. In *Global Governance: Drawing Insights from the Environmental Experience*, ed. O. R. Young. Cambridge, MA: MIT Press, 27–63.

Stripple, J. and E. Lövbrand (2010). Carbon Market Governance beyond the Public-Private Divide. In *Global Climate Governance Post 2012: Architectures, Agency and Adaptation*, ed. F. Biermann, P. Pattberg and F. Zelli. Cambridge: Cambridge University Press, 165–82.

Stripple, J., and P. Pattberg (2010). Agency in Global Climate Governance: Setting the Stage. In *Global Climate Governance beyond 2012: Architectures, Agency and Adaptation*, ed. F. Biermann, P. Pattberg, and F. Zelli. Cambridge: Cambridge University Press, 137–45.

Wapner, P. (1996). *Environmental Activism and World Civic Politics*. Albany: State University of New York Press.

Weiss, T. G. (2011). *Thinking about Global Governance: Why People and Ideas Matter*. Abingdon: Routledge.

Whitman, J. (2009). *The Fundamentals of Global Governance*. Basingstoke; New York: Palgrave Macmillan.

Young, O. R., ed. (1997). *Global Governance: Drawing Insights from the Environmental Experience*. Cambridge: Cambridge University Press.

Yusoff, K. (2010). Biopolitical Economies and the Political Aesthetics of Climate Change. *Theory Culture Society* **27**(2–3), 73–99. doi: 10.1177/0263276410362090.

2

Experimenting on Climate Governmentality with Actor-Network Theory

Anders Blok

Towards an Actor-Network Theory of Climate Governance?

In this chapter, I ask a simple question: What would it mean to make our definitions of climate politics revolve around the specific *issue* – the quasi-object, thing, gathering, concern – rather than have it enter a ready-made political regime of rationalities, practices and powers? Following recent discussions on the political dimensions of actor-network theory (ANT) (Latour 2007; Marres 2007), my overall aim here is to sketch, in broad strokes, an 'issue-centric' theory of climatic assemblies in the making. Secondly, I want to contrast this ANT reading of the climate issue to the body of Foucault-inspired work on climatic governmentalities. Without denying the clear conceptual similarities between these two strands of 'poststructuralist' theorizing, my main contention is that subsuming ANT ontology to an analytics of governmentality risks obscuring what is original in the way ANT reassembles 'politics' along with 'the social' (Latour 2005b). ANT, I want to show, allows us to reassemble climate governance in novel and fruitful ways.

Echoing preeminent ANT scholar Bruno Latour (2005b; 2007), the main question for an issue-centric approach to climate governance is the following: How to devise the 'assemblies' (the forums, the political gatherings) in which the climate 'assemblage' will be shaped, debated and governed? In contrast to Foucauldian studies that seeks to reconstruct the rationalities of power embedded in specific governmental domains, ANT-style politics only starts at the moment when new hybrid associations of humans and nonhumans ('assemblages') render existing political frameworks *problematic*. This view follows closely on that of classical American pragmatists, notably the work of John Dewey on the public and its problems (1927). As Latour notes, pragmatism is the political philosophy dedicated to follow the *consequences* of actions when these are uncertain, overflow the boundaries of administrative routines and render new issues into public concerns (2007: 815). The ANT-pragmatist search for new 'assemblies', then, is quite different from seeking to reconstruct the functioning of 'governmentalities'.

Global warming, I suggest, is a paradigmatic case of what it means for new hybrid assemblages to render existing political assemblies inadequate. In what Ulrich Beck (1998) has aptly dubbed 'world risk society', new scientific and technological connections constantly spur up new public-political matters of concern, from genetically modified organisms (GMOs) to holes in the ozone layer, failing nuclear power plants and – of course – the environmental 'overflows' of carbon (Callon 1998). As Latour argues (2003: 36), Beck's 'risks' are very close to his own 'networks' of uncertain entanglements – or what ANT pragmatists nowadays call an 'issue' (Marres 2007; Callon 2009). In many ways, global warming has today emerged as the ultimate case of new techno-scientific knowledge, as mediated through advanced computer simulations, rendering global economic and public life extremely problematic. By implicating people's seemingly banal everyday actions (driving a car, eating a steak) into prospects of planetary havoc, global warming is fast unsettling most existing social norms of civic responsibility. In short, as a 'risky' issue, global warming means shared uncertainty, ambivalence and political turmoil.

With ANT, governing global spaces will always be a fragile and gradual collective accomplishment of extending actor networks; not a matter of some preexisting, solidified and unified hegemony of global powers, whether corporate or governmental (Kendall 2004). If anything, this fundamental point seems well demonstrated in empirical practice by the many uneven, fragmented and failing attempts at governing the global climate. In the language of ANT cofounder Michel Callon (2009), processes of politicizing climate change start from a situation of initial shock: global warming is an issue that, as defined by authorized science, so far *defies* all attempts to use available political rationalities to answer the questions it raises. As Callon goes on to show, however, this situation does not lead to collective paralysis; rather, it 'stimulates the inventive and creative capacities of actors who are prompted to devise appropriate solutions' (2009: 545). With ANT, politics emerges as ongoing collective experimentation.

In this chapter, having expanded on the theoretical parameters of ANT politics vis-à-vis Foucauldian governmentalities, I illustrate the work of collective experimentation via select domains of climatic assemblies in the making. Specifically, I look briefly at important 'trajectories of problematization' (Callon 2009) by which climatic governance has come to be copractised: through the globalized scientific expertise in the Intergovernmental Panel on Climate Change (IPCC); via the tools of economic commensurability making up transnational carbon markets; and in the emerging authority spheres of transurban sustainability networks. My brief rendering of these 'cases' relies on a combination of primary empirical research and secondary sources.[1] Far from empirical exhaustiveness, the purpose will simply be to illustrate

[1] My own empirical work has so far been primarily on the politics of transnational carbon markets (Blok 2010; 2011; 2012a); in my current research, I look at urban climate sustainability (Blok 2012b).

the core ANT claim, derived mainly from Callon (2009), that climate governance manifests a characteristic topology – one that transforms an unsolvable 'issue' into multiple, possibly solvable, 'problems'.[2]

In conclusion, I return again to the key question: On a backdrop of persistent fragilities in existing forms of global climate governance, what might an ANT-pragmatist sensibility contribute to our collective search for new climatic assemblies? Here, I expand on the notion of collective experimentation by engaging a conception of political assemblies as inclusive, democratic and well-equipped 'hybrid forums' (Callon, Lascoumes and Barthe 2009). As climate politics flickers between a publicly overflowing issue and a meshwork of assemblies in the making, I suggest, there is a clear need for further large-scale political inventiveness in the near future.

Latour on Foucault: Reassembling Climatic Governmentalities?

In the vast literatures on environmental governance, expanding across human geography, sociology, political ecology and kindred domains, governmentality and ANT are often used in close proximity to each other, as complementary approaches to the study of power (e.g. Barry 2001; Rutland and Aylett 2008). Such symbiotic tendencies can be highly productive, given that, as Larner and Le Heron note, the two approaches share a number of substantive and conceptual emphases (2002: 759). Hence, both broadly 'poststructuralist' approaches stress the importance of expertise, techniques, spatiality and subject effects when analysing processes of governance and politics. While ANT foregrounds the role of material objects in governance relative to the more discursive orientation of Foucault (e.g. Nimmo 2008), such nuances can be easily subsumed under the common notion of the 'technologies of politics' (e.g. Asdal 2008).

Nevertheless, as hinted, this symbiotic reading of governmentality and ANT also has its drawbacks, especially from an ANT perspective, because it obscures their rather different understandings of the political. Such drawbacks stand out, in particular, when set against the backdrop of how Foucault's followers, most notably the Anglo-Foucauldians, have put to use his much-cited notion of governmentality.[3] This is complicated analytical territory, and a detailed sorting out is beyond the scope of this chapter (but see Collier 2009). Put briefly, as Bevir observes, 'early' followers of Foucault tended to approach governmentality against the background of an Althusserian structuralist theory of social hegemony (2010: 424). In this vein, the concept has been prone to a certain reification, as though it pointed to a coherent regime of governance dominating an entire epoch. This is particularly noticeable in

[2] As I address later on through the concept of 'problematization', there are important similarities here, as well as differences, between ANT and (some versions of) governmentality studies.

[3] I am thinking here, in particular, of authors such as Graham Burchell, Nicolas Rose and Barry Hindess.

the rather totalizing tendencies of Foucault-inspired scholarship on 'neoliberalism' (see Brenner, Peck and Theodore 2010 for an assessment).

Not surprisingly, such analytical tendencies are also visible (if perhaps not quite as dominant) in the more empirically restricted domain of Foucault-inspired analyses of environmental governance and 'green governmentalities' (Rutherford 2007). One particularly clear illustration is McCarthy and Prudham (2004), who analyse what they depict as a wider transformation, manifesting roughly between the 1960s and 1980s, from a state-led bio-power of environmental regulation to a market-based regime of advanced (neo)liberal government of nature. Concomitant with the neoliberal Reagan-Thatcher years, this analysis suggests, Keynesian-era environmental regulations came under sweeping attacks (McCarthy and Prudham 2004: 278). By implication, environmental subjectivities sensitive to risks and scarcities survive today only as resistance to such hegemonic projects, or in the assimilated ideological form of 'free market' environmentalism (McCarthy and Prudham 2004: 279).

In the domain of climate governance, Oels offers a similar analysis, in claiming that a political rationality of 'advanced liberalism' underlies the main global institutional frameworks through which states render climate change governable, including the United Nations Framework Convention on Climate Change (UNFCCC). This is evinced, she argues, not only in the way carbon markets become prominent organizing principles of governance, but also in a range of new technologies of agency and performance – such as greenhouse gas inventories and the Cities for Climate Protection campaign – by which various nonstate actors are forged as 'calculative' agents in the climate regime. What we witness in global climate regulation, Oels suggests, is a specific version of the wider transition from bio-power to advanced (neo)liberal government, manifesting itself across the Western world since the 1980s (2005: 198).[4]

As I have argued elsewhere in more detail (Blok 2012a), such totalizing narratives of green governmentality carry unfortunate analytical effects in that they downplay the inherent plurality and critical tensions manifest in spaces of environmental politics.[5] As such, they allow us to draw out contrasts to a more ANT-pragmatist reading of climate governance. To paraphrase Kendall, with ANT there is simply too much difference, and too many local solutions that every global actor needs to construct, for there to be any 'hegemony' at all (2004: 62).[6] Instead, networks aiming at the regeneration of power, in the sense of controlling peoples, territories and objects at a distance, always have to be constructed from the ground up. Moreover, such networks

[4] It should be noted that Oels's 2005 article is self-consciously exploratory in its conceptual framework. Her chapter in this volume provides a more recent, and more conceptually fine-grained, statement on governmentalities in the climatic domain, particularly as these pertain to securitization.

[5] Needless to say, a totalizing analytics is not *inevitably* built into Foucauldian readings of the 'topologies of power' (Collier 2009). In this sense, what I present here is a symptomatic reading of analytical tendencies, critiques of which may *also* be found among *other* Foucault-inspired analysts (e.g. Dowling 2010).

[6] I use the term *hegemony* here in the Althusserian structuralist sense implied in some governmentality studies (Bevir 2010). From a strictly 'poststructuralist' starting point, one might arrive at a more 'ANT-compatible' notion of a partial, situated and processual hegemony.

come in the plural; 'big' phenomena, like global governance, is the piling on top of one another, and the active alignment, of myriad specific world-spanning networks. In short, with ANT, ordering the social remains a rare event amidst a sea of irreducible heterogeneity.

One way of capturing this difference is to follow Latour in contrasting the figure of the panopticon – a key metaphor of disciplinary power in Foucault – to his ANT notion of the 'oligopticon' (2005b: 175ff).[7] Whereas the panopticon implies an act of 'totalizing' observation across controlled social spaces, the oligopticon as a network-structuring site allows only for sturdy views of some narrowly connected territory – and only as far as connections extend. Scientific laboratories, financial trading rooms and newspaper editorial offices are oligoptica; that is, well-connected sites exerting some (but not full) control over parts (but not all) of the social territory. The amount of control exercised by any oligopticon is not related to how much power it 'stores', but to the number of actors involved in its composition (Murdoch and Marsden 1995: 372). Power, in this sense, is an *outcome* of collective coordination; more unequivocally than Foucault, ANT embodies an entirely 'productive' notion of the 'powers of association' (Latour 1986).[8]

In the domain of climate governance, the UNFCCC – and the entire conglomerate of scientific, legal and political connections it embodies – has emerged as a central oligopticon, coordinating much of whatever climatic control is exerted within and across nation-states. Returning to Oels's (2005) analysis, she no doubt points to central dimensions of this UNFCCC-led exercise of control, including the way it has increasingly come to rely on technocratic forms of economic expertise and techniques of market calculability. The notion of governmentality, in this context, helpfully allows us to capture a particular *type* of relationship between the governor (UNFCCC) and the governed (nation-states), operating via technologies of shaping and responsibilizing the conduct of others within a matrix of market commensurability. While, in my view, the term *neoliberalism* obscures too much the scientific and political work that goes into constructing new calculative agencies (MacKenzie 2009), such a governmentality analysis does indeed capture important aspects of *how* climate governance is presently practised.

What this analysis of climatic governmentalities neglects, however, is to trace the specific *trajectories* of actor-network construction whereby distinct governance objects – such as the tradable emission permits of carbon markets – emerge and spread over time. This critique mirrors the point Rutland and Aylett make in analysing the gradual constitution of climate change as a local governance priority in Portland,

[7] It should be noted that the figure of the panopticon, and the extent to which it works as a suitable analogy for Foucault's complex (and shifting) conceptions of power, is debated amongst Foucauldians (e.g. Elden 2003; Collier 2009). I thank the editors of this volume for pointing me to these debates.

[8] While much has been made of the constitutive and 'productive' aspects of Foucault's notion of power, in analytical practice, Foucauldians tend to oscillate between this and a more constraining, 'disciplinary' view. Given the all-embracive nature of power in Foucault, this often constitutes a key analytical tension.

Oregon, during the 1990s. In this context, ANT 'supplies a complementary set of tools to help reveal how political priorities and the capacity to achieve them emerge *over time* from the dispersed energies of diverse actants, both human and nonhuman' (2008: 633; italics in original). Specifically, they point to the collective work involved in establishing carbon dioxide emissions as a new object of local concern, through a variety of novel statistical and administrative techniques. Governmentality analytics, they suggest, tells us little about such dynamics of forging new political agencies and interests.

With ANT, climate governance thus presupposes the creation and maintenance of multiple new collective priorities, organized around novel governance objects – or what, following Callon (2009), I later call climatic 'problems'. From this perspective, what needs attention are the historically situated and contested dynamics of 'problematization', whereby specific constellations of actors, material devices and political arenas emerge (or disappear) as authoritative promoters of specific governance objects.[9] Beyond a simple reworking of *existing* political spaces, this process also involves the active *creation* of new spheres of authority and control, as capacities and powers come to be reassembled into new assembly formations. For instance, with the emergence of transnational trade in carbon, the agencies involved in certifying emission permits, including large-scale environmental nongovernmental organizations (NGOs), have *attained* new capacities for co-controlling what counts as 'sustainable development' in the global South (Blok 2011).

In sum, an ANT-pragmatist approach to climate governance tracks how such chains of actors – crisscrossing the governmental, the civic, the scientific, the technological, the economic – come to be tied together by common concerns, as well as organized conflicts, over a multitude of emerging governance objects (Srnicek 2010). Such processes of 'multiform problematization' (Callon 2009: 543), I argue, remain underexplored in those governmentality studies which, like Luke (1999) or McCarthy and Prudham (2004), search for eco-political rationalities *underlying* entire regimes or epochs. By overestimating the coherence of green political rationalities and spatialities, such approaches downplay how *new* governance arrangements come to be proposed, tested and recalibrated in power-inflected but open-ended ways. With ANT, novel assemblies in the making are given their due ontological and political weight; and this sensibility, I suggest, resonates well with global warming as a highly uneven and shape-shifting issue of governance.

The Topology of Climate Governance: A Meshwork of Problematizations

If, as Collier (2009) suggests, we think of Foucault's analyses of power and governmentality as more 'topological' than totalizing – in terms of how they trace

[9] 'Problematization', as noted earlier, is an important case of shared conceptual territory between Foucault and ANT. I return to this in the next section, when setting forth Callon's (2009) usage of the term.

heterogeneous 'patterns of correlation' – this arguably brings us much closer to ANT territory. Transposed from mathematics, the concept of topology signals here an interest in the *variable* organization of sociopolitical relations, in terms of how effects of proximity, distance and authority arise from shifting and particular arrangements of objects and their displacements through space and time.[10] Accordingly, with both Foucault and ANT, we should ask how heterogeneous sociopolitical elements – practices, texts, material forms, techniques of power, ways of problematizing – come to cohere into relatively enduring spatiotemporal configurations of governance. So far, researchers have made few attempts to ask such 'topological' questions of climate governance (Blok 2010; Busch 2011).[11] Beyond the most well-defined institutional spheres, such as the UNFCCC, the variable shapes of global climate governance, I would suggest, are hence still largely unknown.

The one major exception to this rule, I believe, is the brief and somewhat unsubstantiated, but still highly suggestive work on climate governance by ANT cofounder Michel Callon. Amidst an empirical analysis of carbon market construction, Callon makes apposite remarks on the more general problem of how to think about, and hence give intellectual and political shape to, the rather polymorphous issue of global warming. Callon's starting point, as noted, is that global warming registers a situation of initial shock; a situation which he designates by the term *issue*. In line with pragmatist thinking, issues are matters of public concern for which the available political arrangements fail; issues are protean, they overflow institutional boundaries and change as they spread. As Callon summarizes this state of shared uncertainty: 'Global warming in its current state is an issue that is unqualifiable, not in theory but in practice, for no framing is able to embrace it in its entirety' (2009: 542).

Employing a topologically rich metaphor, Callon compares issues to the notion of pluripotent stem cells: like these underdetermined cells, issues have a multiplicity of political fates, specifications and qualifications, all equally possible, but only some of which will materialize. With issues, the crucial moment of trial – and of political inventiveness – is what Callon calls *problematization*: the gradual, contested and incomplete process of jointly dividing the multiform issue of global warming into more well-delineated 'problems', at once distinct from and interdependent on one another. Instead of talking about global warming, Callon observes (2009: 543), people increasingly refer to market efficiency, developing countries' right to development, international politics, technological innovations, new research on scientific models and so on. What emerges then, in topological terms, is a layered 'meshwork' of multi-

[10] Collier's invocation of topology refers explicitly (2009: 100, fn 2) to how this concept has been taken up by Annemarie Mol and John Law (1994), working within ANT. In this sense, like 'problematization', 'topology' may well become a shared conceptual territory between Foucauldians and ANTs. For a deeper discussion of topology and how it may be related to shifting geographies of power, see Allen (2011).

[11] Drawing on the ANT work of Annemarie Mol, Randalls's contribution to this volume shares much of what I dub here a 'topological' interest – without, however, invoking the concept itself.

ple climatic problems, each associated with its own spatiotemporalities, knowledges, techniques and assemblies.[12]

It is interesting, at this point, to note how this ANT concept of problematization resonates with Foucault's own use of this term as set forth in his late interviews. Here, Foucault defines problematization as 'the set of discursive and non-discursive practices that makes something enter into the play of the true and false, and constitutes it as an object for thought ...' (1996 [1984]: 453f). As for Callon and ANT, problematization in Foucault's usage is neither the representation of a preexistent object, nor the creation through discourse of a nonexistent object; rather, it is a process of material-semiotic configuration. More strongly than Foucault's, however, Callon's usage highlights how problematization is not solely a matter of constituting new truth conditions in thought, but also and crucially a process of collective experimentation around fully material and fully practical objects of shared political concern.

What is theoretically crucial here is the way Callon links the notion of collective experimentation to a *substantive* reading of the emerging 'sub-objects' (i.e. problems) of climate governance. In identifying the new (sub)assemblies of climate governance (carbon markets, technological innovation etc.), Callon extends the ANT 'issue-centric' approach to politics into a more nuanced picture of multiform problematizations. The resulting topology of overlapping assemblies in the making is clearly at odds with more homogenizing depictions of a global climate 'polity' made from hierarchically nested political entities (see Bulkeley 2005). New climatic constellations spur the emergence of hybrid problem-solving assemblies, which do not reside at any particular scale or territory, but rather involve their own spatiotemporalities of sociomaterial relations (Blok 2010). Instead of an integrated polity, with ANT, climate governance is a shifting meshwork of superimposed local-global 'problem geometries' (Marres and Rogers 2008).[13]

Likewise, this ANT meshwork of multiple climatic problems also differs from more formalist analytical languages of networked multilevel governance in that it puts more emphasis on the process whereby actors jointly constitute *specific* political objects, capacities and priorities. In the climatic domain, as Callon (2009) points out, such processes of assembly formation always involve the conjoint work of scientific, political and economic actors, knowledges and techniques. This points to another shortcoming of most governmentality approaches, in that they downplay how specific trajectories of climatic problematization *redistribute*, however subtly,

[12] I take here the liberty of improving upon Callon's own topological framing: whereas Callon speaks simply of a 'network of problems', I concur in this case with the critique raised by Ingold (2008), to the effect that the network metaphor is here too orderly and linear for the specific purpose at hand. I therefore opt, in this case, for the more layered, interwoven and rhizomatic notion of a 'meshwork' of problems.

[13] As Marres and Rogers (2008) succinctly note, this also implies that knowing *where* particular forms of climate governance are happening at any one point in time is a matter of huge uncertainty – not only for analysts, but also for political practitioners, who find themselves in need of new tools of mapping and navigation. In this respect, digital platforms and techniques have become crucial tools of politics.

existing boundaries between economics, politics and science. With carbon trading, for instance, esoteric forms of natural science expertise have become integral to the making of new market commensurabilities, for example in terms of calculating the 'carbon equivalences' for greenhouse gases like methane. Likewise, the value of these new market objects depends entirely on political commitments – as witnessed clearly in the European Union carbon trade (MacKenzie 2009).

In sum, Callon's ANT reading of the topology of climate governance serves to raise important new questions for theoretical and empirical research, beyond a Foucauldian governmentality agenda. We should ask: What are the *mechanisms* by which the climate issue transforms into a meshwork of climatic problems – and how do these assemblies in the making, in turn, react to one another? How, indeed, do we *identify* a politically well-entrenched climate problem ('sub-object')? On these points, Callon remains inconclusive. While, as he notes, the formation of problems is 'contingent on the configurations in place when the (stem) issue becomes public', he says little about the specific scientific, political and economic power effects which account for the trajectories of *multiple* climatic problems beyond the case of carbon markets (2009: 543). Indeed, his list of climatic problems seems too restricted; many would say that, for instance, security, urban planning and eco-lifestyles are part of the existing meshwork of climate governance.[14] The question, however, is how to specify this point analytically.

While a thorough answer to this question is beyond the scope of this chapter, Latour's oligopticon provides an initial starting point for thinking about climate assemblies in the making. The oligopticon, as noted, depends for its ordering effects on being a well-connected site that extends sociomaterially across the territory. As such, Callon's meshwork of climate governance may be thought of as a collection of problem-specific oligoptica, each of which emerges as assemblies that tie together specific political actors, material devices and techniques of control. For a climatic problem to become politically well entrenched, then, it requires its own set of oligoptic techniques, which will have to be jointly elaborated by scientific, political and economic actors. This, at least, is my analytical touchstone, when turning now to brief empirical illustrations of such trajectories of joint innovation in climate governance (see Voss 2007).

Trajectories of Climatic Problematization: A Few Empirical Illustrations

In the literature on environmental governance, there is no shortage of work which attempts to capture, in 'panoptic' style, the main tendencies of global-scale climate governance, as this extends across a diverse range of actors, institutions and settings (e.g. Bulkeley and Newell 2010). Likewise, there is no shortage of work dealing with the particularities of specific climatic problems, in Callon's sense, including

[14] Several chapters in this volume provide detailed analyses of the practices and rationalities of these climatic (sub) problems – without, of course, deploying the ANT vocabulary that I develop here.

the constitution of scientific expertise via the IPCC (e.g. Miller 2007); the spread of carbon markets (e.g. Voss 2007); and the strong emergence of cities on national and international climate governance agendas (e.g. Gustavsson, Elander and Lundmark 2009). What is seldom done, however, is to link these styles of enquiry through a 'topological' interest in the trajectories of collective experimentation by which a meshwork of assemblies in the making emerge, composed of problems both distinct from and organically tied to each other. Filling this topological gap, then, is the promise of an ANT approach to climate governance.

In this context, I can only begin to scratch the surface of this empirical agenda, which evidently comes with its own set of methodological challenges. What I will do, then, is simply illustrate Callon's point about multiform climatic formations, by briefly tracking how 'carbon' – the equivalent, in essence, of Callon's stem issue – has been gradually worked into three very *different* yet *interlinked* problem geometries, each associated with its own set of rationalities, practices and ordering powers. The three trajectories of political innovation are: the constitution of the IPCC as a world parliament of experts; the emergence of transnational carbon markets; and the rise of intercity sustainability networks. As compared to the in-depth empirical work needed for a fuller demonstration, these brief illustrations will remain nothing but suggestions for work ahead. Still, bringing the three trajectories together into a single analytical frame is important, I suggest, for starting to think more dynamically, and more topologically, about climate governance.

First Problematization: IPCC as World Parliament of Experts

When searching for the early trajectories of global warming, one may profitably turn to the governance institutions that have hosted and promoted the issue over time, including the 1992 Rio United Nations conference and the successive Conference of Parties (COP) meetings. However, from an ANT perspective, natural science has arguably played *the* crucial role of giving definite shape to climate change as a public-political issue. Since the late 1980s, thousands of scientists from all over the world have been organized in the IPCC, a hybrid science-policy assembly that has played an enormously powerful role of 'ontological kind-making' around climate change (Miller 2007). By mobilizing scientific 'inscription devices' (Latour 1990), including the advanced computer models responsible for authoritative climate predictions, IPCC scientists have managed to write and circulate their 'immutable mobiles' of Assessment Reports, thereby authoritatively aggregating and representing carbon overflows for a global policy audience (Hughes 2011).

As a powerful oligopticon, the IPCC is a science-policy innovation particular to the global warming issue, and its *scale* of worldwide mobilization of scientific expertise for politics is unprecedented in history (Yearley 2009). In this sense, the emergence of the IPCC as a world parliament of experts, mediating between scientific and political communities, exemplifies Callon's sense of how issues spur collective experimentation and creativity. As is well known, this process of politico-scientific consolidation has been and continues to be contested, particularly in the United States, as so-called climate sceptics join forces with conservative political think tanks in attempts to undermine the credibility of the IPCC (McCright and Dunlap 2003). Overall, however, given the relatively short period of time by which climate change has been established as a main issue of worldwide public concern – particularly since around 2006–7 – the IPCC should be seen, in ANT terms, as a major *success* in scientific network extension.

Over the years, the organizational innovation of the IPCC, and its procedures for validating scientific truths across divergences in worldviews, has undergone testing and reworking. Notably, for instance, facing criticisms from Southern governments over the issue of a disproportional Northern expert dominance, the IPCC took steps during the 1990s to expand its geographical reach (Miller 2007). Similarly, in response to socio-epistemic troubles of model predictions and their 'looping effects', since its 2001 report, the IPCC has adopted a novel approach of outlining a variety of future scenarios. Crucially, this change has helped foster a political hardening of the notion of a two-degree 'warming limit' (Yearley 2009). As such, IPCC scientists have been centrally involved in forging those emerging societal norms of future planetary sustainability, around which politicians, business leaders, NGOs and individuals in turn construct their climatic practices – as well as providing a crucial source of criticism against political 'irresponsibility'.

Second Problematization: Transnational Carbon Markets

Many sociopolitical accounts of climate change – ranging from discourses of neoclassical economists to radical NGO critiques of so-called neoliberal ideology – tend unhelpfully to portray carbon markets as coextensive with the whole of climate governance (see Blok 2011). Moreover, such tendencies are mirrored, as noted, in a range of scholarship which depicts carbon markets as simply *expressive* of wider transformations towards (neo)liberal market governance (e.g. Oels 2005; Lohmann 2010). Contrary to such analyses, however, Callon's (2009) ANT take on global warming brings carbon markets down to size, as merely one amongst a multiplicity of politically influential problematizations of the climate issue.[15] More strongly than most governmentality studies, ANT leads to an appreciation of the *specific* tools, techniques and processes involved in constructing 'carbon' as an object of market calculability and commensurability (MacKenzie 2009). Rather than subsuming *all*

[15] To some, this may sound paradoxical, given the way 'critical' scholarship (including Foucauldians) has tended to portray Callon's work as 'ideologically' committed to markets. While true to some limited extent, elaborating the notion of a meshwork of multiple climatic problems, as I have done here, has the added benefit of showing just how far Callon's economics is from any liberal fantasy of the 'free' market.

climatic calculations under the notion of advanced (neo)liberal governmentality, ANT thus stresses the particular trajectory by which 'homo carbonomicus', the hybrid carbon- and price-calculating subject, has come into being (Blok 2012a).

Consistent with a focus on the knowledges and tools of politics, the emergence of carbon trading is contingent on the way economic theories of 'market failure' were gradually elaborated into new technologies of governance (Voss 2007). This process of innovation around emissions trading models took place within networks of American environmental politics of the 1980s, only later to be 'reglobalized' as part of the 1997 Kyoto Protocol compromise, codified as the so-called flexible mechanisms. Somewhat ironically, given this geography of emergence, the main locus of politico-economic experimentation has since become the European Union (EU), whose Emissions Trading Scheme (ETS) represents the primary 'cap-and-trade' carbon market in the world. In time, the ETS has become a key tool of EU climate governance, constituting a volatile space for politically negotiating often contradictory attachments to markets, bureaucratic planning and climatic sustainability as defined by the IPCC (Blok 2012a).

As with the IPCC, collective trials undertaken around European and transnational carbon markets remain contested, as the tools needed for making emission markets work for sustainability are recast over time. In particular, new transnational civil society actors – what Callon (2009) calls 'emerging concerned groups' – have organized themselves around the problem of carbon markets, contesting their specific rules and accountabilities. This is seen, for instance, in the way NGOs like Carbon Trade Watch contest the politics of Clean Development Mechanism (CDM) projects in the global South, thereby reembedding markets into serious public critiques, in places like India, over the injustices of global warming (Blok 2010). If and when deemed persuasive, such NGO practices of witnessing the climatic suffering of the poor may serve to expand civic commitments to climate governance, both in the South and in the North (Blok 2012a). This would be one unexpected consequence of carbon market extension – and a strong illustration as to why ANT emphasizes the ethico-political creativity of publics, in pushing beyond totalizing notions of a 'hegemonic' (neo)liberal governmentality of nature (e.g. McCarthy and Prudham 2004).

Third Problematization: Urban Climatic Sustainability

With successive failures in international climate negotiations – most spectacularly during the COP15 meeting in Copenhagen in 2009 – many analysts, consultants and policy practitioners seem to look towards world cities as the new loci of hope for a low-carbon transition. Saskia Sassen, for instance, recently crystallized this sense of a 'rescaling trend', claiming that 'more than nation-states, cities will be forced into the frontlines by global warming' (2010: 2). Amongst social scientists, however, disagreements exist on how best to interpret such ongoing transformations. Following

Foucault, some analysts (e.g. Slocum 2004) suggest that current urban engagements with climate change primarily manifest an advanced liberal governmentality organized around tools of individualized energy savings. Contesting such depictions, however, other scholars (e.g. Bulkeley 2005) have shown how urban sustainability emerges as a new networked arena of climatic authority, replete with specific knowledges, practices and protocols. This, indeed, is much closer to an ANT topological understanding of climate governance.

As a specific trajectory of climatic problematization, urban concerns with 'carbon' have not been orchestrated (primarily) through the UNFCCC process, nor are nation-states nearly as central as in the problem spaces of the IPCC and carbon markets. Rather, urban climatic sustainability and low-carbon experimentation have largely gained impetus and legitimacy at the junction of Local Agenda 21 procedures born at the 1992 Rio conference, from which a variety of local authority and community-driven initiatives have mushroomed worldwide. Such initiatives are loosely coordinated through intercity networks and oligoptica, including the International Council for Local Environmental Initiative (ICLEI), whose Cities for Climate Protection (CCP) campaign organizes more than 650 cities in more than thirty participating countries (Toly 2008). Among other things, member cities pledge to develop local action plans for achieving specific emission reduction targets. In this sense, the CCP is part of organizing processes whereby local governance is reassembled around shared climatic tools and priorities (Rutland and Aylett 2008).

As important elements in the topological meshwork, cities are thus emerging as so many localized climatic assemblies, each working to rescale the objects and practices of climate governance to fit their particular capacities and interests. While concerns with green-tech innovation and economic competitiveness are important aspects of cities' climatic priorities, local policy commitments often simultaneously extend into new civic identities and solidarities, within and across city boundaries. Hence, as the abstract issue of climate change comes to be translated into more tangible domains of local transport, housing and energy provision, everyday material practices are potentially redefined as globally significant (Sassen 2010). In the city of Copenhagen, for instance, bicycles now serve as important material mediators of a widespread civic sense of global climatic engagement; whereas in Kyoto, a similar sense of attachment is mediated through the reinvention of vernacular wooden eco-houses (Blok 2012b). Here, inhabiting the city itself becomes a site of civic experimentation with new forms of climate governance.

Conclusion: Towards Democratic Climate Experimentations?

To sum up the argument so far: following the pragmatist strand of ANT, this chapter argues for a theoretical reconceptualization of climate politics away from reading political rationalities as what *underlies* administrative rule (e.g. in the UNFCCC) and towards a more open-ended analytics of the *novel trajectories* of hybrid networked assemblies, spurred by the creative problem-solving capacities of actors. Only by adopting this 'issue-centric' perspective, sensitive to how global warming overflows existing political frames and turns public life into a problem, may we begin to trace how 'politics' is slowly being reassembled, in nascent and as yet unpredictable chains of climatic authority and control. Far from having become part of the daily routines of administration – practices to which the term *governmentality* may associate (Latour 2007) – climate politics thus still flickers, I argue with Callon (2009), in between a publicly overflowing *issue* and a meshwork of superimposed *assemblies in the making*.

I illustrate these theoretical claims via three important trajectories of climatic problematization: the hybrid science-policy work of ontological kind-making in the IPCC; the forging of the carbon- and price-calculating subject of transnational carbon markets; and the rescaling trend of urban sustainability networks. While brief, this juxtaposition of *distinct* but *interlinked* problem spaces in the meshwork of global climate governance, I claim, is enough to suggest how a 'topological' ANT sensibility towards experimental politics differs from Foucauldian approaches to climatic governmentalities. Hence, across the three problem spaces, important *heterogeneities* of governance stand out, in terms of spatiotemporal dynamics, material devices and forms of political rationality. For instance, the scientifically codified priorities created via IPCC inscriptions differ markedly from the embodied public engagements enabled via urban sustainability networks – even as both authority chains may traverse the same contested political spaces (Slocum 2004). Overall, I suggest, ANT is here an important vehicle for granting ontological and political weight to the collective inventiveness of diverse climatic publics.

One of the key theoretical benefits of rethinking governance along ANT-pragmatist lines, then, is the way it makes the notion of emerging publics – and, hence, of democracy – integral to the way we think about power and politics in the climate domain (Latour 2005a). In this respect, my analysis of climatic assemblies in the making admittedly raises more questions than it answers. In which sense, and under what conditions, may hybrid assemblies like the IPCC, the CDM Executive Board or the ICLEI network be said to practise *democratic* forms of governance and control? (see Miller 2007 on the IPCC). Following Callon and colleagues' (2009) notion of the 'hybrid forum', one answer – in the abstract – would be: these climatic assemblies are democratic to the extent that they allow for a *pluralistic* and *inclusive* exploration of shared technological and political uncertainties, thereby *empowering* a variety of publics to exercise their creativity in search for new problem-solving arrangements. Democracy, in this view, is not simply a procedural matter; it is also a matter of better *equipping* collectives to deal with their self-imposed problems.

The extent to which existing climatic assemblies, and the wider meshwork of global climate governance, is democratic (and problem-solving) in this ANT-pragmatist sense is a moot point; too little is known, more exploration is needed. What *does* seem certain in the face of persistent fragilities and failures in climate governance, however, is the urgent need for more experimentation and inventiveness in addressing the collective shock of global warming. In terms of the *scale* of the political challenges raised by this risky, protean and shape-shifting issue, we seem to have collectively taken only a few initial steps of problematization; the rest of the work, as they say, still lies ahead of us. Like Callon (2009), I stress that this point applies in practice, not in theory. By following ANT in making climate politics turn around the question of how global warming renders public life problematic, however, I believe we may gain some leeway for broadening the political imagination. This may in turn allow us to start co-elaborating the collective protocols needed for transforming the rather unsolvable issue of global warming into possibly more solvable problems.

References

Allen, J. (2011). Topological Twists: Power's Shifting Geographies. *Dialogues in Human Geography* **1**(3), 283–98.

Asdal, K. (2008). On Politics and the Little Tools of Democracy: A Down-to-Earth Approach. *Distinktion: Scandinavian Journal of Social Theory* **16**, 11–26.

Barry, A. (2001). *Political Machines: Governing a Technological Society*. London: The Athlone Press.

Beck, U. (1998). *World Risk Society*. Cambridge: Polity Press.

Bevir, M. (2010). Rethinking Governmentality: Towards Genealogies of Governance. *European Journal of Social Theory* **13**(4), 423–41.

Blok, A. (2010). Topologies of Climate Change: Actor-Network Theory, Relational-Scalar Analytics, and Carbon-Market Overflows. *Environment and Planning D* **28**(5), 896–912.

Blok, A. (2011). Clash of the Eco-Sciences: Carbon Marketization, Environmental NGOs and Performativity as Politics. *Economy and Society* **40**(3), 451–76.

Blok, A. (2012a). Configuring Homo Carbonomicus: Carbon Markets, Calculative Techniques, and the Green Neoliberal. In *Neoliberalism and Technoscience*, ed. L. Pellizzoni and M. Ylönen. London: Ashgate, 187–208.

Blok, A. (2012b). Greening Cosmopolitan Urbanism? On the Transnational Mobility of Low-Carbon Formats in Northern European and East Asian Cities. *Environment & Planning A* **44**(10), 2327–43.

Brenner, N., J. Peck, and N. Theodore (2010). Variegated Neoliberalization: Geographies, Modalities, Pathways. *Global Networks* **10**(2), 182–222.

Bulkeley, H. (2005). Reconfiguring Environmental Governance: Towards a Politics of Scales and Networks. *Political Geography* **24**, 875–902.

Bulkeley, H. and P. Newell (2010). *Governing Climate Change*. London: Routledge.

Busch, L. (2011). Climate Change: How Debates over Standards Shape the Biophysical, Social, Political and Economic Climate. *International Journal of Sociology of Agriculture and Food* **18**(3), 167–80.

Callon, M. (1998). An Essay on Framing and Overflowing: Economic Externalities Revisited by Sociology. In *The Laws of the Markets*, ed. M. Callon. Oxford: Blackwell, 244–69.

Callon, M. (2009). Civilizing Markets: Carbon Trading between In Vitro and In Vivo Experiments. *Accounting, Organizations and Society* **34**, 535–48.

Callon, M., P. Lascoumes and Y. Barthe (2009). *Acting in an Uncertain World: An Essay on Technical Democracy*. Cambridge, MA: MIT Press.
Collier, S. J. (2009). Topologies of Power: Foucault's Analysis of Political Government beyond 'Governmentality'. *Theory, Culture & Society* **26**(6), 78–108.
Dewey, J. (1927). *The Public and Its Problems*. New York: Holt.
Dowling, R. (2010). Geographies of Identity: Climate Change, Governmentality and Activism. *Progress in Human Geography* **34**(4), 488–95.
Elden, S. (2003). Plague, Panopticon, Police. *Surveillance & Society* **1**(3), 240–53.
Foucault, M. (1996 [1984]). The Concern for Truth, interview by François Ewald, trans. J. Johnston. In *Foucault Live: Collected Interviews, 1961–1984*, ed. S. Lotringer. New York: Semiotext(e), 450–4.
Gustavsson, E., I. Elander, and M. Lundmark (2009). Multilevel Governance, Networking Cities, and the Geography of Climate-Change Mitigation: Two Swedish Examples. *Environment and Planning C: Government and Policy* **27**(1), 59–74.
Hughes, H. (2011). The Practice of Writing and the Intergovernmental Panel on Climate Change (IPCC). Conference paper, *6th International Conference in Interpretive Policy Analysis: Discursive Spaces, Politics, Practices and Power*, Cardiff, June 2011.
Ingold, T. (2008). Bindings against Boundaries: Entanglements of Life in an Open World. *Environment and Planning A* **40**(8), 1796–810.
Kendall, G. (2004). Global Networks, International Networks, Actor Networks. In *Global Governmentality: Governing International Spaces*, ed. W. Larner and W. Walters. London: Routledge, 59–75.
Larner, W. and R. Le Heron (2002). The Spaces and Subjects of a Globalising Economy: A Situated Exploration of Method. *Environment and Planning D* **20**(6), 753–74.
Latour, B. (1986). The Powers of Association. In *Power, Action, Belief: A New Sociology of Knowledge?* ed. J. Law. London: Routledge and Kegan Paul, 264–80.
Latour, B. (1990). Drawing Things Together. In *Representation in Scientific Practice*, ed. M. Lynch and S. Woolgar. Cambridge, MA: MIT Press, 19–68.
Latour, B. (2003). Is Re-modernization Occurring – And If So, How to Prove It? *Theory, Culture & Society* **20**(2), 35–48.
Latour, B. (2005a). From Realpolitik to Dingpolitik, or How to Make Things Public. In *Making Things Public: Atmospheres of Democracy*, ed. B. Latour and P. Weibel. Karlsruhe: ZKM / Cambridge, MA: MIT Press, 14–41.
Latour, B. (2005b). *Reassembling the Social*. Oxford: Oxford University Press.
Latour, B. (2007). Turning around Politics. *Social Studies of Science* **37**(5), 811–20.
Lohmann, L. (2010). Neoliberalism and the Calculable World: The Rise of Carbon Trading. In *The Rise and Fall of Neoliberalism: The Collapse of an Economic Order?* ed. K. Birch and V. Mykhnenko. London: Zed Books, 77–93.
Luke, T. W. (1999). Environmentality as Green Governmentality. In *Discourses of the Environment*, ed. E. Darier. Oxford: Blackwell, 121–51.
MacKenzie, D. (2009). Making Things the Same: Gasses, Emission Rights and the Politics of Carbon Markets. *Accounting, Organization and Society* **34**(3–4), 440–55.
Marres, N. (2007). The Issues Deserve More Credit: Pragmatist Contributions to the Study of Public Involvement in Controversy. *Social Studies of Science* **37**(5), 759–80.
Marres, N. and R. Rogers (2008). Subsuming the Ground: How Local Realities of the Fergana Valley, the Narmada Dams and the BTC Pipeline are Put to Use on the Web. *Economy and Society* **37**(2), 251–81.
McCarthy, J. and S. Prudham (2004). Neoliberal Nature and the Nature of Neoliberalism. *Geoforum* **35**(3), 275–83.
McCright, A. M. and R. E. Dunlap (2003). Defeating Kyoto: The Conservative Movement's Impact on U.S. Climate Change Policy. *Social Problems* **50**(3), 348–73.

Miller, C. A. (2007). Democratization, International Knowledge Institutions, and Global Governance. *Governance* **20**(2), 325–57.
Mol, A. and J. Law (1994). Regions, Networks and Fluids: Anaemia and Social Topology. *Social Studies of Science* **24**(4), 641–71.
Murdoch, J. and T. Marsden, (1995). The Spatialization of Politics: Local and National Actor-Spaces in Environmental Conflict. *Transactions of the Institute of British Geographers* **20**(3), 368–80.
Nimmo, R. (2008). Governing Nonhumans: Knowledge, Sanitation and Discipline in the Late 19th and Early 20th-Century British Milk Trade. *Distinktion: Scandinavian Journal of Social Theory* **16**, 77–97.
Oels, A. (2005). Rendering Climate Change Governable: From Biopower to Advanced Liberal Government? *Journal of Environmental Policy & Planning* **7**(3), 185–207.
Rutherford, S. (2007). Green Governmentality: Insights and Opportunities in the Study of Nature's Rule. *Progress in Human Geography* **31**(3), 291–307.
Rutland, T. and A. Aylett (2008). The Work of Policy: Actor Networks, Governmentality, and Local Action on Climate Change in Portland, Oregon. *Environment and Planning D* **26**, 627–46.
Sassen, S. (2010). Cities are at the Center of our Environmental Future. *Sapiens* **2**(3), 1–8.
Slocum, R. (2004). Polar Bears and Energy-Efficient Lightbulbs: Strategies to Bring Climate Change Home. *Environment and Planning D* **22**, 413–38.
Srnicek, N. (2010). Conflict Networks: Collapsing the Global into the Local. *Journal of Critical Globalization Studies* **2**(2), 30–64.
Toly, N. (2008). Transnational Municipal Networks in Climate Politics: From Global Governance to Global Politics. *Globalisations* **5**(3), 341–56.
Voss, J. P. (2007). Innovation Processes in Governance: The Development of 'Emissions Trading' as a New Policy Instrument. *Science and Public Policy* **34**(5), 329–43.
Yearley, S. (2009). Sociology and Climate Change after Kyoto: What Roles for Social Science in Understanding Climate Change? *Current Sociology* **57**(3), 389–405.

3

Third Side of the Coin: Hegemony and Governmentality in Global Climate Politics

Benjamin Stephan, Delf Rothe and Chris Methmann

Introduction

Even though their ambitions were very different, the theories of Antonio Gramsci and Michel Foucault – 'the imprisoned leader of the Italian communist party and the anticommunist campaigner for reform of the penal system' (Ekers and Loftus 2008: 701) – show a number of commonalities, including a common interest in the study of modern forms of rule, considering elements other than pure government through the state, and, though different, their notions of power go beyond the narrow understanding of power as solely repressive. Hence, it comes as no surprise that over the past years a number of authors have suggested combining Foucault's concept of governmentality with Gramsci's notion of hegemony (Gill 1995; Jessop 2007; Ekers and Loftus 2008; Okereke, Bulkeley and Schroeder 2009; Joseph 2010; Bulkeley and Schroeder 2011).[1] In this chapter, we seek to take this line of thought further by drawing on Ernesto Laclau and Chantal Mouffe's revision of Gramsci's hegemony concept. Their hegemony and discourse theory, we argue, helps to solve crucial problems that surface in the emerging climate governmentality literature – problems identified in the second half of the 1990s in a (self-)critique of governmentality studies in general (Weir, O'Malley and Clifford 1997) and that still haunt the literature on 'green governmentality' (Rutherford 2007): accounts of governmentality often neglect the heterogeneity and 'messiness' of governmental regimes, disregard the role of contestation and resistance and display a lack of critical engagement in the spirit of Foucault's genealogical project (see Chapter 4). This critique obviously does not apply to the

Authors in reverse alphabetical order; The research for this chapter was supported through the Cluster of Excellence 'Climate System Analysis and Prediction' (CliSAP) (EXC177), University of Hamburg, funded by the German Science Foundation (DFG). Furthermore, we would like to thank Anita Engels, Matthew Paterson, Harriet Bulkeley and Johannes Stripple for commenting on earlier versions of this chapter.

[1] Most of these works draw on a historical-materialist reception of Gramsci. As Barnett (2005) rightly states, they herein face certain metatheoretical tensions that cannot be easily reconciled. A number of authors tried to solve this problem by simply subsuming governmentality under a more or less orthodox historical-materialist framework (Gill 1995; Jessop 2007; Joseph 2010). Others apply Foucault's concept of governmentality and Gramsci empirically on equal footing (Ekers and Loftus 2008; Okereke et al. 2009; Bulkeley and Schroeder 2011), but never address the metatheoretical tensions. To overcome this problem, we draw on the poststructuralist reading of Gramsci as developed by Ernesto Laclau and Chantal Mouffe (Laclau 1990; Laclau 1996; Laclau and Mouffe 2001; Laclau 2005).

literature *par tout*. Yet we provide examples from climate governmentality studies showing that some of these critiques are still valid. Furthermore, we demonstrate how a discursive theory of hegemony helps to remedy these problems. Hegemony theory first provides a more profound concept of discourse and articulation that pays attention to heterogeneity and assemblages; second, a more sophisticated understanding of political agency, hegemonic struggle and discursive change; and, third, a basis to critically assess the political effects of regimes of government.

We illustrate the merits of a combined hegemony/governmentality framework with the emergence of Reducing Emissions from Deforestation and Degradation (REDD+) in global climate politics. It is a good example of heterogeneity, change and contestation in governmentality because REDD+ – an increasingly complex governance program – has gained very broad support and has rapidly moved ahead since it was first proposed in 2005, even though the issue of deforestation within international climate negotiations was highly controversial during the 1990s and early 2000s (Stephan 2012a). We focus on this case for reasons of space and simplicity. Yet our suggestions for combining the two approaches have been corroborated in other issue areas (Methmann 2010, 2011; Methmann and Rothe 2012; Stephan 2012a, 2012b, 2013). The next section outlines how REDD+ can be analysed from a governmentality perspective and also discusses how the concept of discursive articulation emerging from the writings of Laclau and Mouffe can enhance such an analysis. The subsequent sections then turn to the role of agency and change in and the political effects of REDD+. Finally, we draw conclusions about combining hegemony and governmentality.

Articulation: Rendering Forests as an Object of Governance

Both the governmentality perspective and hegemony and discourse theory are interested in how particular practices or 'regimes of practices' (Dean 1999: 18; Glynos and Howarth 2007: 125–6) are constituted. The term *regimes of practices* refers to a set of activities and techniques that addresses a particular issue without necessarily constituting a coherent policy programme or following an overall logic. A society's penal system bringing together very diverse practices of surveillance, control, punishment, self-government and so forth would be an example of such regimes of practices. Similarly, the way society is dealing with climate change or addressing deforestation can be conceptualized as a regime of practice. In the following paragraph we briefly point out how the regime of practice concerning deforestation and climate change has evolved historically. Drawing on Laclau and Mouffe's notion of articulation (Torfing 2005: 15) we then sketch the governmental landscape that characterizes this regime of practice.

Even though negotiations under the REDD+ label[2] only began after a submission by Papua New Guinea and Costa Rica to the UN Framework Convention on Climate

[2] The negotiations started as Reducing Emissions from Deforestation (RED). They were subsequently broadened to include degradation (REDD) (UNFCCC 2007: 2CP13) and 'the role of conservation, sustainable management

Change (UNFCCC) in 2005, the issue ties into a forest carbon discourse that goes back to the ninth Conference of Parties (COP) to the UNFCCC in Milan, the seventh COP in Marrakesh and even to the third COP in Kyoto in 1997 (cf. Boyd, Corbera and Estrada 2008). At that point, however, only afforestation and reforestation were accepted as eligible project types under the CDM. Avoiding deforestation in developing countries, by contrast, was excluded and did not play any role under the Kyoto Protocol.

Bäckstrand and Lövbrand (2006) use a governmentality-inspired approach to shed light on this early debate. Starting their analysis by carving out three distinct forms of governmentality[3] – ecological modernization, green governmentality and civic environmentalism – they argue that forest carbon projects can best be understood as a combination of ecological modernization and green governmentality. In short, the economic modernization discourse framed forests as a cost-effective mitigation option. Green governmentality, with its emphasis on global carbon control and management, provided the technologies and practices to render forests governable. Angela Oels arrives at a similar conclusion for the overall Kyoto Protocol (2005).

Applying such *ideal types* of governmentalities is not unproblematic, though (Weir et al. 1997). The problem is twofold. First, many scholars treat governmentalities as rather stable and uniform formations. This seems to originate from a reading of Foucault's lectures that understands sovereignty, biopolitics and liberal and neoliberal governmentality as a historical succession of formations of rule (as, for example, proffered by Dean 1999). Second, using them as schematic templates, scholars tend to isolate and highlight aspects which fit these categories, while discarding others that do not. Rainer Keller thus even talks about a 'depletion' of the governmentality concept (2010: 47). While there is no need to come to such drastic conclusions, we agree that a top-down application of ideal types creates a harmonizing tendency which does not account for contingency and resistance. It obscures the heterogeneity of governmental assemblages and makes it unnecessarily difficult to account for the contingency of social structures. What is more, reading Foucault's governmentality lectures as a succession of independent formations of rule misses the point (Collier 2009). While Foucault put forth such an understanding in his earlier writings, he abandoned them later on in favour of what Collier calls a 'topological approach' to power: mapping how different forms of power and government coexist, are constantly refined, (re)deployed and combined in various contexts (2009: 80). Such a reading of Foucault points to an escape route from the harmonization trap. Not applying preconceived governmentality ideal types but, rather, trying to map and grasp the

of forests and enhancement of forest carbon stocks in developing countries' (REDD+) (UNFCCC 2010: 1CP16 III C).

[3] Even though they draw on the concept of governmentality, Bäckstrand and Lövbrand talk about competing discourses, not governmentalities. In our opinion, however, in this case discourse and governmentality can be equated. This becomes particularly clear if you compare their article with the work of Oels (2005), who in analysing the climate regime as a whole describes similar phenomena but talks about different governmentalities.

case-specific constellation of power configurations and governmental technologies provides the researcher with a detailed understanding of the specific dynamics of each regime of practice.

Enter hegemony theory. The notion of 'articulation' (see Torfing 2005: 15) provides a helpful conceptual tool with which to map the governmental landscape of the regime on deforestation and climate change – or any other regime, for that matter. Articulation is embedded in Laclau and Mouffe's discourse theory, which provides a number of additional helpful insights. First, discourse is not to be reduced to the phenomenon of language. Discourse, in their view, encompasses the relational structuring of words, objects and practices. Every phenomenon in the world, in this perspective, receives its meaning through its relation with other elements. Following this, we should not separate discourse and practice, but look rather at their interplay (Howarth 2009: 311–12). Second, every discourse is founded upon a 'field of discursivity' (Laclau and Mouffe 2001: 135); that is, it is radically contingent and unstable. In principle, all discursive elements are floating, that is social meaning is continuously renegotiated. What turns discursivity into discourse, though, is the practice of 'articulation' (Laclau and Mouffe 2001: 105, 109–13): combining and thereby partially fixing the meaning of several elements. The products of this articulatory practice are 'nodal points' (Laclau and Mouffe 2001: 113) which structure and bring together a broad range of discursive elements. They constitute the basis for temporarily stable and influential, that is hegemonic, discourses. In this sense, the concept of discourse draws attention to the heterogeneity of discursive formations – not only their common 'rules of formation' as in Foucault (1972). It forces scholars to adopt a bottom-up perspective by retracing how the object of governance and the circumjacent regime of practice are articulated in discourse.

The example of REDD+ illustrates that one can hardly speak of one ideal-typical governmentality informing this regime of practice. We subsequently outline how it articulates a variety of different elements into a specific topology of power. We start by sketching out the rationality underwriting REDD+ and by assessing the governmental technologies through which REDD+ is supposed to work. Furthermore, we highlight new forms of subjectivity REDD+ is dependent on and we outline the field of visibility created through this method of addressing deforestation.[4]

With regard to the *rationality* underwriting REDD+ we can distinguish between two central aspects articulated within the discourse on climate change: framing deforestation as a driver of climate change and avoiding deforestation as an efficient mitigation measure. Both articulations have been prominent in the scientific realm for a long time. The first articulation goes back to the 1970s and 1980s when natural scientists – and subsequently the IPCC in its summaries for policy makers in its first

[4] Outlining the regime of practice of deforestation and climate change, we loosely follow Dean's analytics of government. It is a research heuristic based on Foucault's notion of governmentality, comprising the four dimensions of rationality, technologies, subjectivities and field of visibility (Dean 1999: 30–3).

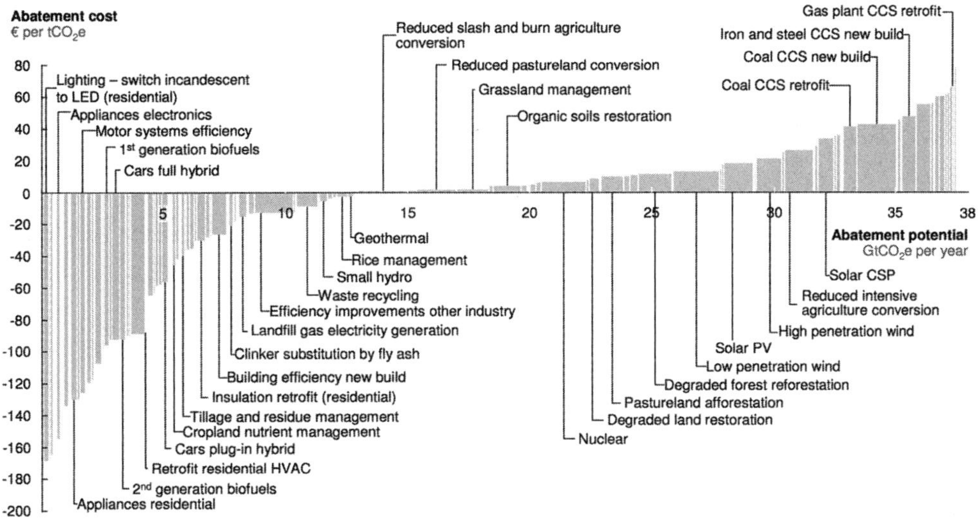

Note: The curve presents an estimate of the maximum potential of all technical GHG abatement measures below €80 per tCO_2e if each lever was pursued aggressively. It is not a forecast of what role different abatement measures and technologies will play.

Figure 3.1 McKinsey's global GHG abatement cost curve V2.1 (McKinsey & Company 2010, 8).

assessment report – began to emphasize the role of deforestation (IPCC 1990a: xxxii; IPCC 1990b: xliii). When avoiding deforestation became an issue for policy makers during the CDM and in the current REDD+ debate it was, however, rearticulated in line with an economic rationale. Ever since, avoiding deforestation is not only a potent mitigation option but also one of the most 'efficient' and 'cost-effective' ones. This has been prominently established by the Stern Review (2006: XIII–XXVI, 215–19). McKinsey's (2010) abatement cost curves – a graphical depiction of the cost and potentials of various mitigation options – provides a similar conclusion (see Figure 3.1). Both received a lot of attention during the early REDD+ debate. Many – particularly developing – countries are using abatement cost curves in assessment exercises to determine their reduction potentials and highlight the most cost-effective ones (see Figure 3.2).

An economic rationale also becomes obvious when we look at the concrete governmental *technologies* proposed to tackle deforestation. They are based on an understanding that deforestation is a problem of missing or false incentives, resulting from a missing valuation of forest ecosystem services in the economic system. Policy makers therefore have to create a price for the carbon stored in forests and compensate agents of deforestation. Within the UNFCCC negotiations countries have agreed that REDD+ should be implemented in phases – starting with capacity-building initiatives and pilot projects – on the basis of results-based payments (UNFCCC 2010: IIIc). Many actors have proposed that REDD+ be integrated into the carbon market once developing countries can produce compliance grade credits. In this case, developing

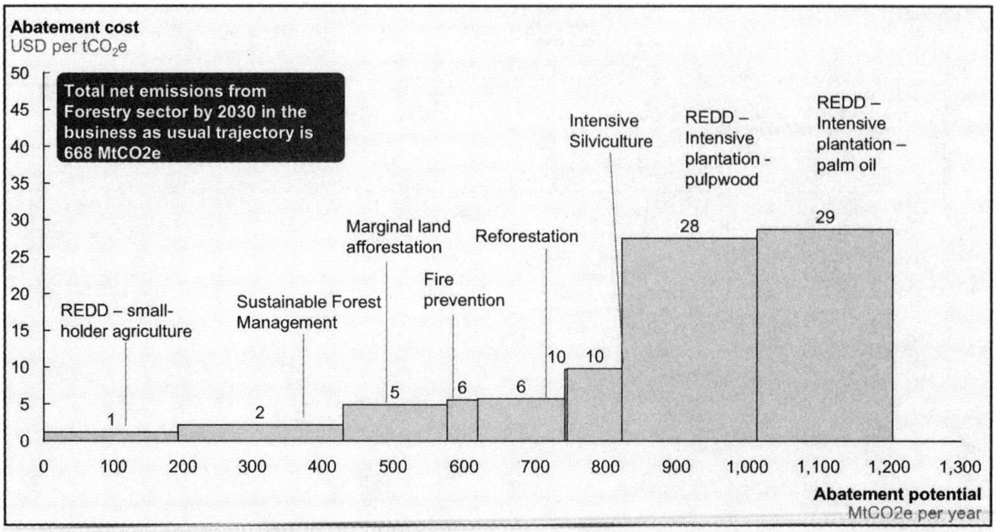

Note: The curve presents an estimate of the maximum potential of all technical GHG abatement measures below EUR 60 per tCO2e if each lever was pursued aggressively. It is not a forecast of what role different abatement measures and technologies will play. Assuming a 4% societal discount rate

Figure 3.2 Abatement cost curve of the Indonesian forestry sector (DNPI Indonesia 2010, 21).

Note:As shown in this figure abatement cost curves are often used for the analysis of reduction potentials within the forest sector, too. This tool produces a very narrow field of visibility, excluding other cost factors or aspects that cannot sufficiently be accounted for in cost analyses at all (for a detailed discussion see Stephan, forthcoming) and thus has become quite controversial (see for example Dyer and Counsell 2010).

countries would receive emissions reduction credits according to the amount of emissions they avoid by reducing their supposed level of deforestation. They can then sell these credits to industrialized countries or companies which can use them to offset reduction commitments they might have under an emissions trading scheme.

How developing countries actually reduce deforestation and implement REDD+ remains a domestic responsibility, however. A broad variety of measures has been discussed. Payments for Ecosystem Services (PES) schemes have especially received a lot of attention. In a PES scheme landowners receive direct payments, the amount of which is set either at the value of the service they provide by protecting a forest or by the opportunity costs they face through not deforesting. Both the carbon market and the PES option heavily rely on the ability of scientists and foresters to determine the amount of carbon stored in forests and to measure and monitor deforestation. They use field inventories, satellite imagery and complex counterfactual modelling to determine the amount of greenhouse gas emissions avoided by protecting a forest. Hence, these scientific and technical practices enable the commodification of avoided deforestation (Stephan 2012a, 2013). It is obvious that REDD+ thus assembles a broad range of technologies of government not easily associated with one particular governmentality.

Governing deforestation in this manner also results in the emergence of new types of *subjectivities*. As Methmann (2011) argues in the case of the CDM, a carbon market-based implementation of REDD+ will be dependent on so-called carbon market professionals (Voß 2007: 340). While policy makers decide to initiate carbon markets, project developers, brokers, verifiers and bankers bring them into being. In the case of REDD+ Stephan (2013: 65–6) highlights the emergence of the carbon forester – actors able to combine a forestry background with a carbon accounting expertise – as an important step for the commodification of avoided deforestation.

Together the linguistic (e.g. the framing of avoiding deforestation as a cost-efficient mitigation measure) and nonlinguistic discursive practices (e.g. the measurement techniques) articulate a *field of visibility* in which forests are reduced to their function as carbon stocks: the measurements and calculations done by carbon foresters present how much carbon is stored in forests, while the carbon market monetizes only this one aspect. All other meanings of forests are being obscured. Stephan (2012a: 632–3, 2013: 62–5) calls this the carbonification of forests.[5]

Even though we have had very limited space to sketch out how deforestation and climate change are being made governable through REDD+, it has become clear that there is a close interaction between heterogeneous science and forestry elements – in governmentality studies usually regarded to be aspects of biopolitics – and market components – part of neoliberalism or advanced liberal government (see e.g. Oels 2005; Bäckstrand and Lövbrand 2006). Other than ideal-typical applications of governmentality, however, a focus on articulation identifies the productive interplay of diverse elements. In the case of REDD+, this diversity generates a particular dynamic. Carbon measurement practices – often perceived as an element of biopolitics – have been a problem in earlier attempts to commodify avoided deforestation. Their levels of uncertainty almost rendered them useless for the market. In other cases, obtaining usable measurement results would have induced transaction costs that exceeded any profits such a system could provide. Using the notion of articulation to assess the regime of practice in a bottom-up manner – as we propose in this chapter – enables us to identify the productive interplay between elements of biopolitics and advanced liberal government that a categorical reading would rather obscure.

Hegemony: REDD+ Becoming the Dominant Approach

In the last section, we demonstrated how the concepts of discourse and articulation as provided by Laclau and Mouffe direct our attention to the heterogeneous elements from which governmental regimes are assembled. This obviously raises a series of questions: How is coherence among these elements established? How is it that such a heterogeneous governmental regime manages to become dominant? And how do

[5] A similar argument is made by Gupta and colleagues, who talk about the 'carbonization of forests' (2012: 729).

we explain change in governmental techniques and rationalities? Here we contend that the concept of hegemony can highlight how regimes of government become hegemonic and are challenged and changed (see also Glynos and Howarth 2007).

Within a Foucauldian approach, the question of change is a difficult one. It is well known that Foucault abandoned his method of archaeology because it was not able to explain discursive change (Dreyfus and Rabinow 1982). His turn to genealogy, thus, puts particular emphasis on the delicate interplay between continuities and discontinuities in different rationalities of government (Foucault 1986). His later works pay attention to the historicity of governmental regimes and trace their evolution through time, but the analysis often remains rather descriptive (see Chapter 4). Foucault himself rarely provided explanations for change. Of course, it would contradict his entire approach to excavate particular 'laws of history', or boil developments down to single causes. Nonetheless, the driving forces for change in governmental rationalities and techniques often remain somewhat obscure. A case in point for this tendency is Foucault's analysis of neoliberalism (2008: 75). He perfectly traces the emergence of a neoliberal governmental rationality but pays only scarce attention to the political agency involved in popularizing it, and the practices of coalition building and consensus formation that eventually made it hegemonic (Mirowski and Plehwe 2009). This argument is also reflected in a number of critical comments regarding Foucault's neglect of collective forms of resistance. Resistance here is bound to the micro level (Kulynych 1997: 328). Foucault claims that 'there is no single locus of great refusal, no soul of revolt, source of all rebellions, or pure law of the revolutionary' (1978: 96). Foucauldian analyses seem to downplay not only the possibility of resistance, but all forms of collective political agency. This tendency is mirrored in some of the Foucauldian analyses of climate change. Angela Oels, for example, in her analysis of climate governmentalities, diagnoses a shift from a biopolitical to an advanced liberal rationality that took place with the adoption of the Kyoto Protocol (2005). While she provides a rich analysis of how climate change is 'rendered governable' through these two rationalities, she does not account for the social and discursive forces that facilitated this shift – for example, the discursive struggles within civil society that eventually turned carbon trading into a legitimate and consensual instrument of climate governance (Stephan 2011). In sum, we would argue that a governmentality perspective is indeed well equipped to map changes in the way social and environmental spaces are governed, but it rarely accounts for the political agency that triggered these changes.

It is here that Laclau and Mouffe's reading of Gramsci's idea of hegemony comes in. Given the fluidity and radical contingency of meaning that marks society, hegemony is that very discursive operation which seeks to tame this field of discursivity. Hegemony seeks to construct a stable and consensual discourse in order 'to arrest the flow of differences, [and] to construct a centre' (Laclau and Mouffe 2001: 112). Laclau and Mouffe have sought to further qualify this discursive operation of hegemony by adding the concepts of antagonism and empty signifier. Any stabilization of discourses depends on the existence of an antagonistic outside – a common enemy (Laclau and Mouffe 2001: 125). And this antagonism comes to be represented in the form of an empty signifier – a signifier that stands for the unity of the discourse vis-à-vis its external negation. Laclau has shown how populist movements constitute around particular empty signifiers which represent the individual demands of its members and unite them in a coherent movement opposing the – antagonistic – established order (Laclau 1996). A case in point is the antiglobalization movement in many countries such as France or Germany that constituted itself around the demand for a Tobin Tax. Thereby it created a chain of equivalence between this particular tax and a broad range of demands such as social justice, peace, economic democracy and the like, as a challenge to the enemy figure of 'globalized capitalism'. In this sense, the creation of a hegemonic discourse depends on the creation of a 'constitutive outside' (Staten 1984) and the articulation of an empty signifier. And these insights obviously also apply to the establishment and maintenance of a particular regime of government. Erik Swyngedouw, for example, has argued that the 'fetishization' of CO_2 represents such a nodal point which holds together all attempts to govern global warming, thereby creating a dangerous or even apocalyptic climate change as a 'constitutive' outside (2010).

Looking at the case of REDD+ we can find some of the discursive structures Swyngedouw identified in the climate change discourse as a whole. As shown in the previous section, the regime of practice implemented through REDD+ also fetishizes carbon. Its narrow focus on emissions obscures the complexity of causes for and consequences of deforestation. It is interesting to note how the threat of deforestation and the threat of climate change are seen to be related: dealing with deforestation is articulated as the prerequisite for addressing climate change and is usually constructed as the more pressing issue (e.g. Eliasch 2008: xx): A recurring narrative argues that while we still have a window of several years within which we can address climate change with various mitigation measures, deforestation needs to be dealt with instantly – otherwise the forests are gone.

Even more interesting in the case of REDD+ is the question of how the issue of avoiding deforestation, very much contested a decade earlier, could suddenly gain such widespread support. Because of the low expected mitigation costs of avoided deforestation projects, critics of including it into the CDM successfully framed it back then as an easy way out for industrialized countries, which would thus prevent domestic reduction efforts and prolong a lock-in into carbon-intensive technologies (e.g. Hare 2000). Some furthermore articulated it as an attack on the sovereignty of developing countries (cf. Laurance 2007: 20–1) while others presented it as a potential biodiversity disaster and as a threat to forest-dwelling communities (cf. Lövbrand 2009: 409).

We can observe significant differences in the narrative structure since the issue has reemerged under the REDD+ label. Avoiding deforestation is still perceived as a cost-efficient mitigation option. With regard to mitigation strategies of industrialized countries, however, this is not perceived as a problem anymore. Cheap REDD+ credits, so the story goes, will enable industrialized countries to take on higher reduction commitments (e.g. Eliasch 2008: xii). Beyond being a solution for mitigating climate change, REDD+ is also portrayed as a remedy for other societal grievances, namely poverty, the increasing destruction of indigenous peoples' livelihoods and biodiversity loss. Through REDD+ the demands to overcome these grievances are combined in a chain of equivalence and REDD+ serves as the signifier that represents them all (Stephan 2012b). Being able to accommodate such a broad spectrum of demands has allowed a variety of very diverse actors to come out in favour of REDD+. Initially put forward by Papua New Guinea and Costa Rica on behalf of a few tropical developing countries and supported by a number of U.S. based conservation nongovernmental organizations (NGOs), REDD+ is now supported by nearly all UNFCCC parties, most big international environmental NGOs, many development organizations, a number of indigenous groups and a small but increasing number of business actors like BP or the multinational bank BNP Paribas (Stephan 2012b). In fact, the emergence of REDD+ is the result of ongoing coalition building forming a coherent discourse.

However, as REDD+ is a rather nascent project one has to wait and see whether it can develop into an empty signifier in its pure form. Thus far, some crucial aspects have not been decided yet – for example the question of whether REDD+ should obtain its funding through integration into the global carbon market or through nonmarket sources. Up until now, this enabled both the actors in favour of market solutions and those that strongly oppose them to support REDD+. Hence, as REDD+ is still partially contested, it represents what Laclau calls a 'floating signifier' (2005: 129–33). If the undecided aspects are resolved in a manner that keeps the chain of equivalence intact and enables subjects with divergent views to continue identifying with it, REDD+ will have become an empty signifier. If this is not the case, the support for REDD+ will crumble (Stephan 2012b).

(De)Politicization: The Political Effects of REDD+

Hegemony theory provides at least one more contribution to the study of governmentality: a more sophisticated understanding of the political. One of the main impacts of Foucault's writings on political thought and analysis is, obviously, a redefinition of the field of the political. Traditional accounts locate politics in a particular sphere of society – the parliament, parties and the government, politics among nations or class struggle. Foucault in contrast highlighted how relations of power traverse the whole society and how government is achieved through a multiplicity of institutions and practices beyond the state (Rose and Miller 1992). In light of his huge influence on political theory and political science, it is quite striking that Foucault was not a very frequent user of the terms *politics* and *the political* as such. In this sense, although he was clearly a political and critical intellectual, Foucault never thoroughly engaged with what he saw as the political. For example, as Weir and colleagues (1997) argue, Foucault tried to separate his explicit political engagement from his mostly historical analyses. As a result, they diagnose that much of the critical ethos of Foucault's genealogical approach is not reflected in present-day governmentality studies (see also Chapter 4). Instead, most of the literature restricts itself to a more or less 'diagnostic' approach and so 'precludes problematizing effects, and thus presumably eliminates the possibility of assigning the costs to existence of any form of governmentality, including neoliberalism' (Weir et al. 1997: 509).

We think that the studies of climate governmentality partly mirror this problem. Recent accounts of carbon accounting, for example, have highlighted the many ways individuals and collectives are rendered governable in terms of their carbon footprint (Lövbrand and Stripple 2006, 2011; Paterson 2010). Again, these analyses provide invaluable insights into the political reason of carbon accounting: how climate change becomes an object of government that can actually be acted upon. However, the evaluation of these effects is sometimes inconclusive. For example, Paterson and Stripple contrast, with regard to individual carbon accounting, the virtues of their diagnostic approach with a more critical perspective. They argue that the critiques widely seen of such 'marketized' governance miss the mark. To operate by shaping and producing individuals as particular types of subjects (managing their carbon budgets, etc.) is precisely how power operates in neoliberalism. Rather than using individual freedom as a depoliticization strategy, it acts through channelling ways that individuals exercise their freedoms (Paterson and Stripple 2010: 359).

We are convinced that the notion of the political implicitly entailed in Foucault can be enriched with what is found in Laclau and Mouffe, allowing the governmentality analyst to evade the choice between fundamental criticism and simple diagnosis. In Foucault's fragmentary notes from his governmentality lectures, Michel Sennelart has unearthed an account that comes close to an explication of the political. There, Foucault remarked that 'nothing is political, everything could be politicised, everything may become political' (quoted in Sennelart 2007). We think that these brief remarks, perhaps not surprisingly, come very close to the concept of the political found in Laclau and Mouffe. Such a 'postfoundational' (Marchart 2007) understanding of the political is not so much interested in the everyday struggle within a separate sphere of 'politics' but rather in the political 'which sets out a particular, historically specific account of what counts as politics and defines other areas of social life as not politics' (Edkins 1999: 2).

Recall that for Laclau and Mouffe all discourses are the radically contingent outcome of a hegemonic struggle. Once a discourse is settled, it excludes some notions and ideas, and thus gives rise to an antagonistic outside as the necessary 'Other' of social life. This assumption allows for distinguishing between two layers within society (Laclau 1990: 33). The 'social' represents the sedimented structures of a given discourse which are taken for granted, where a particular discursive representation has become hegemonic, so that it is not questioned anymore. By contrast, 'the political' refers to those areas of social life where this implicitness has dissolved. This sphere is marked by contestation, instability and hegemonic struggle. In this sense, the political is a latent feature in all areas of social life which comes to the fore when their implicit foundations are called into question. Within the space of the political there are two types of operations. A politicizing treatment of a particular problem brings the underlying antagonisms to the fore and aims for the transformation of sedimented social structures. Depoliticization, by contrast, involves all counter-strategies which seek to conceal the contingency of reality, sew the gaps in hegemonic discourses and channel dislocations in such a way that fundamental social structures remain untouched.

Such an understanding of the political and depoliticization connects well to the Foucauldian concept of governmental power. Not only do both approaches seek to transcend the narrow institutions of the state in their accounts of power and politics, but governmentality can also be understood as a particular form of depoliticization. It is concerned with the 'right disposition of things' (Foucault 2007: 96), which, as long as it remains unchallenged, creates sedimented social structures – what Laclau and Mouffe call the social. Government seeks to manage grievances, problems and demands in a way that does not disturb the dominant order (Howarth 2009: 321). This forms the heart of the liberal doctrine of government. Ever since the emergence of liberal governmentality, it is obsessed with not 'governing too much' (Hindess 2005: 394). Liberalism, hence, 'identifies a domain outside "politics", and seeks to manage it without destroying its existence and autonomy' (Rose and Miller 1992: 180). By constituting this domain as an autonomous and 'natural' entity and managing its disturbances, governmentality depoliticizes it and hence constitutes it as part of the social. In the case of global climate change politics as a whole, Methmann (2011) has shown that, notwithstanding the missing success of the current climate regime – the (weak) reduction commitments of the Kyoto Protocol not being met, a successor agreement not in sight and global emissions trends that further increase despite the urgent need to reverse them – no one challenges the modus operandi. Instead, a 'global carbon governmentality' successfully stabilized the climate regime by reconciling existing antagonisms and creating climate change as the external threat.

As REDD+ is only just being implemented, one cannot say much about its effectiveness yet. But considering the outcome of previous efforts to tackle deforestation and the emerging complexity of the proposed REDD+ mechanism, it is highly questionable whether it will achieve the desired impact. If REDD+ is integrated into the carbon market and hence has the character of an offset mechanism it will be a zero-sum game at best, as the buyers of REDD+ certificates will use them to prolong activities that cause emissions (see Greenpeace et al. 2011). We can already see, however, additional depoliticizing effects REDD+ has on the global climate regime: it seems to soften one of the few antagonisms that occasionally resurfaces in global climate politics – the North-South divide. Ever since anthropogenic climate change has become an issue in international relations, one of the key questions has been who is responsible and hence should pay for mitigating it. Highlighting the 'common but differentiated responsibility' the UN Framework Convention on Climate Change (UNFCCC 1992: §3.1) notes that industrialized countries should take the lead in mitigating climate change. This view was widely shared among developing and developed countries during the 1990s and is also reflected in the Kyoto Protocol, which only includes reduction commitments for industrialized countries.[6] In the face of sharply rising emissions from emerging economies, however, more and more industrialized countries have started to argue that developing countries – particularly emerging economies – have to do their share as well and take on reduction commitments. Thus far, however, they have refused to do so – a central conflict in the current negotiations on a post-Kyoto agreement. REDD+ seems to be a win-win option for everyone and hence dissolves this antagonism: tropical developing countries – among them large emerging economies like Brazil and Indonesia – contribute to mitigating emissions by reducing deforestation. Because of incoming investments and expected co-benefits like poverty reduction, they simultaneously profit from doing so. Industrialized countries, which are expected to provide these investments, benefit from a cost-efficient reduction option to offset some of their own emissions. All that remains to be done, it seems, is to maximize the benefits of REDD+ by making sure that it is carefully designed and managed. Questions on justice and responsibility – the few politicized – that is, highly disputed – aspects left in the climate change discourse – are increasingly buried under discussions on complicated technocratic management issues like nested approaches, Reference Emissions Levels and buffer funds within the REDD+ negotiations.

The hegemony theory of Laclau and Mouffe, however, not only opens a perspective on practices of depoliticization, but also considers possible entry points for a repoliticization of sedimented discourses or governmental practices. As every social or discursive order is marked by a fundamental instability, there is always the possibility of resistance and opposition (Howarth 2009: 317).[7] A starting point for the study of opposition in hegemony is the category of discursive dislocations or dislocative moments. This concept refers to uncontrollable moments that break up the established discursive order, thereby questioning hegemonic knowledge and revealing the contingency of meaning. Dislocations can be found in critical discourse moments,

[6] Not every industrialized country was a strong supporter of the common but differentiated responsibility clause excluding developing countries from reduction commitments. The United States, for example, had been critical of it prior to the agreement of the UNFCCC. And while the United States signed the Kyoto Protocol in 1997, the US. Senate made it clear that it would not ratify the treaty, including reduction commitments, for the United States as long as developing countries are excluded from reduction commitments (U.S. Senate 1997).

[7] This coincides with the notion of productive power in the governmentality concept stating that power – other than domination – can only be exercised over free subjects who always have the possibility to resist (Foucault 1982: 790).

that is unforeseen events like 9/11 or Hurricane Katrina that cannot be represented and explained in given discourses and hence can grow into fundamental challenges.

The failure of COP 15 in Copenhagen presented a critical moment for the discourse on climate change as a whole – and hence for REDD+. However, the modus operandi of the international climate regime has not been substantially questioned or challenged and hence negotiations on REDD+ and other issues continue, and practices remain unchanged. So far REDD+ as a whole is only opposed by a small number of actors[8] whose position has been largely marginalized in the discourse. Among other points of criticism, they challenge claims that REDD+ will help protect biodiversity and provide livelihoods to forest dwellers. Instead, they articulate REDD+ as a threat to forest communities that cannot help protect biodiversity as it is exclusively focussed on carbon emissions (Friends of the Earth International 2008: 10; Cabello and Gilbertson 2010).

Beyond larger critical moments, subjects also experience dislocations in the everyday and routinized practices. The contradictions and irritations that are continuously brought about by such micro-dislocations can be utilized by political projects to challenge the existing regime of practices (Howarth and Glynos 2007: 105). This is the case if political projects are able to articulate the dislocation in a way that links them with their own political demands. Actors that are critical of REDD+ – such as the members of the No-REDD Platform[9] – are carefully monitoring the implementation process, trying to highlight and link cases where REDD+ has not brought about the promised impact. We have to wait for a further implementation of REDD+ to determine whether this results in a broader repoliticization of REDD+. Furthermore, the complicated measuring, monitoring and calculation procedures on which REDD+ is based entail many potential micro-dislocations as well. As Barry (2002) shows, scientific and technical procedures can not only have depoliticizing effects, but calculations that are inaccurate or do not produce clear results and technologies that fail cause dislocations which might then be used to repoliticize an issue.

Conclusion

This volume provides ample evidence that poststructuralist approaches have gained widespread currency in the study of global climate politics. And as many have argued, governmentality and hegemony – two important media of exchange within this conceptual economy – are actually two sides of the same coin. In this chapter we pursued a twofold target: to provide a new perspective on the third side of the

[8] On a country level, Bolivia has been an outspoken critic of REDD+. Among nonstate actors REDD+ is being opposed by some indigenous groups and smaller environmental and development NGOs (see for example Cabello, and Gilbertson 2010). Among the big international environmental NGOs, only Friends of the Earth has voiced fundamental criticism of REDD+.

[9] For details on the No-REDD Platform see http://noredd.makenoise.org.

governmentality-hegemony coin – the edge connecting heads and tails – by exploring how governmentality and hegemony connect in a purely poststructuralist theoretical framework, and to show the empirical value of this novel perspective for the study of REDD+ in global climate politics. This chapter showed that Laclau and Mouffe's reading of hegemony is best equipped for this endeavour as it shares a common meta-theoretical ground with Foucault's notion of governmentality. As the empirical investigation revealed, governmentality and hegemony can best play out their strength when they are not straitjacketed into a single theoretical corset. If applied in a work-sharing fashion, the combination of the two is a valuable tool with which to paint a more elaborate picture of the discursive constitution of climate governance that overcomes important shortcomings in parts of the climate governmentality literature.

First, the concept of hegemonic struggle and articulation provides the missing link between Foucault's everyday and mundane micro-practices of power, and broader macro-structures of governmental power. Using a bottom-up approach, our perspective revealed that current REDD+ discourse draws on a creative articulation of very diverse elements. It brings together an economic efficiency discourse and the idea of a comprehensive carbon management. It relates economic practices (environmental markets and direct payments) to scientific measurement practices that create a narrow focus on carbon (emissions), as opposed to considering the causes for these emissions. And it assembles a whole set of novel subjectivities – from project developers to carbon brokers, verifiers and bankers. Second, drawing on the notion of hegemony we sketched out why REDD+ today receives such broad support despite earlier resistances to address tropical deforestation under the Kyoto Protocol. This success is based on a hegemonic struggle including the rearticulation of avoided deforestation into REDD+, representing a broad range of political demands and the forging of a coalition around this novel, relabelled nodal point. Lastly, Laclau and Mouffe's notion of the political enables the study of how these types of discursive struggles, which render different social and environmental phenomena governable, are intrinsically linked to processes of politicization and depoliticization. We have shown how REDD+ contributes to the depoliticization of global climate politics by replacing debates about equity and justice with technocratic discussions and thus covering up existing antagonisms – such as the divide between North and South in international climate politics. These depoliticizing moments, however, are not set in stone. What might look like a simple technocratic management issue can cause dislocations that may repoliticize how deforestation or climate change are dealt with.

This chapter has provided only a first step to develop this analytical perspective, which tried to drill deeply into the piecemeal and ongoing discursive construction of climate governance. Further research has to follow. An interesting aspect would be a detailed analysis of the different ways oppositional articulations are integrated into and reformulated within hegemonic discourses. In this context, the role of social movements and NGOs in the articulation and stabilization of dominant governmentalities are of particular interest. Finally, more should be done to understand the process and implications of depoliticization in concrete and specific areas of global climate governance beyond the mere identification of a general trend towards a post-political governance.

References

Barnett, C. (2005). The Consolations of 'Neoliberalism'. *Geoforum* **36**(1), 7–12.

Barry, A. (2002). The Anti-Political Economy. *Economy and Society* **31**(2), 268–84.

Bäckstrand, K. and E. Lövbrand (2006). Planting Trees to Mitigate Climate Change: Contested Discourses of Ecological Modernization, Green Governmentality and Civic Environmentalism. *Global Environmental Politics* **6**(1), 50–75.

Boyd, E., E. Corbera and M. Estrada (2008). UNFCCC Negotiations (Pre-Kyoto to COP-9): What the Process Says about the Politics of CDM-Sinks. *International Environmental Agreements: Politics, Law and Economics* **8**(2), 95–112.

Bulkeley, H. and H. Schroeder (2011). Beyond State/Non-State Divides: Global Cities and the Governing of Climate Change. *European Journal of International Relations* **18**(4), 743–66.

Cabello, J. and T. Gilbertson (2010). *NO REDD!* Barcelona: Carbontrade Watch/Indigenous Environmental Network.

Collier, J. (2009). Topologies of Power: Foucault's Analysis of Political Government beyond 'Governmentality'. *Theory, Culture & Society* **26**(6), 78–108.

Dean, M. (1999). *Governmentality: Power and Rule in Modern Society*. London: Sage.

DNPI Indonesia (2010) *Indonesia's Greenhouse Gas Abatement Cost Curve*. Jakarta: Dewan Nasional Perubahan Iklim.

Dreyfus, H. L. and P. Rabinow (1982). *Michel Foucault. Beyond Structuralism and Hermeneutics*. Brighton: Harvester Press.

Dyer, N. and Counsell, S. (2010). *McREDD: How McKinsey 'Cost-Curves' are Distorting REDD*. London: The Rainforest Foundation, UK.

Edkins, J. (1999). *Poststructuralism & International Relations: Bringing the Political Back In*. Boulder, CO: Lynne Rienner Publishers.

Ekers, M. and A. Loftus (2008). The Power of Water: Developing Dialogues between Foucault and Gramsci. *Environment and Planning D: Society and Space* **26**(4), 698–718.

Eliasch, J. (2008). *Climate Change: Financing Global Forests – The Eliasch Review*. London: Earthscan.

Foucault, M. (1972). *The Archaeology of Knowledge*. London: Tavistock Publications.

Foucault, M. (1978). *The History of Sexuality: The Will to Knowledge*. New York: Random House.

Foucault, M. (1982). The Subject and Power. *Critical inquiry* **8**(4), 777–95.

Foucault, M. (1986). Nietzsche, Genealogy, History. In *The Foucault Reader*, ed. P. Rabinow. Hammondsworth: Penguin, 76–100.

Foucault, M. (2007). *Security, Territory, Population: Lectures at the Collège de France, 1977–1978*. Basingstoke: Palgrave Macmillan.

Foucault, M. (2008). *The Birth of Biopolitics: Lectures at the Collège de France, 1978–79*. Basingstoke: Palgrave Macmillan.

Friends of the Earth International (2008). *REDD Myths: A Critical Review of Proposed Mechanisms to Reduce Emissions from Deforestation and Degradation in Developing Countries*. Amsterdam: Friends of the Earth International.

Gill, S. (1995). Globalization, Market Civilization, and Disciplinary Neo-Liberalism. *Millennium-Journal of International Studies* **24**(3), 399–423.
Glynos, J. and D. R. Howarth (2007). *Logics of Critical Explanation in Social and Political Theory*. London: Routledge.
Greenpeace, Friends of the Earth and Rainforest Foundation (2011). *REDD+ and Carbon Markets: Ten Myths Exploded*. Amsterdam: Greenpeace.
Gupta, A., E. Lövbrand, E. Turnhout and M. J. Vijge (2012). In Pursuit of Carbon Accountability: The Politics of REDD+ Measuring, Reporting and Verification systems. *Current Opinion in Environmental Sustainability* **4**(6), 726–31.
Hare, B. (2000). *Should Forests and other Land Use Change Activities be in the CDM?* Amsterdam: Greenpeace International.
Hindess, B. (2005). Politics as Government. Michel Foucault's Analysis of Political Reason. *Alternatives* **30**, 389–413.
Howarth, D. R. (2009). Power, Discourse, and Policy: Articulating a Hegemony Approach to Critical Policy Studies. *Critical Policy Studies* **3**(3), 309–35.
IPCC (1990a). *First Assessment Report –Working Group One 'Scientific Assessment of Climate Change'*. Geneva: IPCC.
IPCC (1990b). *First Assessment Report – Working Group Three 'The IPCC's Response Strategies'*. Geneva: IPCC.
Jessop, B. (2007). From Micro-Powers to Governmentality: Foucault's Work on Statehood, State Formation, Statecraft and State Power. *Political Geography* **26**(1), 34–40.
Joseph, J. (2010). Poverty Reduction and the New Global Governmentality. *Alternatives: Global, Local, Political* **35**(1), 29–51.
Keller, R. (2010). Nach der Gouvernementalitätsforschung und jenseits des Poststrukstrukturalismus? Anmerkungen aus Sicht der wissenssoziologischen Diskursanalyse. In *Diskursanalyse Meets Gouvernementalitätsforschung*, ed. J. Angermüller, and S. van Dyk. Frankfurt a. M.: Campus Verlag.
Kulynych, J. J. (1997). Performing Politics: Foucault, Habermas, and Postmodern Participation. *Polity* **30**(2), 315–46.
Laclau, E. (1990). *New Reflections on the Revolution of Our Time*. London, New York: Verso.
Laclau, E. (1996). *Emancipation(s)*. London: Verso.
Laclau, E. (2005). *On Populist Reason*. London: Verso.
Laclau, E. and C. Mouffe (2001). *Hegemony and Socialist Strategy: Towards a Radical Democratic Politics*. London: Verso.
Laurance, W. F. (2007). A New Initiative to Use Carbon Trading for Tropical Forest Conservation. *Biotropica* **39**(1), 20–4.
Lövbrand, E. (2009) Revisiting the Politics of Expertise in Light of the Kyoto Negotiations on Land Use Change and Forestry. *Forest Policy and Economics* **11**(5–6): 404–12.
Lövbrand, E. and J. Stripple (2006). The Climate as Political Space: On the Territorialization of the Global Carbon Cycle. *Review of International Studies* **32**(2), 217–35.
Lövbrand, E. and J. Stripple (2011). Making Climate Change Governable: Accounting for Carbon as Sinks, Credits and Personal budgets. *Critical Policy Studies* **5**(2), 187–200.
Marchart, O. (2007). *Post-Foundational Political Thought*. Edinburgh: Edinburgh University Press.
McKinsey & Company (2010). *Impact of the Financial Crisis on Carbon Economics. Version 2.1 of the Global Greenhouse Gas Abatement Cost Curve*. London: McKinsey & Company.
Methmann, C. (2010). 'Climate Protection' as Empty Signifier: A Discourse Theoretical Perspective on Climate Mainstreaming in World Politics. *Millenium: Journal of International Studies* **39(2)**, 345–72.
Methmann C. (2011). 'We are all green now' – Hegemony, Governmentality and Fantasy in the Global Climate Polity. Unpublished Ph.D. thesis, University of Hamburg.

Methmann, C. and D. Rothe (2012). Politics for the Day after Tomorrow: The Logic of Apocalypse in Global Climate Politics. *Security Dialogue* **43**(4): 323–44.
Mirowski, P. and D. Plehwe (2009). *The Road from Mont Pèlerin: The Making of the Neoliberal Thought Collective*. Cambridge, MA: Harvard University Press.
Mouffe, C. (1979). Hegemony and Ideology in Gramsci. In *Gramsci and Marxist Theory*, ed. C Mouffe. London: Routledge.
Oels, A. (2005) Rendering Climate Change Governable: From Biopower to Advanced Liberal Government? *Journal of Environmental Policy & Planning* **7**(3), 185–207.
Okereke, C., H. Bulkeley and H. Schroeder (2009) Conceptualizing Climate Governance beyond the International Regime. *Global Environmental Politics* **9**(1), 58–78.
Paterson, M. (2010) Legitimation and Accumulation in Climate Change Governance. *New Political Economy* **15**(3), 345–68.
Paterson, M. and J. Stripple (2010) My Space: Governing Individuals' Carbon Emissions. *Environment and Planning D* **28**, 341–62.
Rose, N. and P. Miller (1992). Political Power beyond the State: Problematics of Government. *British Journal of Sociology* **43**(2) 173–205.
Rutherford, S. (2007). Green Governmentality: Insights and Opportunities in the Study of Nature's Rule. *Progress in Human Geography* **31**(3), 291–307.
Sennelart, M. (2007). Course Context. In *Security, Territory, Population. Lectures at the Collège de France 1977–78*, ed. M. Foucault. Basingstoke: Palgrave Macmillan.
Staten, H. (1984). *Wittgenstein and Derrida*. Lincoln: University of Nebraska Press.
Stephan, B. (2011). The Power in Carbon: A Neo-Gramscian Explanation for the EU's Adoption of Emissions Trading. *Global Transformations towards a Low Carbon Society*, **4**.
Stephan, B. (2012a). Bringing Discourse to the Market: The Commodification of Avoided Deforestation. *Environmental Politics* **21**(4), 621–39.
Stephan, B. (2012b). From Pariah to Messiah: Avoided Deforestation in Global Climate Governance. Paper presented on April 4th, 2012 at the *International Studies Association's Annual Convention*, San Diego, USA.
Stephan, B. (2013). How to Trade Not Cutting Down Trees. In *(De)Constructing the Greenhouse: Interpretative Approaches to Global Climate Governance*, ed. C. Methmann, D. Rothe and B. Stephan. London: Routledge, 57–71.
Stephan, B. (forthcoming). *Governing the Forest Frontier*.
Stern, N. (2006). *The Economics of Climate Change: The Stern Review*. Cambridge: Cambridge University Press.
Swyngedouw, E. (2010). Apocalypse Forever? Post-political Populism and the Spectre of Climate Change. *Theory, Culture & Society* **27**(2–3), 213–32.
Torfing, J. (2005). Discourse Theory: Achievements, Arguments, and Challenges. In *Discourse Theory in European Politics: Identity, Policy, and Governance*, ed. D. Howarth and J. Torfing. New York: Palgrave Macmillan, 1–32.
UNFCCC (1992). *The Framework Convention on Climate Change*. Rio de Janeiro: United Nations Conference on Environment and Development.
UNFCCC (2007). *FCCC/CP/2007/6/Add.1 – Report of the Conference of the Parties on its thirteenth session, held in Bali from 3 to 15 December 2007*. Bonn: UNFCCC.
UNFCCC (2010). *FCCC/CP/2010/7/Add.1 – The Cancun Agreements*. Bonn: UNFCCC.
U.S. Senate (1997). *105th CONGRESS. 1st Session. S. RES. 98 – Byrd-Hagel Resolution*. Washington, DC: U.S. Congress.
Voß J-P. (2007). Innovation Processes in Governance: The Development of Emissions Trading as a New Policy Instrument. *Science and Public Policy* **34**(5), 329–43.
Weir, L., P. O'Malley and S. Clifford (1997). Governmentality, Criticism, Politics. *Economy and Society* **26**(4), 501–17.

4

The Limits of Climate Governmentality

Carl Death

Introduction

As this book shows, the Foucauldian-inspired governmentality literature is becoming an increasingly popular and productive resource for attempts to theorize, analyse and critique contemporary environmental and climate governance. By drawing attention to forms of power that work through 'rule at a distance' and the 'conduct of conduct', the eco-governmentality literature has persuasively illustrated the interconnectedness of power and freedom; the importance of fields of visibility, regimes of knowledge, techniques and technologies; and the production of subjects in contemporary environmental governance. In the field of climate change, for example, recent analyses have powerfully drawn attention to the constitution of 'the climate' as a field of visibility and the rationalities and practices of government within that field (Oels 2005; Bäckstrand and Lövbrand 2006; Lövbrand, Stripple and Wiman 2009; Okereke, Bulkeley and Schroeder 2009; Whitehead 2009; Bulkeley and Newell 2010: 14; Paterson and Stripple 2010; Swyngedouw 2010; Methmann 2011).

However, the burgeoning literature on environmental governmentality has now reached the stage where (a) critiques are beginning to accumulate, and (b) differences of approach and analysis are increasingly evident amongst those drawing on the concept. For these reasons, theorists of climate governmentality need to be more explicit about their usage of key terms, and aware of some of the limits of these approaches. Of course, as any good student of Foucault is well aware, attempts to inscribe boundaries of knowledge and acceptable and authoritative ways of speaking 'in the true' are deeply political moves, and this chapter is not an attempt to define some usages of governmentality as correct and others as illegitimate, but rather an attempt to make reflexive conversations about theoretical and methodological practices and assumptions more explicit and visible.

To this end the chapter begins by briefly sketching one view of the way the eco-governmentality field has developed over the past fifteen years. There has been a shift from a focus on statist and disciplinary governmentalities by what might be termed broadly 'post-Marxist' scholars; towards an emphasis on the production of self-governing advanced liberal subjects; followed by a renewed interest in practices of sovereignty and violence. The next section identifies three prominent critiques or limits of the applicability of governmentality as it has been applied to climate governance: regarding the *globality* of climate governmentality; regarding the *agency* of climate governmentality; and regarding the *ethics* of climate governmentality. This chapter argues that in these three areas climate governmentality faces certain challenges. Yet the chapter concludes that an 'analytics of government' approach can be a very flexible tool, and with greater theoretical precision it remains one of the most exciting and illuminating approaches to climate politics.

Eco-Governmentality: Statism, Neoliberalism and Liberal Violence

Michel Foucault's work on governmentality – in which he showed how the population emerged as the focus of government in fields like political economy, urban planning, grain supplies and crime (Foucault 2007; see also Gordon 1991; Dean 1999; Miller and Rose 2008) – was first applied to the field of environmental politics in the 1990s, at a time when interest in global environmental governance and issues like climate change was sharply rising (Ferguson 1994; Escobar 1995: 192–211; Kuehls 1996). Whilst the majority of political scientists working on environmental issues were primarily concerned with solving environmental problems, either through technical and institutional means or through more radical diagnoses of the underlying structural production of environmental harms, the concept of eco-governmentality was deployed to critically interrogate the political risks and dangers of new forms of environmental governance.[1]

One of the trail blazers of this new literature was Timothy Luke. His work, along with the work of other early eco-governmentality scholars, was animated by the post-Marxist concern that modern techniques of government had brought an active 'nature' under human management as the passive 'environment', and had reincorporated the environmental movement within capitalist and statist power relations (Luke 1995, 1997, 1999). For example, Luke read the work of a prominent environmental nongovernmental organization (NGO), the WorldWatch Institute, as a 'continuous attempt to reinvent the forces of Nature in the economic exploitation of advanced technologies' (1995: 58). For Luke, such NGOs 'operate as a green panopticon, enclosing Nature in rings of centered normalizing super-vision', in order to dominate, exclude and repress people and the environment (1995: 77). In a similar vein, he suggests elsewhere that the environmental movement is the latest in a 'solid series of statist practices beginning in the eighteenth century' (1999: 122), through which consumers 'must be forced to be functional in accord with the regulatory goals of geo-environmentalizing bureau-

[1] For a snapshot of the field at this time see Vogler and Imber (1996), and contrast their work with that of Darier (1999a).

cracies' (1999: 134). He argues that, under such a global capitalist governmentality, 'the health of global populations as well as the survival of the planet itself necessitate that a green spreadsheet be draped over Nature, generating an elaborate ecomarket of global reach and scope' (1997: 94). The model of eco-governmentality Luke exposes here is a repressive and exploitative one which dominates people and nature in the interests of political elites, science and capital.

In contrast to this quite particular view of eco-governmentality as a statist, repressive, dominating form of power, which is based on a conception of power, the state and capitalism drawn from the Marxist tradition, other invocations of eco-governmentality have rather emphasized how environmental governance regimes operate through the conduct of conduct, at a distance from traditional centres of authority, through the production of particular kinds of free subjects. Akhil Gupta, for example, argued that new 'postcolonial spaces' are being created between the global and the local, where 'the "unbundled" space in which these forms of governance are exercised creates its own possibilities for opposition to coalesce' (1998: 326). Gupta drew attention to the biopolitical and disciplinary tendencies within the environmental governance – the bringing of 'life and its mechanisms into the realm of explicit calculations' (Foucault 1998: 143) – but other accounts have seen a clearer distinction between disciplinary and advanced liberal governmentality. For example, Arun Agrawal explains 'why, when, how, and in what measure people come to develop an environmentally orientated subject position' through community-based environmental management schemes in rural India (2005: 2–3). He describes in detail how forests and local communities have been governed in Kumaon between 1850 and 2000 through the creation of environmental subjects, 'governmentalized localities' and 'regulatory communities' (2005: 2–3). As such, rather than presenting the disciplinary and repressive effects of eco-governmentality, he is interested in the ways the 'very individuality that is supposed to be constrained by the exercise of power may actually be its effect' (2005: 217).

Angela Oels displays a similar sensitivity to the processes through which actors are empowered and subjects created in her analysis of the changing ways the climate has been rendered governable. She suggests that 'shifts in the discourses and practices of climate change should be understood in terms of a shift of governmentality from bio-power to advanced liberal government in Western industrialized countries' (2005: 186).

> Climate change as framed by bio-power creates the basis for justifying far-ranging policy interventions and even the extension of state power in the name of 'survival' of life on planet Earth. Advanced liberal government, on the other hand, renders climate change governable as an issue of state failure requiring market-based solutions or the creation of markets. (2005: 201)

The form of governance represented by the Kyoto Protocol and carbon trading mechanisms, Oels argues, relies on 'market-based solutions that spur technological innovation and economic growth' (2005: 201), thereby constructing the broader population as active participants in governance via markets, rather than passive objects of discipline or regulation (see also Bäckstrand and Lövbrand 2006).

Whilst these usages of eco-governmentality have focussed on the neoliberal commitment to governing through specific practices of freedom, and have generally sought to draw a distinction between governmental and sovereign power relations, more recent work on governmentality has reemphasized the integral role of violence within liberal power relations (Peluso and Watts 2001: 6; Watts 2002; Hindess 2005; Dean 2007; Nadesan 2008). Majia Nadesan's analysis of governmentality and biopower in a range of contexts from health care to the U.S. war on terror indicates that 'although liberal governmentalities profess resistance to centralized and repressive power, they simultaneously rely on them to securitize everyday life and to extend the liberal freedoms of the market' (2008: 210). In another example, the 2002 Johannesburg World Summit on Sustainable Development was a site where techniques of environmental governmentality both promoted voluntary public-private partnerships and self-governing notions of environmental and social responsibility, as well as the division of civil society participants into 'responsible' and 'irresponsible' actors. Those deemed uncivil and irresponsible were overtly marginalized, repressed and coerced (Death 2010). In the field of climate governmentality, Chris Methmann argues that 'sovereignty and governmentality must not be thought of as mutually exclusive', instead highlighting the ways climate governmentality works through sovereign states and the ever-present threat of force and coercion (2011: 3).

These trends and developments within the eco-governmentality literature suggest a pluralist field of analysis with many different nuances. Some authors have used governmentality almost synonymously with neoliberalism, whereas others see it as more of a framework for analysis of different power relations. The relationships between specific governmentalities and capitalism, the state, resistance and freedom have also been framed in different ways. In part this pluralism stems from Foucault's own ambiguous usage of governmentality. In his famous lecture of 1 February 1978, Foucault describes governmentality in three ways: as a specific form of power 'that has the population as its target, political economy as its major form of knowledge, and apparatuses of security as its essential technical instrument'; as the historical trend towards the predominance of 'governmental' over sovereign and disciplinary forms of power; and finally as what he terms the 'governmentalization of the state' from medieval to early modern and Enlightenment Europe (2007: 108–9). In the following year's lecture series, moreover, his view of governmentality became even broader, and suggests that it should 'be considered simply as a point of view, a method of decipherment' (2008: 186). This difference between viewing governmentality as either a specific form of power relation or 'a proposed analytical grid for these relations of power' (2008: 186) is an important distinction which may further refine the precision and utility of the concept of climate governmentality. This distinction can also help respond to prominent critiques of the concept. The rest of this chapter discusses three alleged limitations of the field organized around critiques regarding the globality, agency and ethics of a climate governmentality approach.

Global Climate Governmentality

The first critique of the climate governmentality literature to be addressed emerges from a broader set of concerns regarding the notion of 'global governmentality' (Larner and Walters 2004; Lipschutz and Rowe 2005; Neumann and Sending 2010). The claim that Foucault's concept of governmentality, which he primarily used in the context of domestic politics in Western Europe, can be extended both to non-Western parts of the world and to the realm of global governance, has attracted the criticism that, rather than describing genuinely global power relations, it actually only describes a rather limited set of Western neoliberal power relations (Joseph 2009). Given the emphasis in much of the governmentality literature on free markets and the construction of liberal civil societies, critics have suggested that governmentality analyses risk producing a rather anodyne and liberal reading of 'flat' global politics, missing out on the continued violence, coercion and materiality of power relations in most of the rest of the world (Hindess 2005; Dean 2007; Selby 2007; Joseph 2010). If governmentality is used to refer to neoliberal power relations, then the limits of neoliberalism also represent the limits of a governmentality analysis.

In one respect, the specific field of climate governmentality has quite a strong riposte to this critique: climate change is a quintessentially global problem, as greenhouse gas emissions wherever they are produced contribute to the problem (although not all parts of the world have equal emissions), all parts of the world will be affected by the phenomenon of climate change (although not equally), and the emerging architecture of climate governance has attempted to extend 'control to the entire planet' (Bäckstrand and Lövbrand 2006: 54; see also Methmann 2011: 3). This globality can sometimes be overemphasized, however, and there is considerable room within the critical literature on climate change for more nuanced readings of the scales and spaces of climate change, and attention to the spatial and social limits of climate governmentalities.[2] Climate governmentality research tends to focus on 'global' institutions and discourses – such as the UNFCCC and Kyoto mechanisms or the geosciences (Lövbrand et al. 2009; Lövbrand and Stripple 2011) – which are dominated by Western centres of power and knowledge, or individualized practices and discourses – such as carbon offsetting or consumption monitoring (Paterson and Stripple 2010) – with material largely confined to advanced liberal societies. In contrast, as an article by Eva Lövbrand and Johannes Stripple has persuasively argued, one of the effects of discourses and practices of global carbon cycle governance is

[2] See Methmann's discussion of CDM projects in China in direct response to Joseph's critique (2011: 15–16).

a renewed interest in land use, forestry and national sinks, which 'signifies not just the importance of sovereign political spaces, but also the importance of territoriality as such, that is as a geographical expression of social power' (Lövbrand and Stripple 2006: 218).

One of the dangers of too quick proclamations of the globality of climate governmentality is what might be termed adopting the 'programmer's view' (Dean 2007: 83). Whilst dominant climate institutions and discourses promote global fields of visibility, rationalities and practices, these are invoked and articulated and reproduced in highly varied and sometimes contradictory manners (Rydin 2007). They have their silences, invisibilities and shadows, as well as the smooth extension of global governmentality. For example, the language of environmental responsibility and good citizenship has become, in the Bolivian context, the language of *buen vivir* and the rights of Mother Earth (Gudynas 2011). In another example, the practices of civic engagement, participatory democracy and multistakeholder dialogues have become, for many environmental activists at major summits, a set of repertoires of mass marches, sit-ins and denunciations of political leadership (Death 2011: 12).

However, by regarding a governmentality approach as more of an analytical lens for examining the practices and rationalities of different forms of government, rather than as a specific form of neoliberal or global power relations, it can have considerable potential in mapping the different scales and alternative articulations of local climate governmentalities. An excellent example of this potential is Arun Agrawal's insightful account of the creation of governmentalized localities in village forest management schemes in India, which reveals how environmental subjects and new conceptions of nature itself have been produced through forms of environmental management. Crucially, this sensitivity to local politics and alternative rationalities is enabled by what he describes as a 'selective' use of Foucault 'as a source of possible provocations leading toward a framework for environmental political analyses' (2005: 216). Rather than using governmentality as a synonym for neoliberalism (cf. Joseph 2009: 416, 2010: 227), and thereby limiting its usefulness to those parts of the world in which neoliberal power relations are hegemonic, Agrawal argues that 'treating governmentality as an analytical construct exposes its potential usefulness in investigating the nature of institutionalized power outside of Western modernity' (2005: 219).

Viewing governmentality as an analytical construct draws upon the work of Mitchell Dean, who sets out an 'analytics of government' which charts the fields of visibility, regimes of knowledge, techniques and technologies and the production of subjectivity that various mentalities and rationalities of government employ (1999: 23). For Agrawal, this leads to a study of environmentality defined as 'the knowledges, politics, institutions and subjectivities that come to be linked together with the emergence of the environment as a domain that requires regulation and protection' (2005: 226). Locally grounded, empirically rich and theoretically informed studies such as this – the Clayoquot Sound project is another prominent example (Magnusson and Shaw 2003) – which explore how global discourses and institutions reshape and are reshaped by local politics, are one of the greatest strengths of the governmentality lens. These can help to put 'traditional distinctions between local and global, small and large, domestic and international – and much else – into serious question' (Magnusson 2003: 1).

Agency and Climate Governmentality

Another frequent line of critique of the governmentality literature, and an area in which greater clarification of climate governmentality arguments would be useful, relates to questions of agency and subjectivity. For example, David Chandler (2006: 15–18) argues that governmentality analyses of global governance have misinterpreted the interests and motivations of Western agencies when it comes to development. The response is often that governmentality theorists seek to focus on the *how* of power rather than the *why* (Dean 1999: 2; Miller and Rose 2008: 6). Yet this raises its own problems, as it is frequently argued that focussing on the 'descriptive how' to the neglect of the 'causal why' means that important questions about the constitution and shape of global politics remain unanswered by governmentality theorists (Selby 2007: 337; Joseph 2010: 242). McCarthy and Prudham, for example, suggest that it is essential 'to remain focused on the need to identify specific winners and losers' in terms of class coalitions and interest-based politics in the politics of environmental governance (2004: 280).

Theorists of climate governmentality need not shy away from such challenges. Causal conversations are *not* one of the limits of a governmentality analysis. Analysis of how forms of governance work in practice are an important contribution to showing why world politics looks the way it does. On the other hand, governmentality theorists are right to be more wary of discussions of interests, motivations and unproblematic ascriptions of agency. Foucault's reluctance to frame his work in terms of 'why' questions can be interpreted, at least in part, as a refusal of the notion of an unproblematic subject which preexists power relations (Foucault 2000a,b). Rather, one of the central analytical insights of the climate governmentality literature has been the way power relations produce self-governing subjects and new forms of agency (Agrawal 2005). Foucault's conception of liberal governmentality noted how the population became both the subject and object of government. This has been exceptionally well illustrated by Paterson and Stripple with regard to new technologies for governing the carbon emissions of individuals. They assert that 'there is an emergent governmentality that entails the "conduct of carbon conduct" through moulding and mobilizing a certain subjectivity (the individual as concerned carbon emitter) to govern his or her own emissions in various ways – as counters, displacers, dieters, communitarians, or citizens' (2010: 342).

This poses an interesting additional problem in the area of climate governmentality, however, as well as for environmental governmentality more broadly. Whilst it is easier to see how the poor, the criminal, the unemployed or the sick, for example, can become active subjects through techniques of liberal governmentality, it is less easy to see how 'the climate' or 'the environment' becomes an active self-governing subject. To what degree do we regard the climate or the natural environment as self-governing – and does this take us into the realm of Gaia theory, self-regulating ecosystems or actor-network theory (ANT) (Cudworth and Hobden 2011: 145)? It is notable that most work on environmental governmentalities has concentrated on showing how local communities, indigenous groups, NGOs and others have been integrated into self-government, rather than nonhuman nature (Bryant 2002).

Yet work by Eva Lövbrand and others on the rise of earth system science provides one way of resolving this tension. Particularly important here is the concept of the Anthropocene: 'a new geological époque dominated by human activity' (Lövbrand et al. 2009: 10). According to this argument, the rise of 'Earth System Science is involved in the making of a new kind of population, namely humankind for the first time understood as a major geological force' (Lövbrand et al. 2009: 8). This suggests that the Foucauldian emphasis on self-government and the production of new forms of subjectivity that is at the heart of the governmentality literature might be applicable to an understanding of the carbon cycle which does not rely on a human-nature dualism. Methmann reaches a similar conclusion, and suggests that it is the way that this earth system science pictures the carbon cycle in terms of a 'coupled human and ecological system' which enables us to properly apply the concept of governmentality to the climate (2011: 9–10).

> The key epistemic figure that makes it possible to think of human life in terms of carbon is that of the 'carbon footprint'. It integrates individual activities (e.g. flying) and lifestyles into the visible space of the carbon cycle, and can be extended to the impact of firms, communities, NGOs or entire nation states. Governing carbon, in this sense, is also governing *through* carbon. (2011: 10)

This ascription of subjectivity and even agency to the carbon cycle, rather than solely human actors, reveals why there has been so much interest in the intersections between ANT and governmentality. Future work on climate governmentality can therefore learn much from complexity theory and posthuman critical theory (Cudworth and Hobden 2011). Importantly, this illustrates how questions of agency and causality are not a limit of governmentality approaches, but should rather be seen as an area where such approaches might make some of their most important contributions.

The Ethics of Climate Governmentality

The third area in which governmentality analyses have attracted significant criticisms relates to the ambiguous ethical stance of Foucauldian political theory in general, and the governmentality literature in particular. Without a clear normative foundation, some have argued, Foucault's work can be used to justify almost any ethical position and most often ends up supporting a rather apathetic conservatism (Fraser 1985; Cohen and Arato 1994: 270–98; McCarthy and Prudham 2004: 280). Many progressives and leftists remain therefore deeply frustrated by Foucault's critique of Enlightenment ideals and emancipatory projects. Moreover, as Neil Levy points out, 'from the perspective of the Green movement … poststructuralism must appear dangerously relativistic and abstract, absorbed in arguments about representation while forests fall' (1999: 203). Foucault's hostility towards appeals to ethical foundations, especially those grounded on some notion of *nature* (human or otherwise), makes him an unlikely theorist to be at the forefront of a new climate politics (Darier 1999b).

This ethical ambiguity can be seen quite clearly in the divisions within the eco-governmentality literature as to whether the forms of power they address are positive or negative. Whilst authors in the post-Marxist tradition like Timothy Luke (1999), Ronnie Lipschutz and James K. Rowe (2005) and Erik Swyngedouw (2010) are stridently critical of new forms of governmentality, many others remain more ambiguous. Eric Darier's discussion of the Canadian 'Green Plan' in the 1990s, for example, concludes that 'one cannot answer the abstract question whether the Green Plan is good or bad' (1996: 603). By showing how the Green Plan sought to produce environmentally responsible citizens who would take greater ownership of their actions, he suggests that 'the Green Plan could be one of the conditions for the subsequent emergence of radically different green subjectivities and a step toward the greater disciplining of the population on global scale' (1996: 603). Likewise, Paterson and Stripple conclude that in the context of their analysis of the individualization techniques represented by management of personal carbon emissions, 'it is difficult to envisage how limiting global warming to 2°C (emission reductions by 50–75% below 1990 levels by 2050) might be achieved without such an intensive, managerial (and self-managerial) effort' (2010: 359).

Whilst Foucault's own work – and much of the climate governmentality literature inspired by him – might be ethically ambiguous (in the sense of eschewing firm ethical foundations), it is certainly not normatively neutral (Brown 1994). Indeed much of Foucault's later work explicitly focussed on ethical relations, and there is much here that could be more explicitly introduced into climate governmentality studies (Foucault 2010). Although there is not space in this chapter to develop these ideas more fully, this section concludes by noting three aspects of Foucault's work on ethics which could be productive for reflecting on the normative politics of climate change.

The first aspect is the broadly critical stance of Foucauldian theory in general and governmentality in particular. This entails a scepticism towards established power relations, a cynicism of the intellect and 'an ethic of permanent resistance' (Simons 1995: 87). Foucault observed that all institutions operate on the basis of some kind of rationality, and all forms of government have some kind of thought behind them. 'Criticism', he argued, 'consists in uncovering that thought and trying to change it: showing that things are not as obvious as people believe, making it so that what is taken for granted is no longer taken for granted. To do criticism is to make harder those acts which are now too easy' (2000c: 456). As Louise Amoore argues, for Foucault 'critique is not that which seeks out resolution, reconciliation or the smoothing out of difficulty, but rather that which discomforts and unsettles one's sense of certainty' (2008: 274). Colin Gordon suggested that Foucault's political stance was a kind of 'moral judo' or agonism, where government and ethics and politics are in continual tussle and interplay (1991: 48). Much of the climate governmentality literature indeed reflects the implications of this: a scepticism towards the mechanisms, institutions and power relations of climate change that stands in marked contrast to more problem-solving literatures. Intergovernmental summit negotiations (Death 2011), the Clean Development Mechanism (Methmann 2011) and personal carbon offsetting (Paterson and Stripple 2010) are all examples of where governmentality analyses have sought to actively destabilize the accepted wisdom within climate governance circles; an accepted wisdom which usually concludes that we need many more of these mechanisms and on a much larger scale.

The second aspect stems from Foucault's contribution to the study of power relations, and in particular his description of a spectrum ranging from strategic relations which are fluid, reversible and ever present; through techniques of government for regulating the conduct of others; to states of domination, in which power relations have become sedimented (1997: 283; see also Dean 1999: 35). Whilst it is important not to try and elevate this section of his work into a general normative framework, Foucault tended to be broadly in favour of the attempt to pursue structures of power relations that were flexible and reversible (1997: 299; see also Hindess 1996: 102). He clarified that whilst he did not believe it was possible to escape power relations altogether, we should attempt 'to acquire the rules of law, the management techniques, and also the morality, the ethos, the practice of the self, that will allow us to play these games of power with as little domination as possible' (Foucault 1997: 298). In the field of climate governance, this could be reflected in a willingness to consider and advocate forms of governance outside multilateral legal regimes, and encompassing a wide range of individual, community, civic, regional, national and continental initiatives (Okereke et al. 2009; Bulkeley and Newell 2010).

The third and final aspect noted here draws on the themes of Foucault's later lecture series, *The Government of Self and Others*, which was far more explicitly focussed on the possibility of the care of the self as an ethical project, and the role of practices of truth-telling and free speech (2010). The care of the self and the aesthetics of working on the self were a significant theme of this later work, where he noted that 'to constitute oneself as a governing subject implies that one has constituted oneself as a subject who cares for oneself' (1997: 293). In this light Darier proposes that Foucault can help us towards a possible green 'aesthetics of existence' (1999b: 219). This would be, according to Darier, a perpetual process of self-reflection and self-examination of one's place in relation to 'the environment' (1999b: 227). This is the most underexplored dimension of the climate governmentality literature, but there are exciting possibilities of drawing on long traditions of green thought, as well as new philosophies of 'posthumanism' (Cudworth and Hobden 2011) and 'becoming' (Connolly 2011), in exploring alternative climate governmental rationalities which might present new forms of identity and subjectivity and new ways to work on the self.

Each of these aspects provides productive and fertile soil for cultivating reflections about the ethics of climate governmentality, even if the connections and links are always contingent and fragile rather than robust and declamatory. On which note, it is important to observe that Foucault's ethical cynicism should not be read as political passivity (Edkins 1999: 45). Foucault was very clear that he advocated a hyper and optimistic activism in political life, and this too is worthy of reflection from a climate governmentality perspective. Nikolas Rose declares that the 'political function' of governmentality studies must be 'to strengthen the resources available to those who, because of their constitution as subjects of government, have the right to contest the practices that govern them in the name of their freedom' (1999: 60). How well we succeed or fail in achieving this must be one of the yardsticks by which we assess the field of climate governmentality studies.

Conclusion

The literature on climate governmentality now constitutes a significant and vibrant subfield within global environmental politics. As it has grown over the past two decades, significant differences have emerged within this literature, and important critiques have been posed from outside. As such it is important for climate governmentality scholars to begin to address more systematically some of their principles, assumptions and limitations. This chapter has aimed to contribute to this debate by highlighting a key conceptual difference between those who treat governmentality synonymously with neoliberalism, and those who regard governmentality as more of an 'analytical perspective for relations of power in general' (Foucault 2007: 388).

These two uses of governmentality are very closely linked in both Foucault's work and in the work of those who have drawn upon him. Indeed, it is quite plausible that whilst governmentality provides broader analytical tools for power relations in general, these are more adept at dissecting neoliberal governmentalities which work primarily through free markets and the creation of economically rational individuals. On the other hand, the ways governmentality is defined and used has consequences for some of the critiques directed at it. I have argued here that a broader approach to governmentality as an analytical construct is better suited to addressing both the global dimensions of climate governmentality, as well as the local specificities of how climate politics plays out in different locations at different times. Secondly, treating governmentality in this manner rather than as a synonym for neoliberalism allows a more explicit focus on some of the most important insights of the field: the creation of new forms of subjectivity and new types of agency. Rather than the frequently alleged limitation of governmentality in only addressing 'how' questions, this focus on the production of identities is an important part of causal conversations about why climate politics works the way it does. Finally, a common criticism of Foucauldian approaches has been their ambiguous ethical and normative stance. In this respect adopting an 'analytics of government' approach may be less clearly normative than the condemnation of neoliberalism associated with post-Marxist authors like Luke, Lipschutz and Swyngedouw, but there are perhaps greater opportunities for drawing upon the more nuanced ethical discussions associated with Michel Foucault's later work.

These ethical questions pose possibilities for the future directions of climate governmentality research. If the lack of an explicitly ethical dimension has been one of the limitations of the literature so far, then perhaps integrating Foucault's notion of the 'care of the self' into an analysis of climate governmentalities might be one productive avenue. In this regard there is much work to be done in bringing the diverse ways climate change is experienced and governed around the world, through myriad different conducts of conduct, and counter-conducts, into a climate governmentality approach. As William Connolly reflects, perhaps the point of new forms of climate conduct might be to:

> Make a direct difference through our conduct, to open us to new experiences that might alter our relational sensibilities further, to unscramble role assumptions assumed by others, to form operational connections with others from which larger political movements might be generated, and to make new connections with noble role warriors in other regions and walks of life to enlarge the space and visibility of positive action. (2011: 145)

Reflecting on these questions may provide ways for the climate governmentality literature to transcend some of its critiques and limitations, as well as to chart new futures for progressive climate politics.

References

Agrawal, A. (2005). *Environmentality: Technologies of Government and the Making of Subjects*. Durham, NC: Duke University Press.

Amoore, L. (2008). Foucault against the Grain. *International Political Sociology* **2**(3), 274–6.

Brown, C. (1994). "Turtles all the way down": Anti-Foundationalism, Critical Theory and International Relations. *Millennium* **23**(2), 213–36.

Bryant, R. L. (2002). Non-Governmental Organisations and Governmentality: 'Consuming' Biodiversity and Indigenous People in the Philippines. *Political Studies* **50**(2), 268–92.

Bulkeley, H. and P. Newell (2010). *Governing Climate Change*. Abingdon: Routledge.

Bäckstrand, K. and E. Lövbrand (2006). Planting Trees to Mitigate Climate Change: Contested Discourses of Ecological Modernization, Green Governmentality, and Civic Environmentalism. *Global Environmental Politics* **6**(1), 50–75.

Chandler, D. (2006). *Empire in Denial: The Politics of State-Building*. London: Pluto Press.
Cohen, J. L. and A. Arato (1994). *Civil Society and Political Theory*. Cambridge, MA: MIT Press.
Connolly, W. E. (2011). *A World of Becoming*. Durham, NC: Duke University Press.
Cudworth, E. and S. Hobden (2011). *Posthuman International Relations: Complexity, Ecologism and Global Politics*. London: Zed Books.
Darier, E. (1996). Environmental Governmentality: The Case of Canada's Green Plan. *Environmental Politics* **5**(4), 585–606.
Darier, E. (1999a). *Discourses of the Environment*. Oxford: Blackwell.
Darier, E. (1999b). Foucault against Environmental Ethics. In *Discourses of the Environment*, ed. E. Darier. Oxford: Blackwell, 217–40.
Dean, M. (1999). *Governmentality: Power and Rule in Modern Society*. London: Sage.
Dean, M. (2007). *Governing Societies: Political Perspectives on Domestic and International Rule*. Maidenhead: Open University Press.
Death, C. (2010). *Governing Sustainable Development: Partnerships, Protests and Power at the World Summit*. Abingdon: Routledge.
Death, C. (2011). Summit Theatre: Exemplary Governmentality and Environmental Diplomacy in Johannesburg and Copenhagen. *Environmental Politics* **20**(1), 1–19.
Edkins, J. (1999). *Poststructuralism and International Relations: Bringing the Political Back In*. London: Lynne Rienner.
Escobar, A. (1995). *Encountering Development: The Making and Unmaking of the Third World*. Princeton, NJ: Princeton University Press.
Ferguson, J. (1994). *The Anti-Politics Machine: 'Development', Depoliticization and Bureaucratic Power in Lesotho*. Minneapolis: Minnesota Press.
Foucault, M. (1997). On the Ethics of Concern for Self as a Practice of Freedom. In *Ethics, Subjectivity and Truth: Essential Works of Foucault 1954–1984*, ed. P. Rabinow, tr. R. Hurley et al. New York: The New Press, 281–301.
Foucault, M. (1998). *The Will to Knowledge: The History of Sexuality 1*. Tr. R. Hurley. London: Penguin.
Foucault, M. (2000a). Truth and Power. In *Power: Essential Works of Foucault 1954–1984*, ed. J. D. Faubion, tr. R. Hurley. New York: The New Press, 111–33.
Foucault, M. (2000b). The Subject and Power. In *Power: Essential Works of Foucault 1954–1984*, ed. J. D. Faubion, tr. R. Hurley. New York: The New Press, 326–48.
Foucault, M. (2000c). So Is It Important to Think? In *Power: Essential Works of Foucault 1954– 1984*, ed. J. D. Faubion, tr. R. Hurley. New York: The New Press, 454–8.
Foucault, M. (2007). *Security, Territory, Population: Lectures at the Collège de France 1977–1978*. Ed. M. Senellart, tr. G. Burchell. Basingstoke: Palgrave Macmillan.
Foucault, M. (2008). *The Birth of Biopolitics: Lectures at the Collège de France 1978–1979* Ed. M. Senellart, tr. G. Burchell. Basingstoke: Palgrave Macmillan.
Foucault, M. (2010). *The Government of Self and Others: Lectures at the Collège de France 1982–1983*. Ed. F. Gros, tr. G. Burchell. Basingstoke: Palgrave Macmillan.
Fraser, N. (1985). Michel Foucault: A 'Young Conservative'? *Ethics* **96**(1), 165–84.
Gordon, C. (1991). Governmental Rationality: An Introduction. In *The Foucault Effect: Studies in Governmentality*, eds. G. Burchell, C. Gordon and P. Miller. Hemel Hempstead: Harvester Wheatsheaf, 1–51.
Gudynas, E. (2011). Buen Vivir: Tomorrow's Today. *Development* **54**(4), 441–7.
Gupta, A. (1998). *Postcolonial Developments: Agriculture in the Making of Modern India*. London: Duke University Press.
Hindess, B. (1996). *Discourses of Power: From Hobbes to Foucault*. Oxford: Blackwell.
Hindess, B. (2005). Politics as Government: Michel Foucault's Analysis of Political Reason. *Alternatives* **30**(4), 389–413.

Joseph, J. (2009). Governmentality of What? Populations, States and International Organisations. *Global Society* **23**(4), 413–27.
Joseph, J. (2010). The Limits of Governmentality: Social Theory and the International. *European Journal of International Relations* **16**(2), 223–46.
Kuehls, T. (1996). *Beyond Sovereign Territory: The Space of Ecopolitics*. Minneapolis: University of Minnesota Press.
Larner, W. and W. Walters (2004). *Global Governmentality: Governing International Spaces*. London: Routledge.
Levy, N. (1999). Foucault's Unnatural Ecology. In *Discourses of the Environment*, ed. E. Darier. Oxford: Blackwell, 203–16.
Lipschutz, R. D. and J. K. Rowe (2005). *Globalization, Governmentality and Global Politics: Regulation for the Rest of Us?* London: Routledge.
Luke, T. W. (1995). On Environmentality: Geo-Power and Eco-Knowledge in the Discourses of Contemporary Environmentalism. *Cultural Critique* **31**, 57–81.
Luke, T. W. (1997). *Ecocritique: Contesting the Politics of Nature, Economy and Culture*. Minneapolis: University of Minnesota Press.
Luke, T. W. (1999). Environmentality as Green Governmentality. In *Discourses of the Environment*, ed. E. Darier. Oxford: Blackwell, 121–51.
Lövbrand, E. and J. Stripple (2006). The Climate as Political Space: On the Territorialisation of the Global Carbon Cycle. *Review of International Studies* **32**(2), 217–35.
Lövbrand, E. and J. Stripple (2011). Making Climate Change Governable: Accounting for Carbon as Sinks, Credits and Personal Budgets. *Critical Policy Studies* **5**(2), 187–200.
Lövbrand, E., J. Stripple and B. Wiman (2009). Earth System Governmentality: Reflections on Science in the Anthropocene. *Global Environmental Change* **19**(1), 7–13.
Magnusson, W. (2003). The Puzzle of the Political. In *A Political Space: Reading the Global through Clayoquot Sound*, eds. W. Magnusson and K. Shaw. Minneapolis: University of Minnesota Press, 1–20.
Magnusson, W. and K. Shaw (2003). *A Political Space: Reading the Global through Clayoquot Sound*. Minneapolis: University of Minnesota Press.
McCarthy, J. and S. Prudham (2004). Neoliberal Nature and the Nature of Neoliberalism. *Geoforum* **35**(3), 275–83.
Methmann, C. (2011). The Sky is the Limit: Global Warming as Global Governmentality. *European Journal of International Relations*, published online, 1–23.
Miller, P. and N. Rose (2008). *Governing the Present: Administering Economic, Social and Personal Life*. Cambridge: Polity.
Nadesan, M. H. (2008). *Governmentality, Biopower and Everyday Life*. Abingdon: Routledge.
Neumann, I. B. and O. J. Sending (2010). *Governing the Global Polity: Practice, Mentality, Rationality*. Ann Arbor: University of Michigan Press.
Oels, A. (2005). Rendering Climate Change Governable: From Biopower to Advanced Liberal Governmentality. *Journal of Environmental Policy and Planning* **7**(3), 185–207.
Okereke, C., H. Bulkeley and H. Schroeder (2009). Conceptualising Climate Governance beyond the International Regime. *Global Environmental Politics* **9**(1), 58–78.
Paterson, M. and J. Stripple (2010). My Space: Governing Individuals' Carbon Emissions. *Environment and Planning D: Society and Space* **28**(2), 341–62.
Peluso, N. L. and M. Watts (2001). Violent Environments. In *Violent Environments*, eds. N. L. Peluso and M. Watts, Ithaca, NY: Cornell University Press, 3–38.
Rose, N. (1999). *Powers of Freedom: Reframing Political Thought*. Cambridge: Cambridge University Press.
Rydin, Y. (2007). Indicators as a Governmental Technology? The Lessons of Community-Based Sustainability Indicator Projects. *Environment and Planning D: Society and Space* **25**(4), 610–24.

Selby, J. (2007). Engaging Foucault: Discourse, Liberal Governance, and the Limits of Foucauldian IR. *International Relations* **21**(3), 324–45.

Simons, J. (1995). *Foucault and the Political*. London: Routledge.

Swyngedouw, E. (2010). Apocalypse Forever? Post-Political Populism and the Spectre of Climate Change. *Theory, Culture, Society* **27**(2–3), 213–32.

Vogler, J. and M. F. Imber (1996). *The Environment and International Relations*. London: Routledge.

Watts, M. (2002). Preface: Green Capitalism; Green Governmentality. *American Behavioural Scientist* **45**(9), 1313–17.

Whitehead, M. (2009). *State, Science and the Skies: Governmentalities of the British Atmosphere*. Chichester: Wiley-Blackwell.

Part II

Cases of Climate Government: Theorizing Practice

5

Neuroliberal Climatic Governmentalities

Mark Whitehead, Rhys Jones and Jessica Pykett

Introduction: Thinking Globally, Acting Neurologically

Recent developments within the behavioural and neurological sciences are beginning to cast valuable light on the drivers of human behaviour towards the environment (see Gertner 2009). At the simplest level, research suggests that the nature of human impacts on the environment present significant challenges to the so-called *greening of the brain* (see Centre for Research on Environmental Decisions 2009). At one level, the green orientation of the brain appears to be inhibited by the fact that many of the consequences of our environmental actions (particularly those associated with climate change) tend to be displaced in both time and space (Jones et al. 2013). Consequently, our carbon dioxide emissions on any given day are only likely to have a discernable climatic effect in the near future. Furthermore, if we live in a relatively advanced economy, the worst impacts of our greenhouse gas emissions are most likely to afflict people in distant lands. On these terms, the lack of relevant feedback on our environmental actions works to inhibit rational responses to our current climate predicament. On another level, the very fact that many people will not directly experience the worst consequences of climatic actions tends to inhibit our more emotional prompts to pro-environmental behaviour. In essence, the nature of climate change tends to inhibit both our rational and emotionally based drivers to behaviour change.

If climate change presents a particular set of neurological challenges to pro-environmental behaviour, recent developments within behavioural psychology and behavioural economics are beginning to suggest policy-based solutions to these problems. Collectively referred to as the Behaviour Change Agenda, this fusion of psychology and public policy making is now having an impact on climate change polities in states throughout the world. This chapter charts and critically analyses the emergence of the so-called Behaviour Change Agenda in the United Kingdom and its impacts on policies for climate protection. Taken together, this emerging policy agenda draws particular attention to the importance of understanding the more-than-rational context for human decision making (see Jones, Pykett and Whitehead 2011; Whitehead, Jones and Pykett 2011). In charting this experimental system of governance (which in addition to the environment is also affecting the design of public health and financial policies), this chapter argues that despite appearing to constitute a more interventionist and post-neoliberal approach to climate protection, such policies are actually creating a form of neuroliberalism, which recreates the market-based approaches to climate government at a molecular level. Notwithstanding this assertion, this chapter goes on to claim that the Behaviour Change Agenda provides the potential basis for a more radical, progressive, but as yet unrealized, system of neurologically sensitive climate governance.

This chapter begins with a brief introduction to the intellectual and political history of the Behaviour Change Agenda. The following section explains the particular connections that exist between this agenda and climate change policies. In this context, analysis reveals an important set of natural synergies that appear to exist between the principles of behaviour change and the problematics of climate change. Through reference to theories of the neurotic citizen and governmentality, Section 3 develops a critical interpretation of the underlying assumptions of behaviour-changing policies. The final section explores a series of alternative interpretations of behaviour change theory, which appear to suggest a series of more radical policy designs and goals. In relation to this volume's attempts to explore the value of the governmentality approach within the study of global climate governance, this chapter makes two assertions. At one level, it argues that despite appearing to diverge from the rationally oriented systems of self-conduct identified within work on governmentality, the Behaviour Change Agenda actually reflects the extension of governmentality into the more-than-rational registers of human behaviour. At a second level, this chapter asserts that governmentality provides a valuable interpretive framework within which the Behaviour Change Agenda can be critically contextualized in relation to the broader history of liberal government and biopolitical desire.

Origins – Bounded Rationality and the (Partial) Death of Homo Economicus

The scientific and political history of the Behaviour Change Agenda is inextricably tied to changing perceptions of the paradigmatic figure of homo economicus. Since its nineteenth-century invention (see Persky 1995), the figure of homo economicus has stood for an idealized trope of human behaviour primarily defined by rational self-interest. Notwithstanding the perennial critiques of the theory of the economic human (see Sen 1977), it appears that the concept has never been as deeply embedded within classic economic theory as is often claimed. On the terms of John Gray, homo economicus may have been nothing more than a *necessary myth* for a modern society that clings to the vision of 'unitary, conscious subjects' controlling their own destinies (2002: 70). Or as Žižek, following Kierkegaard, observes, perhaps the classical economists only 'believed that they believed' in the empirical veracity of the assumption of perpetual human rationality (2009: 3). It is within such contexts that the ideas that undergird the Behaviour Change Agenda come to the fore. They do so not because they confirm the *always already* realization of the absurdity of homo economicus, but because they offer the world something else to believe in.

The figure that preoccupies the Behaviour Change Agenda is that of the *Real Human*. The idea of the Real Human is essentially a shorthand term for a subject that is characterized by systematic patterns of suboptimal socioeconomic behaviour. This is a figure firmly grounded in the time constraints and cognitive liabilities of everyday life. On these terms, the idea of the Real Human is a depiction of behaviour with which we are all only too familiar. But the notion of the Real Human is more than merely a statement of what everyday experience reveals to be all too obvious. The emergence and study of the Real Human is actually the product of a hard-fought battle within economics between a neoclassical tradition, which remains wedded to the rationality assumption, and a strand of behavioural economics built on the theory of *bounded rationality*.

Herbert Simon, an American political scientist and polymath, first proposed the idea of bounded rationality (see 1945, 1957). In essence, the idea of bounded rationality proposes a foundational challenge to the rationality assumption that is associated with classical economics and its avatar homo economicus. The rationality assumption depicts human decision making (by the consumer and entrepreneur) as the outcome of the consistent application of relevant knowledge, critical reflection and deliberation. For Simon rationality was bounded because of inadequate knowledge, suboptimal cognitive practices and the lack of time available to the average decision maker. Simon utilized his theory of bounded rationality to explain why people favour short-term need over long-term gain, make impulsive judgements, allow the behavioural norms of those around them to influence their decisions, and see their choices arbitrarily shaped by the decisions they have habitually made in the past. While such behavioural aberrations were seen as beyond the purview of classical economics (embodying unwanted remainders of an equation that began with the theory of homo economicus), the behavioural economics tradition Simon initiated embraced them as their primary target of analysis.

To gain more analytical purchase on the irrational/intuitive composition of Real Human decision making, early behavioural economists turned to psychology. While Simon's attempts to develop a creative dialogue between economics and psychology saw him ostracized from the classical mainstream of economic enquiry, by the early 1970s this interdisciplinary relationship was beginning to flourish. At the centre of this *Psychological Spring* within economics were prominent psychologists Daniel Kahneman and Amos Tversky. In their pioneering work (for which, and following Tversky's death, Kahneman would eventually win the Nobel Memorial Prize in Economic Sciences), Kahneman and Tversky sought to uncover the actual processes and practices that produce the bounded rationality Simon identified (see Kahneman and Tversky

1974; see also Kahneman, Slavic and Tversky 1982). Drawing on detailed empirical studies, Kahneman and Tversky isolated a series of heuristics (or necessary shortcuts), which are routinely used to guide human decision making and behaviour, and which ultimately limit the supposed rationality of action. The heuristics identified by Kahneman and Tversky include the *representative heuristic* (where certain characteristics of an object, subject or situation are falsely equated with their underlying nature), *anchoring* (the tendency to base judgements on available relative measures – like upgrading to leather seats when buying an expensive car because the perceived cost of the upgrade is conditioned by the relatively high price of the car), the *availability heuristic* (when behaviour is guided by the available information on the basis that its accessibility equates to its quality and relevance) and the *validity heuristic* (when the arbitrary patterning of events is interpreted as a sign of underlying trends) (Kahneman et al. 1982).

What is most relevant about the work of Kahneman and Tversky for this chapter, and the broader Behaviour Change Agenda today, is that their interpretation of cognitive biases saw them position heuristics as pernicious survival techniques that lead rationally inclined individuals into bad decision making. Consequently, while Simon suggested that the identification of bounded rationality should see economics embrace the study of new more-than-rational subjects, Kahneman and Tversky's work partially resuscitated homo economicus as an enduring super ego that had merely been undermined by the irrational push of everyday life. This crucial disjuncture in the early iterations of behavioural economics has led Sent to identify old and new schools within the subdiscipline (2004). Sent suggests that the old school of behavioural economics was characterized by Simon and his desire to forge a strong break with classical economics' rationality doctrine. The new school was led by Kahneman and Tversky, and rather than breaking with classical economics attempted to identify the ontological barriers that act to inhibit the enduring rational impulses of the economic subject. The distinction between the old and new schools of behavioural economics is important in practical policy terms because it suggests two very different ways of approaching the regulation of human behaviour. On one hand, the old behavioural school suggests that policies should be synced within the complex emotive drivers and values of the more-than-rational subject. The new school of behavioural economics, on the other hand, suggests that policies should be designed to correct the irrational tendencies of human behaviour and to return them to market-oriented norms. To these ends, new behavioural economists appear to see irrationality as the failure of the analytical subject to process knowledge, as opposed to an indicator of an emotionally driven actor.

The most prominent contemporary expression of new behavioural economics is libertarian (or soft) paternalism. Inspired by the work of Chicago economist Richard Thaler, and famously encoded within his volume *Nudge: Improving Decisions about Health, Wealth and Happiness* (Thaler and Sunstein 2008), libertarian paternalism maps out the techniques in and through which the insights of behavioural economics can be applied to behaviour-changing policies. At the heart of libertarian paternalism is: 1) a recognition of the need for governmental intervention within the field of human decision making (let us call this the *paternalist imperative*); and 2) an appreciation of the methods governmental intervention within human decision making can be achieved in a way that does not inhibit personal freedom, and can actually increase personal choice (the *libertarian imperative*).

These seemingly incompatible imperatives are pursued within libertarian paternalism through the reconfiguration of 'choice architectures' or environments. Choice architectures are shorthand for the various frameworks within which everyday decisions are made. Choice architectures can include household environments (the design of thermostats, kitchen appliances and electricity meters), public spaces (like high streets, school canteens and hospitals) and bureaucratic procedures (including the design of energy bills, pension enrolment schemes and organ donation programmes). Libertarian paternalism, as with many behaviour-changing policies, operates on the basis that the governmental or corporate redesigning of choice architectures can help to promote improved long-term decision making without the use of coercion. A classic example of libertarian paternalist policy, which has direct relevance for climatic change governance, is the publication of comparative energy use statistics on domestic household energy bills. The idea behind this policy is that, if individuals can see that their household energy use is in excess of a social average, a new psychological incentive to reduce energy use and related carbon emissions will be generated (the reverse could, of course, occur for households falling below the average level). This policy intervention actually achieves two things: 1) it provides relevant, comparative statistical information in and through which people can assess whether their household energy practices require reform (addressing the heuristic biases identified by Kahneman and Tversky); and 2) it generates a form of peer pressure within which a household may feel more compelled to change its daily energy use practices to conform with broader social norms (reflecting the more emotional component of decision making identified by Simon). This policy can then be seen as paternalistic to the extent that it encourages forms of environmentally beneficial behaviours, and libertarian to the extent that such behaviours are promoted only on the basis of increased information flow and more informed systems of choice.

Although libertarian paternalism is one of the most influential contemporary frameworks within which the Behaviour Change Agenda operates, it is important to note that it actually has very diverse sources and associated modes of operation (see Table 5.1). Uniting this broad governmental movement, however, is a recognition of the limits to rational human thought and action and an appreciation of how various psychological initiatives can be deployed to direct human behaviour in more socially and ecologically sustainable directions. One policy executive working in the UK government's Cabinet Office Strategy Unit described the impacts

Table 5.1 *Origins and modes of operation of the behaviour change agenda*

Intellectual Influences	Modes of Operation
Behavioural Economics	Nonconscious Priming
Behavioural Psychology	Intelligent Assignment
Cognitive Design	Presumed Consent
Engineering Psychology	Mandated Choices
Ethology	Anchoring
Intuitive Judgement Theory	Culture Change
Material Psychology	Channelling Factors
Neuroeconomics	Collaborative Filtering
Neuropsychology Preference Theory	Disclosure
Psychographics	Feedback
Social Cognition	Self-Registered Control Strategies
Social Influence Theory	Peer-to-Peer Pressure
Social Marketing	Norm Formation
Theories of Affect	Choice Editing
Time Preference Theories	Default positions
User-Centred Design	
Visual Perception Theory	

of behavioural economics and the broader behaviour change agenda to us on the following terms:

> [b]ehavioural economics sort of does that compared to classical economics, in relation to policy design. It moves you from rather rough, crude, stick-man type clothing, into one which is really fitted, to change the metaphor, which goes with the grain of human cognition. (Interview 2010)[1]

At its heart then, this new era of behavioural government appears to be about moving beyond the stick-man that is homo economicus and designing policies that work within the complex analytical and emotional grains of real human decision making.

Connections – Barriers to Governing Climate Conducts

Having established the background and nature of the Behaviour Change Agenda, this section considers the ways this agenda is connected to emerging systems of climate change governance. According to Thaler and Sunstein, 'It helps to think about

[1] This interview was one of a series of interviews carried out with senior policy executives in the United Kingdom who were responsible for developing behaviour-change policies.

the environment as a global choice architecture system' (2008: 185). Thinking of contemporary forms of climate change as a product of a global choice environment is helpful on a series of counts. To appreciate why this is helpful, however, it is necessary to outline the nature of this global choice architecture. The first thing to note is that despite recent attempts to generate carbon markets and trading schemes, the historical emergence of climate change is actually the product of a world within which the emission of greenhouse gases has been seen as an uncosted externality. In other words, there has been, and continues to be for many, an incentive deficit when it comes to tackling climate change. This incentive deficit means that even if we all conformed to the rational expectations of homo economicus, the lack of market-based signals would inhibit any significant behavioural shifts in response to climate change. But even when we acknowledge the folly of homo economicus, climate change exhibits specific qualities that make it a particularly difficult context within which to promote behavioural shifts (see Gertner 2009).

One of the peculiar features of climate change is the fact that the processes with which it is associated tend to provide individual actors with very little immediate feedback on the nature and consequences of their actions (see Centre for Research on Environmental Decisions 2009). This lack of feedback begins at the points of production of greenhouses gases (the car exhaust, the central heating duct and even the beef sandwich), and the fact that the very acts that produce such emissions tend to obscure the nature and extent of the emissions they produce. Furthermore, the very nature of climate change also inhibits behavioural feedback as the consequences of actions related to climate change tend to be displaced in time (to future generations) and in space (to more vulnerable geographical locations). These two forms of disconnect from climate change relate to different forms of human decision making: the first an analytical disconnect (pertaining to a lack of knowledge), the second a more emotional form of disconnect (concerning a lack of experiential connection to the actual effects of climate change) (see Gertner 2009). Climate change's association with uncertainties of various kinds generates further problems for both analytical and more emotionally based decision making. These uncertainties range from debates about the very existence of anthropogenic climate change to the precise future impacts it may have. What is clear is that such uncertainties can often delay the onset of rational and more-than-rational responses to the threats of climate change.

In addition to presenting policy problems that behaviour-changing techniques appeared to be designed to directly address, climate change has a further quality that has contributed to its close association with the Behaviour Change Agenda. One of the primary limitations on the emergence of behaviour-changing techniques in government, in liberal societies at least, has been the enduring principle of 'harm to others'. The 'harm to others' principle states that government interventions within the everyday life of citizens are only justified if they are designed to prevent that individual causing harm (broadly defined as physical, political and emotional) to other citizens. On these terms, and despite their apparent suitability, behaviour-changing policies have found it hard to gain traction in fields relating to healthy eating and financial planning. In contrast to this, issues of climate change (and environmental policy more generally) have been associated with strong evidence of the myriad ways greenhouse gas emissions in one place can have deleterious affects on peoples and communities in other locations. Because of the particular nature of the climate change problematic, climate protection policies have been associated with one of the most sustained programmes of policy experimentation relating to behavioural-changing techniques (see Shove 2010; Whitmarsh and O'Neill 2010; Whitmarsh, O'Neill and Lorenzoni 2010).

In the UK context, this period of policy experimentation began in 2006 when the Department for Environment, Food and Rural Affairs (hereafter Defra) published a scoping study on environment-related behaviour-change strategies (Defra 2006). This scoping study involved policy executives in Defra consulting with a range of behaviour change experts to develop a more systematic way of thinking and approaching so-called pro-environmental behaviours. In addition to drawing on the insights of behavioural economics and science, this scoping study also developed a provisional segmentation model in and through which particular strategies could be specifically designed to address the behavioural tropes of particular groups within society. Building on this initial scoping study, in 2008 Defra published a definitive strategy for environment-related behavioural policy: *A Framework for Pro-Environmental Behaviours*. This report was based on a wide range of research projects on sustainable behaviours and consumption practices, which Defra had commissioned, and sought to enable the government to move beyond a relatively simplistic model of behaviour change based on the communication of relevant information. The report focussed on a series of *Headline* environmental behaviours, including home insulation, energy management, reductions in nonessential flying, reduced food waste production and lower car usage. To address these behavioural priorities the report suggested the use of classic behaviour-change techniques. At one level these techniques involved using peer-to-peer networks to promote behaviour change, which deliberately tapped into people's competitive instincts and emotional responses to the actions of those around them. To overcome key cognitive biases in climate change behaviours, the report also recommended the use of household environmental audits and domestic visual units to analytically sensitize people to their levels of energy use of greenhouse gas production (see Chapter 8 of this volume for a broader discussion of the use of smart meters and the broader process of energy governance in the home).

These early studies and reports culminated in the 2011 *Behaviour Change and Energy Use* strategy (Behavioural Insights Team 2011). Produced by the Cabinet Office's Behavioural Insights Team (a unit established by the Coalition Government to promote the application of behavioural science within British public policy), in conjunction with the Department of Energy and Climate Change, and the Department for Communities and Local Government, this strategy set out the government's vision for a Green Deal. From the outset, the strategy acknowledges that the government's Green Deal is based on the insights of behavioural economics and psychology. A key dimension of the strategy was the promotion of up-front incentives (including tax breaks and pay-as-you-go investments) for the installation of low-carbon technologies in the home. In typical behavioural economics-inspired fashion, this incentives system sought to overcome the perennial problem associated with green technologies: namely that they involve significant short-term investment which yield returns only in the long term. Echoing previous Defra reports, the strategy also supported the more effective production and display of environmental information: from home monitoring technologies that measure energy use in real time, to improvements in the format of Energy Performance Certificates so that homebuyers could more effectively assess the environmental efficiencies of the properties they were viewing.

When these varied policy initiates are taken together they appear to reflect a definitive shift in systems of climatic governance. The key shift relates to how the subjects at the centre of climate governance system are conceived. Instead of the idealized homo economicus motivated purely by rational price incentives and undifferentiated environmental data, we find environmental subjects more responsive to social vectors of emotional pressure who can be nudged by the effective framing and presentation of environmental information. Perhaps what characterizes these policies most, however, is their reliance on established corporate techniques of persuasion and manipulation. These corporate techniques are most clearly expressed in the use of segmented marketing strategies and subconscious prompts to action. Thaler and Sunstein argue that the corporate-inspired nature of much of the Behaviour Change Agenda means that it reflects a kind of *Real Third Way* which utilizes corporate techniques to pursue collective governmental goals (2008: 252–3). The agenda also appears to reflect a facet of what Rose has described as *Advanced Liberalism*: a condition under which the state not only steers away from unnecessary intervention within the marketplace, but also becomes subject to the practices, values and assumptions of the corporate world itself (1999: 137–66).

Interpretations – The Neurotic Citizen and the Marketization of Existence

A series of critical interpretative perspectives have already been applied to the Behaviour Change Agenda (see Hausman and Welch 2010; Wells 2010; Jones et al. 2011; Whitehead et al. 2011). Drawing on the analysis set out in this chapter so far, this section contributes to this endeavour by asserting that the Behaviour Change Agenda does not reflect a rejection of neoliberal assumptions about human conduct, but instead involves the remaking of market-based society by other means. This section concludes by considering the potential implications of this for the efficacy of behaviourally oriented climate change governance.

A key starting point for critically analysing the Behaviour Change Agenda is to recognize that abandoning the rationality assumption and recognizing the irrational drivers of human action does not necessarily involve a rejection of neoliberal orthodoxy. The true heart of the neoliberal project is not homo economicus, but a belief in the unrivalled ability of markets to efficiently organize society, preserve freedom and harness entrepreneurial energies (Hayek 1944; Friedman 1982). To these ends, homo economicus has never been an ontological necessity within the neoliberal canon – but rather a vision of what an ideal subject would be like within a market-based society. In his 1979 lecture series at the Collège de France, *The Birth of Biopolitics*, Michel Foucault takes up this point as he considers the connections between neoliberal theory and notions of rationality (2008). Reflecting on the work of prominent Chicago School economist Gary Becker, Foucault observes

> The most radical of American neo-liberals [namely Becker] … says that it is still not sufficient, that the object of economic analysis can be extended beyond rational conduct … and that economic laws and analysis can perfectly well be applied to non-rational conduct … If an individual's conduct reacts to reality in a non-random way. (2008: 269)

The key proviso for more-than-rational considerations to enter the field of economic analysis is that nonrational conduct reflects a nonrandomized reaction to real-world circumstances. The notion that individuals are *predictably irrational* has, of course, been an underlying trait of work within new behavioural economics (see Ariely 2008). In this context, it is possible to interpret the collective work of behavioural economists, cognitive psychologists and neuroscience as essentially exposing the more-than-rational dimensions of human conduct to neoliberal scrutiny and correction. For Foucault, this moment, which has actually been an emerging feature of neoliberal society throughout the postwar period, is historically significant because it marks a key shift in ideologies of government. If the idealized homo economicus was essentially ungovernable (or at least best left to her/his own self-interested, rational pursuits), the Real Humans of behavioural economics are not only in need of governance (to correct their suboptimal decision making), but are also susceptible to government (because of their predictably systematic responses to changes in their choice environment) (Foucault 2008: 270). For Foucault then, the subject that has been forged out of the assiduous work of behavioural economics marks a kind of threshold point after which it starts to become possible to imagine, and legitimately develop, a neoliberal system of governmentality.

The idea that neoliberal governmentality is born at the point when psychology begins to enter the realm of economic analysis is an important one. It is important because it suggests that Rose's notion of advanced liberalism may not convey effectively all that there is to the Behaviour Change Agenda. While it may be true that the Behaviour Change Agenda marks the deepening impact of corporate practices within state actions, this process is only one facet of the broader application of neurological power within the practices of governance. This point is directly addressed by Isin in his analysis of the rise of the so-called *neurotic citizen*. In a move that appears to, in part, reflect what is happening in the Behaviour Change polity, Isin observes:

> The figure that also occupies a central role in our times is the neurotic citizen who governs itself through responses to anxieties and uncertainties. The rise of the neurotic citizen signals a new type of politics (neuropolitics) and power (neuropower). I suggest a new concept, neuroliberalism – a rationality of government that takes as its subject the neurotic citizen – as an object of analysis. (2004: 220)

Isin's neurotic citizen is a figure essentially governed through the irrational registers of fear, anxiety and neurosis. While the idea of 'governing through neurosis' may only convey part of the palette of policy options deployed within contemporary behavioural policy, Isin's broader deployment of neuroliberalism has perhaps a broader utility. The notion of neuroliberalism is redolent of a form of market-oriented society within which the neurological is not only positioned as a progressive plan of therapeutic intervention and care, but as a further context within which market failures can be remedied and contradictions resolved.

The use of neurological power to more deeply marketize human existence raises two significant issues in relation to climate governance: (1) to what extent can climate protection actions be connected to market-compatible goals (such as those typically espoused within ecological modernization; and (2) what if it is these market-based goals and their associated values which are at the psychological root of the contemporary climate change problematic? In relation to the second issue, there already exists a substantial body of work that has charted the co-evolution of liberal society, capitalism and the psychology of market-based individualism (see Elias 2011; Ophuls 2011; Quilley 2013). If *bigger-than-self* environmental issues, like climate change, are a product of an individualizing, self-interested and highly competitive psychological substrata within society, a contradiction becomes apparent in contemporary behavioural policies. This contradiction stems from the fact that the Behaviour Change Agenda in the United Kingdom is using psychological techniques to exploit an already marketized subconscious (with its positive emotional connections to consumption, accumulation and competition), and to further establish market-oriented values as sociocultural norms. Some would argue that such techniques may ultimately only reinforce the psychological processes that have led us to this particular juncture in human-climate relations (Crompton 2010).

A helpful exposition on this conundrum is provided within a prominent climate protection initiative currently being developed in the United Kingdom by Italian car manufacturer Fiat. *Eco-Drive* is a joint initiative between Fiat and Microsoft that attempts to provide a framework within which participants can make changes to their driving behaviours and reduce the production of greenhouse gases. Through the use of an onboard computer and a Universal Serial Bus flash drive, the *eco-Drive* provides driving-style analysis, statistics breakdowns of carbon dioxide produced and automatic prompts concerning how to ecologically regulate driving patterns. *Eco-Drive* has an online community (the *Fiat Ecoville*) of some sixty-thousand participants able to compare statistics and engage in various eco-competitions. In its use of automatic prompts, peer-to-peer comparison and various rationalization techniques, *eco-Drive* is, in many ways, emblematic of behaviour-changing policies in the United Kingdom. The aforementioned psychological contradiction is manifest in the case of *eco-Drive* to this extent that although this behavioural initiative attempts to pursue short-term pro-climate gains, it also promotes car-ownership, interpersonal competition and market savings. These are ultimately values that can work against long-term action on climate change.

Alternatives – Climate Conducts and Counter Conducts

In the United Kingdom, there have already been a series of critical responses to the pro-climate behaviour change strategies being developed within central government (see House of Lords 2011). These critical responses have been paralleled by the production of a series of alternative, psychologically inspired approaches to environmental behaviours. These alternative strategies have been predominantly produced by environmental nongovernmental organizations such as the WWF, Oxfam and the Climate Outreach and Information Network. These alternative strategies appear to have in common a desire to draw on the complex insights of behavioural economics and psychology in the construction of climate change policy. In keeping more with the old, as opposed to the new, school of behavioural economics, such strategies appear concerned more with recognizing and facilitating nonmarket-oriented climatic subjects. Arguably the most influential of these counter-strategies has been the 2010 *Common Cause: The Case for Working with Our Cultural Values* (Crompton 2010). *Common Cause* was produced in partnership between the Climate Outreach and Information Network, the Campaign to Protect Rural England, Friends of the Earth, Oxfam, and the WWF, and was written by WWF change strategist Tom Crompton.

Focusing predominantly on the potential impacts of behavioural sciences on environmental campaigning, Common Cause draws particular attention to the relationship between our emotional registers of more-than-rational decision making and our cultural values. The report observes:

> It seems that individuals are often predisposed to reject information when accepting it would challenge their identity and values. Campaigning approaches that rely on the provision of information may well work for people whose existing values are confirmed through accepting, and acting upon, that information. But for others, the same information (for example, about the scale of the challenge that climate change presents) may simply serve to harden resistance to accepting new government policies or adopting new private sphere behaviours. (Crompton 2010: 9)

In recognizing the limits of rational argumentation when trying to persuade people to reform their climatic behaviours, Common Cause has much in common with the UK government's Behaviour Change Agenda. Where it differs from the mainstream behavioural agenda, however, is the connection it makes between more-than-rational/emotional responses and particular forms of values.

According to the Common Cause report, cultural values can be divided, admittedly crudely, into two broad categories: (1) intrinsic values (which are self-transcendent and include empathy and community orientation); and (2) extrinsic values (concerned with self-betterment and social status). While neoliberal orthodoxies and the paradigm of homo economicus suggest that extrinsic values are a natural part of the human condition, a significant body of scientific research now suggests that humans are an inherently empathic species (see Rifkin 2009). Placing biological fundamentalism to one side, however, Common Cause argues that the prevalence of extrinsic value systems within industrialized societies means that many people have a kind of 'protective cognition' against the fundamental messages of personal restraint and global care that are part of the climate protection change agenda (Crompton 2010: 19). The implications of this perspective are that psychological prompts to pro-climate behaviours are likely to be unsuccessful if they do not address the underlying values which inform contemporary patterns of climate-related actions. Furthermore, even if such actions are successful in the short term, they are unlikely to change long-term patterns of behaviour. Drawing on the work of cognitive linguist George Lakoff, Common Cause argues that, to succeed in the long-term, behaviour-change policies must target the *deep cognitive frames* that condition human behaviour (see Lakoff 1996). It would appear that only when these frames are oriented towards more egalitarian and empathic values are behaviour-change programmes likely to make a significant contribution to addressing the climate change problematic.

The Common Cause report has clear limitations. First, it provides very little guidance as to precisely how values can be accessed in order for them to be realigned. Second, it fails to acknowledge that the very act of shifting the fundamental values associated with modern industrial society may actually be a greater challenge than addressing climate change itself. Finally, it says little about the nature of the potential gap that exists between values and actions and whether values are a reliable way of trying to understand and shape collective forms of human behaviour (see Shove 2010). Notwithstanding these issues, it is clear that Common Cause sets out an alternative, potentially more progressive way, in which it is possible to apply the insights of the behavioural sciences to climate change government. Its message is predicated on the assertion that merely recognizing that human decision making is a complex amalgam of rational and more-than-rational registers of action is not going to solve the complex problems associated with climate change. What is also required is an appreciation of the political and economic systems that have historically conditioned and framed human values in ways that have perpetuated a range of harmful environmental relations.

Conclusions

This chapter has drawn attention to the emerging neurological strategies that are beginning to shape the governance of the global climate. Although this chapter has focussed on the so-called Behaviour Change Agenda in the United Kingdom, the forms of policies described in this chapter are now having a significant influence on environmental policy development and governance in the United States, France, Australia, Canada, Brazil, Germany and the Netherlands. At the heart of this new policy agenda is a desire to move beyond the narrowly defined interpretations encoded within the figure of homo economicus and to develop a more complex picture of the rational and more-than-rational prompts to human behaviour. Having established the reasons the Behaviour Change Agenda has particular salience to a personalized form of climate governance, this chapter described how many climate-change-related behavioural policies continue to direct human conduct in heavily market-oriented directions. In identifying a form of neuroliberalism that appears to infuse contemporary behavioural governance in the United Kingdom, the analysis presented here has explored the extent to which such policies may ultimately recreate the kinds of extrinsic values that facilitated the emergence of the climate change problematic in the first place. This chapter culminated with a discussion of the recently produced *Common Cause* report. Drawing attention to the emphasis that this report places on the underlying values that inform various forms of rational and more-than-rational conduct, this chapter argues that this strategy may embody a more progressive and potentially efficacious response to the problems of climate change. Ultimately, it would appear that emerging systems of climate governance operating at global, national, urban and neurological scales are defined by ongoing struggles between neoclassic calls for deeper forms of marketization and more progressive programmes for building eco-egalitarian futures. The real problem with the current Behaviour Change Agenda in the United Kingdom is that it appears to be pursuing a psychologically expended neoliberal agenda in the guise of more progressive neuroecological mandate. Theories of governmentality provide a context within which both the neoliberal undertones of the Behaviour Change Agenda can be exposed and the contingency of the neurological present (and the possibilities that exist to transform it) becomes clear.

References

Ariely, D. (2008). *Predictably Irrational. The Hidden Forces that Shape Our Decisions.* London: HarperCollins Publishers.

Behavioural Insights Team [with the Departments of Energy and Climate Change and Communities and Local Government] (2011b) *Behaviour Change and Energy Use* (Cabinet Office, London).

Centre for Research on Environmental Decisions (2009). *The Psychology of Climate Change: A Guide for Scientists, Journalists, Educators, Political Aides, and the Interested Public.* New York: CRED.

Crompton, T. (2010). *Common Cause: The Case for Working with Our Cultural Values*. London: UK.

Defra (2006). *An Environmental Behaviours Strategy for Defra: Scoping Report*. London: Defra.

Defra (2008). *A Framework for Pro-Environmental Behaviours*. London: Defra.

Elias, N. (2011). The Society of Individuals. In *Collected Works Norbert Elias*. Ed. Robert van Krieken. Dublin: UCD Press.

Friedman, M. (1982). *Capitalism and Freedom*. Chicago, IL: University of Chicago Press.

Foucault, M. (2008). *The Birth of Biopolitics: Lectures at the Collège De France 1978–1979*. Tr. Burchell, G. Basingstoke: Palgrave/MacMillan.

Gertner, J. (2009). Why Isn't the Brain Green? *New York Times*, 19 April.

Gray J. (2002) *Straw Dogs: Thoughts on Humans and Other Animals*. London: Granta Books.

Hausman, D. M. and B. Welch (2010). Debate: To Nudge or Not to Nudge. *The Journal of Political Philosophy* **18**, 123–36.

Hayek, F. A. (1944). *The Road to Serfdom*. London: Routledge.

House of Lords (2011). *Behaviour Change Report* [HL paper 179]. London: Stationary Office.

Isin, E. F. (2004). The Neurotic Citizen. *Citizenship Studies* **8**, 217–35.

Jones, R., J. Pykett and M. Whitehead (2011). Governing Temptation: On the Rise of Libertarian Paternalism in the UK. *Progress in Human Geography* **35**, 483–501.

Jones, R., J. Pykett and M. Whitehead (2013). *Changing Behaviours: On the Rise of the Psychological State*. Cheltenham: Edward Elgar.

Kahneman, D. and A. Tversky, (1974). Judgment under Uncertainty: Heuristics and Biases. *Science* **185**, 1124–31.

Kahneman, D., P. Slavic and A. Tversky, ed. (1982). *Judgment under Uncertainty: Heuristics and Biases*. Cambridge: Cambridge University Press.

Lakoff, G. (1996). *Moral Politics: What Conservatives Know that Liberals Don't*. Chicago, IL: University of Chicago Press.

Ophuls, W. (2011). *Plato's Revenge. Politics in the Age of Ecology*. London: MIT.

Quilley, S. (2013). De-Growth Is Not a Liberal Agenda: Relocalization, and the Limits to Low Energy Cosmopolitanism, *Environmental Values* **22**, 261–85.

Persky, J. (1995). Retrospectives: The Ethology of Homo Economicus. *The Journal of Economic Perspectives* **9**, 221–31.

Rifkin, J. (2009). *The Empathic Civilization: The Race to Global Consciousness in a World in Crisis*. Cambridge: Polity Press.

Rose, N. (1999). *Powers of Freedom: Reframing Political Thought*. Cambridge: Cambridge University Press.

Sen, A. K. (1977). Rational Fools: A Critique of the Behavioural Foundations of Economic Theory. *Philosophy and Public Affairs* **6**, 317–44.

Sent, E. M. (2004). Behavioural Economics: How Psychology Made Its (Limited) Way Back into Economics. *History of Political Economy* **36**(4), 735–60.

Shove, E. (2010). Beyond the ABC: Climate Change Policy and Theories of Social Change. *Environment and Planning A* **42**, 1273–85.

Simon, H. (1945). *Administrative Behaviour: A Study of Decision-making Processes in Administrative Organization*. New York: The Free Press.

Simon, H. (1957). *Models of Man: Social and Rational*. London: John Wiley and Sons.

Thaler, R. and C. Sunstein (2008). *Nudge: Improving Decisions about Health, Wealth and Happiness*. London: Yale University Press.

Wells, P. (2010). A Nudge One Way, a Nudge the Other: LP as Political Strategy. *People, Place and Policy Online* **4**, 111–18.

Whitmarsh, L. and S. O'Neill (2010). Green Identity, Green Living? The Role of Pro-Environmental Self-Identity in Determining Consistency across Diverse Pro-Environmental Behaviours. *Journal of Environmental Psychology* **30**, 305–14.

Whitmarsh, L., S. O'Neill and Lorenzoni, I., ed. (2010). *Engaging the Public with Climate Change: Behaviour Change and Communication*. London: Earthscan.

Whitehead, M., R. Jones and J. Pykett (2011). Governing Irrationality, or a More-Than-Rational Government: Reflections on the Re-scientization of Decision-making in British Public Policy. *Environment and Planning A* **43**, 2819–37.

Zizek, S. (2009). *First as Tragedy, Then as Farce*. London: Verso.

6

Making Carbon Calculations

Sally Eden

Introduction

As part of my Environmental Management Principles class in 2011, my twenty-eight students and I tried out a variety of online 'carbon calculators' to measure our individual (or household) 'carbon footprint'. Given that these students had deliberately chosen to take an environmental management module, I had assumed that they would be more inclined towards 'green' issues and would thus find the exercise interesting. But when we compiled the greatly divergent set of results, what became more intriguing to me than the overall result in terms of CO_2 measurement or the students' levels of 'greenness' was the process of calculating carbon and how this select – and in many ways privileged – group of environmental citizens failed to make the calculators 'work'. The class turned out to be not merely an exercise in measuring carbon but also an exercise in the public understanding of environmental science and the ways environmental measurements are made through complex assemblages and multiple practices.

How can we analyse such complex calculations? Carbon calculators have often been read as attempts to govern consumption outside regulatory regimes of control, through voluntary participation and self-surveillance, especially when promoted by governmental agencies. However, many carbon calculators are set up by nongovernmental actors, such as environmental pressure groups or private-sector consultants, which complicates this reading, not least where these offer alternatives or even resistance to those set up by the state. Moreover, carbon calculators are also examples of how environments and humans are coproduced through varied calculative strategies that depend very strongly on the state and its subjects, as well as on a variety of (often highly variable) objects and devices.

In this chapter, I want to use ideas from actor-network theory (ANT) to analyse how carbon calculations are coproduced as complex assemblages and as imperfect systems for measuring carbon – what we might call 'carbon metrologies'. This means that the implications for climate policy governance need to go beyond a reading of control because, in order for carbon calculators to work as technologies of government, they must first be co-constructed as socio-technological assemblages, making the subjects of government also active users and co-creators of environmental calculation.

Theoretical Tools for Analysing Carbon Policy and Calculations

Let me begin by outlining some theoretical tools that we might use to analyse carbon calculations. In this volume (e.g. Chapter 7 and Chapter 1) and elsewhere (e.g. Luke 1999; Rutherford 2007; Lovell and Liverman 2010), concepts of 'environmental' or 'green' governmentality have been used to analyse how environmental conduct is governed, that is, how processes of environmental policy making, management and measurement shape individual subjects, their sense of self and their actions, as well as shaping environments and organizational practices. However, as Rutherford noted, such work 'often focuses on state regulation and its impacts and effects' (2007: 292). Hence, the ecopolitical project of reducing carbon under a governmentality analysis is often dominated by the will or agenda of the state – Rutherford (2007: 296) refers to its 'goals' and Blok (Chapter 2) to the persistence of 'hegemony' – even where this will is manifest only implicitly and across multiple, dispersed sites.

For example, Luke's (1999) 'green governmentality' or 'environmentality' is very much state-centric in analysing ecological means by which the state seeks both national security and geopolitical power – in this sense of 'green geopolitics', it is geopolitics first and green very definitely second (if at all). Like Rose in describing statistics as 'the science of the state', Luke (1999: 149) refers to 'statist ecology' as the particular rationalities and technologies used by the state (albeit sometimes in ways not necessarily explicitly linked to state agencies) in its various power projects, and he also cast the Brundtland Commission (a United Nations (UN) and therefore nonstate body) as a statist form of 'environmental protection agency' (Luke 1999: 138; see Chapter 4 for a more detailed discussion of Luke's work).

This is not to say that green governmentality is restricted to state power; indeed, authors have frequently emphasized that the practices and technologies of government increasingly reach out 'beyond the state' (Rutherford 2007: 293; also Chapter 1) to households (e.g. Paterson and Stripple 2010; Chapter 7) and the psyche (e.g. Rose 1991). However, the central project of governing people (and things) remains dominant, even in this dispersed, seemingly flatter form, implicit in notions of the origins of governmentality as 'the governmentalization of government' (Dean 1999). This, in my reading, problematically reinforces the importance of the state's will as the key driver/design of environmental relations, even if those relations are played out in areas notionally organized around and by nonstate actors and with an emphasis more on how people are governed rather than from where and by whom (Chapter 1).

Further, although notions of governmentality have been used successfully to critique and thus resist forms of oppression, such as of sexuality and the body, in support of a more liberal academic agenda, some writers find it more difficult to use such notions to critique and yet support environmental positions. Rutherford is a case in point when she critiques the push to 'be a better environmental citizen' but then supports such a push as something that 'people in the North *must incorporate* into their lives' (2007: 299, my emphasis). Focussing entirely on environmental governmentality also prevents a more positive consideration of forms of environmental conduct that are themselves constructed as resistance to others that are sponsored (in both senses of the word) by the state or neoliberal capital; it also prevents a consideration of where governmentality fails or is ineffectual (Rutherford 2007: 300), such as in many environmental fields. Death (Chapter 4) shows in more detail the difficulties of drawing ethics from governmentality analyses and the often ambivalent consequences.

Other perspectives take a less top-down view of how environmental practices are understood and made and have attempted to treat successful and unsuccessful attempts at ordering the world more symmetrically. Here, I draw on theoretical tools from ANT, which has become well known for emphasizing that power is produced through human-nonhuman interactions, rather than residing in particular actors or solely in the human or nonhuman realm (e.g. Latour 1986). But using ANT also increasingly reflects a more general turn to hybridity, assemblages and associational theory in some areas of social science to attend explicitly to the heterogeneous array of actors (or 'actants', to avoid implicit assumptions about these being necessarily human, see Latour 2005) involved in making social and environmental worlds. This turn also seeks to shed light on the ways knowledge, practice and power are coproduced and travel through ongoing (but often precariously unstable) networking, rather than through preexisting patterns of agency and ability. And in this networking, all these diverse actants can be (and indeed are likely to be) themselves transformed.

Under this kind of associational thinking, power does not originate at the top and flow down, but is produced through interaction and association (Latour 1986, 2005). Hence, state-led processes of governing become problematic – and often ineffectual and ultimately unsuccessful – where the internalization of state values by individuals is resisted (Chapter 7) or where shifts in governance towards a nonstate agenda need to be explained. Indeed, state values themselves – the imperative to govern – can also be seen as a product of associational power; states in recent years have gradually shifted towards low-carbon policies because of the influence of others, such as those in scientific and nongovernmental organizations (NGOs).

So, the theoretical tools offered by governmentality and ANT work differently to analyse carbon policy and practices (see Chapter 2 for more details). Some have argued that the two can be combined. For example, Lovell and Liverman argued that analyses using governmentality and analyses using ideas from science and technology studies (such as ANT) can be 'complementary', in that both analyse technologies, but the first in terms of technologies primarily as ideas by which people's conduct is controlled in the interests of governance, and the second in terms of technologies also as material devices which have agency (2010: 271). This distinction between conceptual and material technologies somewhat simplifies the underlying analytical frameworks and their notions of how the world develops and how power is produced. However, Lovell and Liverman rather undermine the usefulness of governmentality by concluding that two sorts of carbon offset markets can be distinguished precisely by the applicability of governmentality: the compliance market is seen as amenable to 'a governmentality analysis', whereas the voluntary market is less amenable to such an analysis (2010: 270). Indeed, Lovell and Liverman imply that the green credentials of the voluntary market are far more credible and (perhaps surprisingly for a paper in *New Political Economy*) see such offsetting not as a project driven by government or neoliberal capital, but as morally grounded.

The difficulties in adding together or combining governmentality and ANT is also nicely illustrated in the contrast between two geographical papers. First, Rutland and Aylett used ANT to analyse how an environmental policy develops in Portland, Oregon, through 'the interlinked and ongoing processes of constituting the local environment', but then switched to governmentality to analyse how that policy is implemented subsequently through 'the making of political subjects' in relation to the local state, although they do insist that the two sets of processes are 'interwoven' and overlapping in time (2008: 629). Second and in contrast, Higgins and Larner reverse this sequence, seeing governmentality more readily applied to the development of state policy and ANT more applicable to the implementation of that policy, arguing that 'many of the empirical analyses in the governmentality literature have been criticized for tending to study how technologies of governing are constituted discursively from the perspective of those "programmers" seeking to govern, rather than investigate how they are put into place and the consequences of this at the level of everyday practice. ANT provides a conceptually coherent, and complementary, approach that addresses these limitations' (2010: 5).

In practice, both works problematically blur the distinction between policy development and policy implementation; in practice, policy continues to change even as it is implemented (or fails to be implemented). A scalar distinction also lurks beneath both, in that ANT tends to be seen as more applicable to 'localized processes through which standards are implemented, rendered workable and shape identities' (Higgins and Larner 2010: 8) or individual-level processes and practices, whereas governmentality is seen as more applicable to 'the programmatic level' (Higgins and Larner 2010: 5), the strategic agenda of local government or 'the worldview of the state' (Rutland and Aylett 2008: 643).[1] Again, in practice these two scales link and merge as policy develops and cascades down and up scalar hierarchies and as individual choices effectively make national policy work (or fail).

[1] Even Dean struggles to explain how a governmentality approach such as his 'analytics of government' is different from a theory of the state, especially as his 'series' of sovereignty-discipline-governmentality brackets these together (1999: 23–6). To distinguish governmentality from sovereignty, he argues for seeing the state not as a

ANT has not been applied very much to carbon calculations, although it has been applied to other forms of calculation. For example, Callon argued that 'calculating tools' (such as from accounting practices) create both calculativeness and 'spaces of calculability' for economic markets and that, although a governmentality perspective would see this accounting as strongly disciplining people's behaviour and decisions, such disciplining is not 'mechanical, irreversible or irrevocable' but changes as the (calculating) tools that enact it are themselves reconfigured (1998: 23f). As Barry and Slater put it in their overview of Callon's work on economics, 'the capacity to calculate depends on a set of technical devices and discursive idioms that make calculation possible' and the calculative processes themselves create 'new realities', that is, new objects for trade or management that involve further processes of calculation (2002: 181).

Although online carbon calculators designed for domestic users are not strictly economic calculations, they connect with economic processes of purchasing and using consumer goods (including domestic energy) and thus collectively construct spaces in which consumption can occur. Hence, calculation is not performed by people alone (as atomistic, rational agents under classical economic theory), but collectively through 'complex socio-material arrangements' (Barry and Slater 2002: 182) of humans and nonhumans. Calculativeness itself, under this understanding, is not a property or power possessed (innately, solely or at all) by humans, but is relationally produced (performed) through these interactive arrangements.

More recently, Callon has applied these ideas explicitly to carbon markets, casting them as 'carbon experiments' for which an open-ended, flexible 'experimental, agnostic approach' to monitoring and improvement is preferable (Chapter 2, for a more detailed discussion of Callon's paper) (2009: 536). Callon argues that, unlike some economic experiments which result from the ivory towers of universities and 'the outcome of prior intense theoretical reflection', developing carbon markets are 'fruits of the imagination of innovators in the wild … truly collective, distributed experimentation deployed in time and space, more or less chaotically' (2009: 538). So these experiments are not only unusual in that they are happening *in vivo* (live/in the world), rather than *in vitro* (in the lab), but also because, by being in the world, they include not only traditional 'economic agents' for example companies and consumers, but also other heterogeneous sorts of agents, knowledge producers and influences, for example scientists, NGOs and European Union (EU) bureaucrats. Hence, the initial problematization of 'global warming' or 'climate change' as an issue proliferates in the world into other sorts of problematizations (Callon 2009: 543), which include measurement problems.

unified actor that holds power, but as a composite of multiple agencies or 'assemblages' (page 26) – this last word in particular, although also the whole argument, resonates with ideas from ANT. Rose (1991) also occasionally draws on Latour through notions of inscription and the state 'acting at a distance' on individuals, again suggesting resonances between theories at the time, although perhaps the gulf between these theories has since increased with the years.

Another reason ANT is useful in analysing carbon calculations is its origins in the sociology of science and particularly the way that knowledge practices develop, are shared or contested, work with devices, are inscribed (or not) in different ways and become authoritative (or not) (e.g. Callon 1998, 2009; Law 2002). Moreover, in relation to public consumption, this work has also focussed on how interactions between humans and technology shape humans and technology, making both malleable through reciprocal and symmetrical influence (in principle, if not necessarily in terms of degree). For example, 'public engagement exercises' have been shown to be themselves technologies that make publics in distinctive ways (Felt and Fochler 2010), rather than merely reach out to some neutral or preexisting group of people.

> Publics are not merely being encouraged to express their citizenly concerns, they are also being 'made' as particular types of citizens by virtue of the models of the public that inform public engagement. (Michael 2009: 619)

These publics may particularly be shaped by how they are engaged with as either 'consumers' (with assumptions that they are self-interested, hedonistic and driven by price, instant gratification and individualism) or as 'citizens' (with assumptions that they are driven mainly by a sense of public duty and responsibility to others, as well as long-term goals within an ongoing relationship with the state through voting, participating in planning and so forth). However, in practice, this boundary between publics behaving as consumers and publics behaving as citizens is blurred (Michael 1998), as both constructs are idealized rather than realized.

I now want to apply some of these theoretical ideas to the messy reality of carbon calculation. This has two elements, which continually co-evolve – designing and using carbon calculations – and I take these in that order in what follows.

Making Generic Calculative Devices

Designing carbon calculators is a messy process. As Paterson and Stripple note, their designers might include or exclude multiple possibilities, and 'the vast majority of calculators resolve them in the direction of the easier to calculate' (2010: 350). But for these kinds of public engagement technologies, the decision to include an 'easier to calculate' calculation is then buried behind the 'user-friendly' online front, which often deploys animation, cartoons and visual aids in an attempt to make the user's experience more entertaining and less confusing.

Hence, these carbon calculators are a good example of public engagement 'black boxing' (Latour 1986) where a host of (mostly government-sponsored) environmental science is bundled up and buried in algorithms. Some black boxing is clearly done before the calculators even encounter users on the Internet. Each calculator has its own scope and set of questions – WWF's, for example, asks immediately about the user's diet (others do not), recycling habits and what type of heating system is used in the home (e.g. oil or electricity), but does not ask for details such as how much fuel (e.g. in litres or kilowatt hours) is used to power that heating system. Although these in/exclusions can be analysed if the different calculators are compared by consumers, this itself requires extra (and unlikely) work, because the underlying metrological choices remain implicit.

There is a division both of labour and of calculation here. First, government commissions[2] (and sometimes itself conducts) research to establish commensurability between different sorts of data, for example to agree upon conversion factors for energy types, to support diverse policy aims. In the United Kingdom, this has mainly been done by Department of the Environment and Rural Affairs (Defra). Second, these factors or data are also deployed specifically in carbon calculators. In the United Kingdom, the government-sponsored online carbon calculator is 'Act on CO_2' and its design was overseen by the Department for the Environment and Climate Change (DECC 2009: 12), but incorporating Defra's conversion factors.

It is notable that, when DECC revised the 'Act on CO_2' calculator after two years of online existence, it sought to make 'the back-end engine and all data and assumptions "open source". This allows anyone with an interest in carbon calculations or footprinting tools to see exactly what has been considered ... and to freely copy the method and, potentially, improve on it' (2009: 4). The details of the revisions were published in a long (156 pages) methodological document detailing the assumptions built into the 'Act on CO_2' calculator. This demonstrates the messy and potentially controversial choices made in (re)designing the carbon calculator and also offers an information subsidy for other (nongovernmental) actors wishing to develop their own versions, effectively to recalculate the calculator.

Making Individual Calculations

Hence, carbon calculators are presented online with multiple political choices already embedded and obscured, that is, 'black boxed' so that they may not be modified or even detected by the ordinary user/consumer. But even where the design of the calculator seems to have been finished and it is presented online for public use, in practice it remains unfinished *until it is used*. And how it is used changes what it is and how it works because, like all technology (if we take an ANT perspective), carbon calculators are part of an assemblage that makes carbon calculations, an assemblage that includes not only the designers, but also the users and their other devices, including computers and paper documents.

Let me illustrate this using examples from the class exercise that I mentioned at the beginning of this chapter. To use the UK government's 'Act on CO_2' calculator, users

[2] Other, nonstate bodies have also commissioned research in different carbon and environmental metrologies, such as the Welsh Assembly's work with the WWF about developing an ecological footprint, as mentioned later.

need to do extra work, such as looking up their energy bills and entering the figures either in pounds sterling or kilowatt hours of electricity or litres of oil and perhaps converting across time periods (from quarterly bills to years or months, for example). Most students seemed to have completed the exercise fairly quickly and so, when they did not have the answers to hand or did not want to spend the time looking for the answers (such as energy bills) or 'working it out exactly', they simply guessed, because this speeded up the calculation.

This was especially the case when students were renting their accommodation and paying energy bills bundled up into an all-inclusive monthly rent, rather than as separate amounts. In other cases, students were not familiar with the structure and fabric of their accommodation – whether it was insulated and how. One online calculator asked what the average temperature of the house was and the student felt unable to judge this because different areas of the student's (shared) house were cooler or hotter than others, not least depending on the different predilections of the occupants.

Users therefore can choose to forego the extra effort required to retrieve bills, to convert monthly amounts into yearly amounts and to perform other ad hoc calculations – they resist, guess, estimate and fudge to speed calculation up. Anticipating this, online calculators are often designed to build some of this work into their own subroutines by providing shortcuts for users: where users do not have (or do not wish to find or use) their bills with measures in units that physically measure volume or flow, such as litres of oil or kilowatt hours, calculators use proxy measures by matching users to generic profiles or categories of consumption based on their answers to behaviour questions such as 'Do you often leave your appliances on standby?'

A good example is when the user can choose a national average to stand in as a proxy for some part of his or her household or individual consumption, instead of working out what he or she actually uses/spends. For the United Kingdom, the proxy represents a mythical middle ground, black boxing those taken-for-granted practices that represent the British standard, the fictional norm abiding within the traditional political territory of the nation-state, reflecting also the calculator's sponsorship by the nation-state. This also obscures the ways being in Britain itself shapes how people consume carbon, compared to, for example, being in Nepal or Nigeria, in that the (infra)structural influences on individual consumption, both social and technological (e.g. Shove 2003), go unexamined.

I shall come back to the problems with national proxies a little later, when I discuss benchmarking. For now, I will briefly note that choosing this national figure means that the individual user will not input the data; instead, a standardized value will be incorporated without the user having to do any work of his or her own. The calculation thus becomes generic and modelled:

> Where no bill data information is given, the calculating engine will take the information provided through the questions and model an approximate carbon footprint for both the heating

and lighting of a home.... This method is based on in-built, generalised assumptions about average emissions for each type of home, appliance or travel mode and hence has limitations, but is a good approximation in the absence of the bill data. (DECC 2009: 5)

As mentioned earlier, DECC revised the 'Act on CO_2' calculator after two years of operation online and the main reason given was to improve estimates: 'to add more accuracy to the inferred modelled calculations (i.e. those made without bill data available)' (2009: 5). To put this another way, DECC wanted to support user guessing more effectively by providing both 'a bill data and a non-bill data journey' through the carbon calculations, depending on the user's choice. This builds multiplicity into calculative practice by relying on algorithms to convert, for example, pounds (as a proxy measure) to kilowatt hours or litres of oil and then converting these in further algorithms to CO_2.

Users can also choose the form of their data in both time and physical measures. In terms of time, they can input bill data under the headings 'Monthly', 'Quarterly', 'Yearly' or 'Pre-payment meter', depending on their household circumstances. For inputs using the 'Monthly' format, 'the calculator assumes the costs are evened out to accommodate the lower bills in summer and higher in winter. Hence the calculator multiplies the amount by 12 and applies the appropriate calculation' (DECC 2009: 10). For 'Quarterly' inputs, 'the user must indicate which quarter the input pertains to. As both bill data and energy usage are seasonally sensitive, the calculator will make an adjustment depending on which quarter is chosen' (DECC 2009: 11). Here, rather than users being standardized, the seasons are standardized and the climate averaged out – ironically, stability and predictability of climate are assumed, despite the policy imperative for the calculations being climate *change*.

In terms of physical measures, there is often a choice of scientific units (litres of oil or gas, kilowatt hours of electricity, kilograms of coal or biomass) or of monetary units of pounds and pence. Students (and I) often found monetary units easier to work out in terms of how much we paid each month, as did the householders studied by Hargreaves (Chapter 7), although monetary units were also less accurate, not least because 'the user's particular energy tariff [is needed] to exactly determine what they pay for a unit of energy' (DECC 2009: 5). Here, the user and the conversion factors built into the calculator together turn a motley of monetary figures into a singularity of carbon. As Law suggests, the calculations are fluid, blurred and incomplete and 'when fluid objects or subjects do become visible in network space they tend to look dangerously *elusive, vague and sloppy*' (2002: 101f, italics in original).

And, unlike the physical conversion of electricity and oil to carbon dioxide (which is unaffected by recession or politics or undersupply), the socioeconomic conversion of money to carbon is itself a complex assemblage that shifts in and out of accuracy with the changing market. For example, DECC's conversion factors from converting pounds paid for domestic oil into kilograms of CO_2 in 2009 have dated rapidly: it converts at thirty-nine pence per litre of oil, while by 2012 domestic oil in England had reached nearer fifty-nine pence per litre (2009: 13). So, although the online results seem fresh each time, the carbon calculator is always historical. This emphasizes the disconnection between the appearance of immediacy online and the actuality of metrological history that is bundled up behind/into the calculation.[3]

Similarly, technology always seems to be ahead of the calculators: for example, LED televisions are now purchased and used in UK homes, but they are too new for Defra to have yet produced energy conversion factors for them (DECC 2009), meaning that users must call them something else (probably LCD, which are more carbon consuming), using another proxy to complete their calculations. In this case, the hard technology, rather than the soft money, has changed since the calculators were last revised – the fossil record here does not reach as far as the latest speciation.

In general, we could say that time itself is poorly accommodated in the calculators, which do not seem to be directly linked in any 'live' way to prices and are sometimes unclear about the time period for which users should enter information. However, the Carbon Account's calculator[4] provided a way to log data on a weekly or monthly basis and thus updated the calculations over longer time periods, providing some monitoring over time of the user's consumption, but also requiring more commitment by the user to keep on gathering and inputting data.

Sometimes users abandon the attempt to calculate, giving up and perhaps going elsewhere on the Internet, tired of the extra effort involved in making individual calculations, looking for something more interesting to do or disconnected by computer problems. As Hargreaves (Chapter 7) notes, the effects of such domestic carbon calculations are uneven across diverse populations. Several of the students noted that the online system was frustrating in producing errors or wiping data and one commented that they would have given up after the first couple of attempts if they had not been asked to do this by me. DECC's (2009) monitoring shows that, in its first two years, the 'Act on CO_2' Carbon Calculator received over 1.6 million unique Web visits but with a 'web conversion rate' of only 40 per cent, that is, 40 per cent of users who began the calculation actually completed it, meaning that over half did not.

Benchmarking Calculations

So far, I have outlined the calculations performed through designing a carbon calculator, then the calculations (and sometimes deliberate noncalculations) performed through using one. There is then often a third calculative stage in which many calculators offer a comparison of the user's carbon footprint against some form of calculated benchmark. Table 6.1 gives some examples of such benchmarks that demonstrate

[3] This reflects a digital/digitized world where time is increasingly erased, history is lost and temporal sequencing disappears as old information is recirculated and reposted online as fresh.

[4] Via http://www.thecarbonaccount.com/.

Table 6.1 *Benchmarking examples from online carbon calculators*

	Benchmarking examples
Act on CO_2 http://carboncalculator.direct.gov.uk/index.html	National Average: 4.46 tonnes of CO_2 per year National Average in household of same type: 14.09 tonnes per year
Carbon Footprint http://www.carbonfootprint.com/calculator.aspx	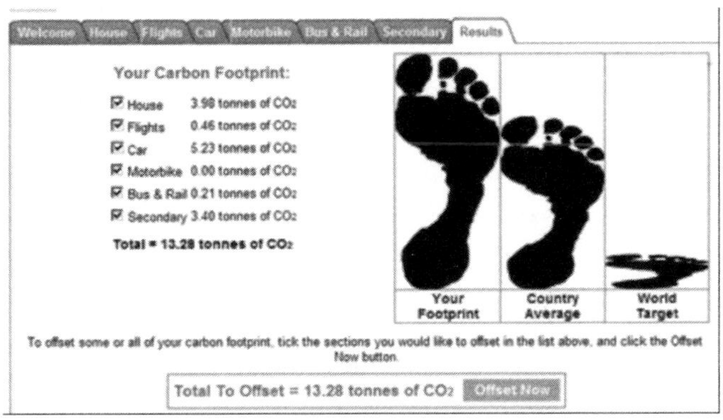 (Copyright Carbon Footprint Ltd. www.carbonfootprint.com) 'The average footprint for people in United Kingdom is 9.80 tonnes The average for the industrial nations is about 11 tonnes The average worldwide carbon footprint is about 4 tonnes The worldwide target to combat climate change is 2 tonnes'
WWF http://footprint.wwf.org.uk/	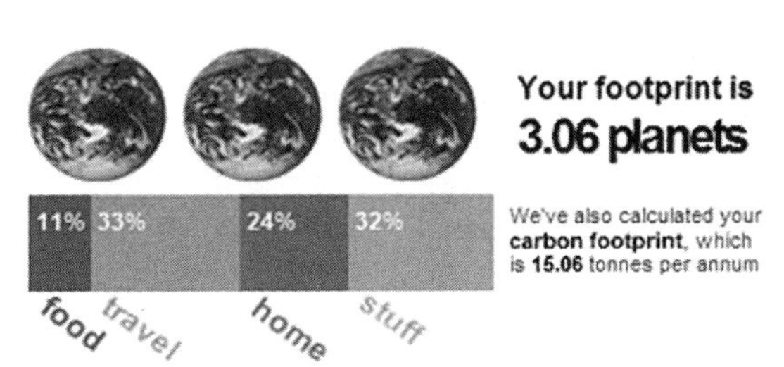 (Photos & graphics © WWF)

the different measurement strategies and units chosen by the designers. The UK government-sponsored 'Act on CO_2' calculator uses 'National Average in household of same type' as its benchmark, suggesting a form of 'peer review', whereas WWF, the international environmental NGO, uses 'planets' as units and as visual representations. The Carbon Footprint company uses both tonnes of carbon dioxide as units and black footprints of various sizes as visual representations of impact benchmarked against both national and global averages.

By connecting different scales of environmental behaviour and reference, from individual/household through regional or genre to national and global, these carbon knowledge practices 'simultaneously totalise (aggregating social practices, overall greenhouse gas emissions) and individualise (producing reflexive subjects actively managing their greenhouse gas practices)' (Paterson and Stripple 2010: 359). However, the two vectors of aggregation and individualization are themselves calculatively connected. This is because benchmark calculations are themselves built upon the messy calculations and assemblages already outlined as performed by/through use of online calculators by individuals. Using ANT, this analysis sees both devices (online calculators that are both digital and material) and humans (online users that are also both digital and material) as actants, coproducing both knowledge about and powers to shape carbon consumption in a hybrid world.

For example, when upgrading the 'Act on CO_2' calculator in 2009, DECC's designers used the data users inputted over the first two years of the calculator's online operation to calculate average footprints according to both national profiles and profiles of groups 'who have similar circumstances in terms of house type and size; heating type; and number of people in the home' (DECC 2009: 6). Hence, the calculator is reflexive and directly shaped by the users' choices about the data they input, data which is used to benchmark other (or possibly the same) users. Again, the calculations and noncalculations created by the online devices are both buried deep within this metrological process and also accumulated and rebroadcast into the world through recalculation.

The calculators are also and perhaps more obviously reflexive through using peer pressure and group surveillance through social networking (Figure 6.1), where users' carbon footprint would be visible to other users invited to share the results through a

Figure 6.1 Social networking as peer pressure from WWF's carbon calculator (Photos and graphics © WWF).

form of reflexive monitoring of consumption (Giddens 1991). Here, the moral ordering of consumption is facilitated not through the visibility of fashion brands on clothing and other products, but through the virtual visibility of carbon calculations. Similarly, Whitehead, Jones and Pykett (Chapter 5) argue that publishing average usage figures on household energy bills to increase peer pressure can be understood as 'libertarian paternalist policy' and part of the behavioural politics of 'nudge', although they also argue that such a rationality-based approach to behavioural change may prove ineffectual because the barriers to pro-environmental behaviour change come from much deeper, neurologically based values and ideas.

Offsetting Calculations

And finally, there is often a fourth calculative stage, in which many calculators calculate not only the carbon footprint of the user, but also how much users would have to spend on carbon offsetting to neutralize their footprint and thus buy a (net) carbonless lifestyle.

Much has already been written about carbon commodification through trading and offsetting and the neoliberalization of nature. Pieces of information in the form of tonnes of CO_2 equivalent become the 'commodity form' of carbon through complex metrologies (Bumpus and Liverman 2008) and thus contribute to capitalist accumulation and flows of finance. There has been particularly strident criticism of offsetting as (falsely) buying legitimacy or an ethical position and as an embodiment of neoliberal entrepreneurialism, but Newell and Paterson (2011) also see such mechanisms more positively as ways to 'experiment' in carbon markets.

Calculating for Climate Change – Practical and Analytical Considerations

In this chapter, I have used ideas from ANT to analyse how carbon calculations are coproduced as complex assemblages and as imperfect carbon metrologies, emphasizing how they depend on messy assumptions, black boxing, noncalculation and users' guesses. I do this to highlight how carbon calculators are co-constructed as socio-technological assemblages, rather than being straightforward technologies or instruments of government, and how users act as co-creators of environmental calculation, rather than merely subjects of these governmental technologies. As Blok (Chapter 2, page 55) notes, ANT helps us in 'granting ontological and political weight to the collective inventiveness of diverse climatic publics'. The choices online users make, including to ignore, lie about or fail to provide information, bring very different results and also undermine any notion of straightforward truth beyond the collectivity in which they participate.

Emphasizing co-creation also helps to underline the work that the user is required to do to obtain a carbon footprint and how each co-creation event can produce different footprints. This may arise because different calculators are used (one student reported that her carbon footprint ranged from 1.8 tonnes to 8 tonnes on different calculators, a diversity of results that made students distrust all the calculators on principle) or because the choices made by users as to input data changed over time (I used one calculator three times over a period of days and got different results each time, not least because I made greater efforts each time to find real figures, for example energy bills).

I have therefore emphasized these knowledge practices of assembling, guessing and questioning figures to suggest that some of the analyses of carbon calculators as technologies of governmentality overclaim the potential influence in terms of changing consumption. In a governmentality approach, statistics become a device, a technology wielded by the state to shape the thinking and doing of its citizens and to push them towards reducing their carbon dioxide emissions. For example, Rutland and Aylett see the success of reducing carbon emissions in Portland as an example of the success of the (local) state in coercing individuals to internalize its agenda and thus critique ANT for not providing 'an adequate understanding of precisely how actants like the state attempt to shape the subjectivities and behaviours of other actants' (2008: 643).

However, from an ANT perspective, we can instead see Portland's policy/agenda as a product of networking individuals and other actants. This perspective sees the state as the product of associations of actors, as an assemblage rather than some clearly defined and overly powerful actor with a clear will or agenda which it attempts to force upon its subjects, albeit through more dispersed sites and methods. Hence, carbon calculation does not merely produce carbon subjects by 'shaping not only their behaviour but their internal rationalities, identities, what they fundamentally regard as "normal" behaviour' (Paterson and Stripple 2010: 359), but is also partly produced by those subjects.

Another key point is the role of the state in environmental regulation, a role frequently overstated in recent years in that carbon policy more generally is often pushed forward not by state interests (indeed, may often be against state interests, as has been the claim of the United States for over a decade), but by nongovernmental interests in the form of environmental pressure groups. Carbon footprinting (and its predecessor, ecological footprinting) are thus technologies nongovernmental actors use to push both individuals *and states* to consider their carbon (and environmental) impacts more generally. For example, WWF Wales has been collaborating with the Welsh Assembly for nearly ten years in using the idea of ecological footprinting to raise environmental priorities in assembly decision making. Environmental pressure groups have therefore developed their own carbon calculators, such as WWF's, and also pushed government at various scales to invest in measuring and managing carbon in domestic and commercial settings. This again suggests that carbon calculators do not solely serve as technologies of governmentality by/for the state, nor as ways to govern and produce environmental citizens. Rather, they are also contesting technologies that compete to influence not only consumers, but also state agencies, NGOs and even each other. In this sense, ANT seems more ready to acknowledge the ineffectual operations of carbon calculations and their failure to influence carbon consumption, thus addressing one area in which governmentality analysis has been lacking (e.g. Rutherford 2007: 300).

Finally, from an ANT perspective, how much do calculations of consumption, benchmarks and offsetting themselves cause changes in consumer behaviour, become themselves actors? How influential are such devices against the barrage of other actors (advertising, social conventions, hedonistic desires) in continually pushing up energy use and thus carbon emissions? The failure of carbon calculations to persuade my students again underlines the wider failure of the state to persuade its citizenry to change its habits through such technologies of governmentality. Hargreaves (Chapter 7) provides further evidence that some citizens are influenced far less than others by carbon metering and, even where householders initially change behaviour by reducing their energy consumption in response to more (and more visible) information about carbon usage, these effects may quickly fade with time. The power to shape carbon behaviour therefore resides neither in the state nor the individual consumer, but is distributed through the complex assemblage of humans and nonhumans. In this way, the ability of an ANT perspective to track both success and failure symmetrically is analytically helpful, if not very optimistic about the future for carbon control.

Acknowledgements

I would like to thank my second year students at the University of Hull, especially Laura Jordan and Tris Terry, for their participation and helpful comments when we discussed and tested carbon calculators.

References

Barry, A. and D. Slater (2002). Introduction: The Technological Economy. *Economy and Society* **31**(2), 175–93.

Bridge, G. (2010). *Heading 'Downstream': Towards a Cultural Political Economy of Energy Consumption*. [Online.] http://www2.le.ac.uk/departments/geography/documents/research/seminar_bradshaw_UK_ESRC_Energy_151010_summary.pdf. Accessed 22 January 2013.

Bumpus, A. G. and D. M. Liverman (2008). Accumulation by Decarbonization and the Governance of Carbon Offsets. *Economic Geography* **84**(2), 127–55.

Callon, M. (1998). Introduction: The Embeddedness of Economic Markets in economics. In *The Laws of the Markets*, ed. M. Callon. Oxford: Blackwell/The Sociological Review, 1–57.

Callon, M. (2009). Civilizing Markets: Carbon Trading between *in vitro* and *in vivo* Experiments. *Accounting, Organizations and Society* **34**, 535–48.

Dean, M. (1999). *Governmentality: Power and Rule in Modern Society*. London: SAGE.

DECC (2009). *Act on CO2 Calculator Version 2.0: Data, Methodology and Assumptions Paper*. [Online.] http://www.decc.gov.uk/assets/decc/what%20we%20do/global%20climate%20change%20and%20energy/tackling%20climate%20change/ind_com_action/calculator/1_20091120174357_E_@@_actonco2calculatormethodology.pdf. Accessed 22 January 2013.

Felt, U. and M. Fochler (2010). Machineries for Making Publics: Inscribing and De-Scribing Publics in Public Engagement. *Minerva* **48**, 219–38.

Giddens, A. (1991). *Modernity and Self-Identity: Self and Society in the Late Modern Age*. Cambridge: Polity Press.

Higgins, V. and W. Larner (2010). Standards and Standardization as a Social Scientific Problem. In *Calculating the Social Standards and the Reconfiguration of Governing*, ed. V. Higgins and W. Larner. Basingstoke: Palgrave Macmillan, 1–17.

Latour, B. (1986). The Power of Association. In *Power, Action, Belief: A New Sociology of Knowledge*, ed. J. Law. London: Routledge & Kegan Paul, 264–80.

Latour, B. (2005). *Reassembling the Social: An Introduction to Actor-Network Theory*. Oxford: Oxford University Press.

Law, J. (2002). Objects and Spaces. *Theory, Culture & Society* **19**, 91–105.

Lovell, H. and D. Liverman (2010). Understanding Carbon Offset Technologies. *New Political Economy* **15**(2), 255–73.

Luke, T. W. (1999). Environmentality as Green Governmentality. In *Discourses of the Environment*, ed. É. Darier. Oxford: Blackwell, 121–51.

Michael, M. (1998). Between Citizen and Consumer: Multiplying the Meanings of the 'Public Understanding of Science'. *Public Understanding of Science* **7**, 313–27.

Michael, M. (2009). Publics Performing Publics: Of PiGs, PiPs and Politics. *Public Understanding of Science* **18**, 617–31.

Newell, P. and M. Paterson (2011). *Climate Capitalism*. Cambridge: Cambridge University Press.

Paterson, M. and J. Stripple (2010). My Space: Governing Individuals' Carbon Emissions. *Environment and Planning D: Society and Space* **28**, 341–62.

Rose, N. (1991). *Governing the Soul: The Shaping of the Private Self*. London: Routledge.

Rutherford, S. (2007). Green Governmentality: Insights and Opportunities in the Study of Nature's Rule. *Progress in Human Geography* **31**(3), 291–307.

Rutland, T. and A. Aylett (2008). The Work of Policy: Actor Networks, Governmentality, and Local Action on Climate Change in Portland, Oregon. *Environment and Planning D: Society and Space* **26**, 627–46.

Shove, E. (2003). *Comfort, Cleanliness and Convenience: The Social Organization of Normal*. London: Berg.

7

Smart Meters and the Governance of Energy Use in the Household

Tom Hargreaves

Introduction

The development of a 'smart' electricity grid that is able to accommodate and balance decentralized forms of electricity generation is a central plank of the UK government's 'Low Carbon Transition Plan' (Department of Energy and Climate Change [DECC 2009b]). The first step towards achieving this is the roll-out of smart meters to all UK households by 2020 (DECC 2009a). At an estimated cost of £11.3bn (National Audit Office 2011), this is not only a major investment in the energy infrastructure, but also a key means by which the UK government seeks to engage the entire UK population, household by household, in the effort to tackle climate change. Each smart meter will be accompanied by a real-time display (RTD) monitor that, the DECC claims, 'will provide consumers with real-time information on their electricity use to help them control consumption, save money and reduce emissions' (2009a: 7). A key aim of the roll-out is to address the 'double invisibility' (Burgess and Nye 2008) of energy in everyday life. Smart meters and RTDs introduce a new field of visibility into homes, enabling householders to 'see' their energy use and therefore perform new kinds of calculation about which kinds of usage are desirable or otherwise, and thus to introduce new forms of self-government of their energy use and carbon emissions (cf. Wilhite and Ling 1995). Whilst this rather simplistic linear, information-deficit model of behaviour change has been widely criticized (e.g. Burgess et al. 2003; Shove 2010), previous trials of RTD monitors have suggested they can generate savings of anywhere between 3 per cent (AECOM 2011) and 15 per cent (Darby 2006).

It is clear from these sorts of policy statements and research findings that the smart meter roll-out represents an effort to introduce a particular expression of carbon governmentality right into the heart of people's everyday lives. Nonetheless, as Darby (2006) observes, the overwhelming majority of energy feedback studies focus solely

The author would like to thank the UK Engineering and Physical Sciences Research Council (EPSRC) and E.ON UK-funded 'Transition Pathways to a Low Carbon Economy' project (Grant: EP/F022832/1); Carbon Connections (Grant: CC29); Green Energy Options; SYS Consulting Ltd.; British Gas and all participants in the Visible Energy Trial for making this research possible.

on what level savings are or are not achieved and say nothing of *how* such savings are realized. In this respect, the household is rendered as something of a 'black box'. Here, a Foucauldian focus on the *how* of power and the quotidian technologies and rationalities of government is perhaps especially insightful (Miller and Rose 2008). Such an analysis focuses attention on the ways new programmes of government problematize particular fields of reality (in this case households) and seek to govern conduct within them by introducing new forms of knowledge in an attempt to structure action and practice in new ways. The smart meter roll-out can thus be usefully viewed through such a governmental lens as it is based on a clear set of assumptions about everyday energy use and how it might be changed by the introduction of new knowledge through instantaneous feedback. Whilst there is some evidence to suggest that these assumptions may have some purchase in certain situations, as of yet, no studies have examined *how* the particular form of carbon governmentality that smart meters introduce to homes actually functions in practice.

Drawing on repeat interviews with twelve households participating in a trial of RTD monitors in the east of England, this chapter examines the ways householders understand, interact with and respond to the form of carbon governmentality that smart meters extend. Specifically, after a brief review of the relevant literature on carbon and energy governmentality and an overview of the Visible Energy Trial, it focuses on three core themes that emerged from the interview analysis. First, that carbon appears to be a relatively weak rationality of government. Second, that the monitors' carbon governmentality did not operate in the same way on all household members, but was filtered through preexisting relationships between household members. Third, that the monitors were actively resisted by householders both through appeals to longer-standing household ethics and aesthetics, and through a partial rejection of the territorialization of climate change at the level of the household. This chapter concludes by identifying a need for more research on the 'messy actualities' (O'Malley, Weir and Shearing 1997) of carbon governmentality. Specifically, it calls for further examination of how newly introduced forms of carbon government are necessarily superimposed upon, intersect with and become embedded inside other, preexisting modalities of power, and on the hybrid forms of government and mongrel identities that emerge as a result.

The Messy Actualities of Carbon Governmentality

The analytical approach of governmentality focuses on the alignment and convergence of agendas between governmental programmes and their subjects by examining the techniques and tactics through which individual subjectivities are constantly made and remade to (try and) achieve particular ends (e.g. Dean 1999; Lövbrand and Stripple Chapter 1, this volume). Miller and Rose distinguish between 'rationalities of government' to refer to the various ways particular issues are made to appear problematic and in need of governing, and 'technologies of government' to refer to the many and various methods through which this governing is attempted. One of the crucial distinctions between governmentality and other analyses of government is that, as Miller and Rose show, these rationalities and techniques of government do not exist in grand political ideologies or shows of military-industrial might, but rather in a heterogeneous cast of 'humble and mundane mechanisms' such as: 'techniques of notation, computation and calculation; procedures of examination and assessment; the invention of devices such as surveys and presentational forms such as tables; the standardization of systems for training and the inculcation of habits; the inauguration of professional specialisms and vocabularies; building designs and architectural forms' (2008: 63). As is clear from this list, these mechanisms do not act by imposing power *over* practices, but rather work *through* practices by attempting to structure 'the field of possible action of individuals' (Paterson and Stripple 2010: 346).

Whilst Foucault himself is said to have turned his back on the environment (Darier 1999), since the mid-1990s a growing strand of governmentality studies has turned its attention to the environmental problematique to examine how 'the environment' has been made into a set of problems for which particular kinds of intervention and management are possible, desirable and even seemingly 'natural' (e.g. Darier 1996, 1999; Rutherford 2007). Within this general move towards understanding forms of eco-governmentality or environmentality (Agrawal 2005), and as this volume illustrates, there has been a growing focus on how climate change is understood and acted upon as a governable concern (e.g. Slocum 2004; Rutland and Aylett 2008). Whilst much of this field focuses on the configuration of programmes of carbon governmentality at the national or even global scale (e.g. Oels 2005; Lövbrand and Stripple 2006), Paterson and Stripple have begun the important challenge of analysing how climate change, or more specifically carbon dioxide, is increasingly made into an individualized concern to be acted upon in the course of everyone's normal everyday life.

In their paper 'My Space: Governing Individual Carbon Emissions', Paterson and Stripple (2010) examine how carbon emissions have been individualized – such that people think of 'my emissions' or 'your emissions' (hence the 'My Space' metaphor) – in a range of increasingly common governance programmes, for example carbon footprinting, carbon offsetting, carbon dieting, carbon rationing action groups (CRAGs) and Personal Carbon Allowances. Their analysis persuasively outlines the particular rationalities and technologies of government deployed in these programmes to create different kinds of 'self-regulating, carbon-conscious citizens' (2010: 345). For example, where carbon offsetting initiatives attempt to produce 'the carbon displacer' motivated by guilt to 'do-good' by the climate, CRAGs try to make up a solidaristic subjectivity of 'carbon communitarians' focussed on self-sacrifice and collective learning to reduce emissions.

As energy and climate change have increasingly converged on the UK policy agenda (Lovell, Bulkeley and Owens 2009), so Paterson and Stripple's (2010) analysis has been mirrored by several recent studies within energy social science focussed on the kinds of 'energy public' envisaged within transitions to a low-carbon economy. For example, Van Vliet, Chappells and Shove (2005) identify four distinct energy consumer identities including the captive consumer, the customer consumer, the citizen consumer and the co-provider. Similarly, Walker and Cass (2007) outline no fewer than ten distinct public identities (as captive consumers, active customers, service users, financial investors, local beneficiaries, project protestors, project supporters, project participants, technology hosts and energy producers) in renewable energy futures. In each case, the concern of these studies has been to expand on representations of passive energy consumers who simply respond to centralized concerns and move towards more active imaginaries of 'energy citizens' (Devine-Wright 2007) able to participate more fully in critical energy debates.

It is here, at the intersection of debates around self-regulating carbon-conscious citizens and active energy citizens, that UK policy interest in smart meters, real-time display monitors (RTDs) and the impacts of energy feedback on domestic energy use is perhaps best situated. In its hope that the provision of real-time information to householders through the smart meter roll-out will 'help them control consumption, save money and reduce emissions' (DECC 2009a: 7), the DECC is clearly introducing a particular form of carbon governmentality further beyond the state and right into the heart of citizens' kitchens, lounges and even bedrooms. As Paterson and Stripple (2010) observe, the broad governmental programme of carbon governmentality, at least when applied to individual citizens and consumers, can take a number of specific practical forms. This chapter addresses the particular form that carbon governmentality takes through the introduction of RTDs to people's homes. Specifically, RTDs introduce new rationalities and technologies of government into households, isolating electricity and gas consumption as the specific issue requiring control, rendering energy consumption visible in a more pervasive way and through wholly new presentational forms (forms that are far more precise, interactive and insistent than traditional energy bills, for example – and see Wilhite and Ling 1995), and quietly demanding new forms of calculation and action from householders (for example, is the kettle or the tumble drier worse for the climate?).

Following Paterson and Stripple's analysis, however, it is not (yet) clear what new kinds of energy and carbon subjectivities the smart meter roll-out may produce. All of the forms Paterson and Stripple outline – carbon counters, displacers, dieters, communitarians and citizens – as well as many more, or even none at all, are possible. What will eventually emerge, however, depends fundamentally not on the rationalities and technologies of government that smart meters embody, but rather on whether and how smart meters are 'domesticated' (Silverstone, Hirsch and Morley 1992) into people's everyday lives and homes. In short, and as will be the focus for the rest of this chapter, how does 'the conduct of carbon conduct' (Paterson and Stripple 2010: 342) actually function amid the complexities of everyday life?

Before addressing this question through empirical analysis, it is worth highlighting two of Foucault's early observations about governmentality that have not always been remembered in subsequent studies. First, Foucault observed that new formations of power never completely replace preexisting ones, but always link in and mesh together with them in ways that are themselves generative of new and alternative subjectivities (see Chapter 2 for more on the many hybrids created by the emerging meshwork of climate governance). Second, and relatedly, Foucault observed that 'there are no relations of power without resistances' (1980: 142). Examining the form of carbon governmentality that smart meters introduce to homes in isolation would therefore risk both overplaying their relative strengths and neglecting the fact that governmental programmes are always invoked, articulated and reproduced in varied ways. Instead, empirical analysis should explore the limits, or the edges, of particular governmental programmes to see how they mesh together to co-operate and/or compete with others, as well as how they are resisted by different people in different times and places. The rest of this chapter attempts to begin this challenge by turning attention to the messy actualities involved in a trial of RTD monitors.

The Visible Energy Trial

Throughout 2009 and 2010, 275 households in the east of England were recruited to trial a range of RTD monitors, designed and manufactured by a company called Green Energy Options (GEO) for a period of twelve months. As Figure 7.1 shows, three different monitors were used in the Trial, from the *Solo* which was the most basic and monitored only overall levels of household electricity use, to the *Trio*, an advanced system capable of monitoring up to 100 individual appliances separately. Whilst there are differences in the level of detail and specificity with which each monitor introduced carbon governmentality into the trial households, for the purposes of this chapter the crucial point is that all of the monitors offered near instantaneous feedback on current electricity use, an indication of how much energy had been used so far in any twenty-four-hour period, and an historical comparison with a previous twenty-four-hour period. In each case the monitors could display the information as kilowatt-hours used, carbon dioxide emitted or pounds sterling spent. In these respects, all of the monitors met the minimum display requirements for, and are thus comparable with, the RTDs that will be distributed as part of the national-level smart meter roll-out (Ofgem 2010).

From the 275 participating households, a sample of 12 households (stratified to ensure four of each type of monitor) was selected to take part in two semi-structured interviews – the first conducted a few months after the monitors were first installed, the second conducted twelve months later. The interview protocol focussed on exploring how the monitors had been used within the different households, how they had fitted in with or disrupted existing household routines and practices, how different

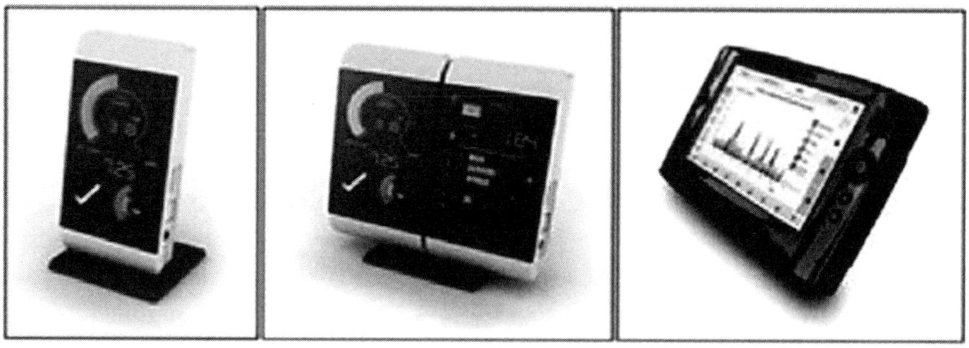

Figure 7.1 GEO's real-time display monitors (showing from left to right: the Solo; the Duet; the Trio).

household members had used the devices, and how, if at all, they had influenced household dynamics, relationships and decision-making processes. Full details of the trial are available in Hargreaves (2010) and Hargreaves, Nye and Burgess (2010, 2013).

The rest of this chapter draws on this interview data to highlight how the particular expression of carbon governmentality the monitors introduced played out in the real-life contexts of the twelve interviewed households. Specifically, it highlights three core themes that challenge and add nuance to the notion that the broad programme of carbon governmentality is becoming increasingly common and powerful, at least at the level of the household. First, that carbon is in many ways a weak rationality that made little sense to many householders and was often simply ignored. Second, that a focus on the moulding of *individual* carbon subjectivities neglects that individuals do not exist in a social vacuum. Third, that household members offered several forms of resistance to the form of carbon governmentality the RTDs carried, significantly lessening its impact.

Carbon as a Weak Rationality of Government

As part of the UK government's broad efforts to transition to a low-carbon economy, smart meters and the RTDs used in this trial represent a single new technology of government that simultaneously extends three kinds of governmental rationality into households: financial, energy and carbon. Through these three rationalities, the monitors render new areas of everyday domestic practice almost constantly visible and newly problematic by introducing forms of knowledge into households that, if their programme of government is to work, should encourage householders to consider the energy, financial and/or carbon implications of their energy-consuming routines and practices (cf. Paterson and Stripple 2010). Whilst this is no doubt true, interviews with householders suggested that the manner in which these new rationalities of government functioned in the course of everyday life differed considerably. Perhaps most interestingly, given the monitors are (at least marketed as) part of a broader effort to cut carbon dioxide emissions, all but a few interviewees stated that understanding their household energy use in terms of carbon was extremely confusing. Not only did they not understand how much a kilo or a tonne of carbon actually was, in the absence of something meaningful to compare it with, they also found it difficult to grasp whether this was good or bad or what effect it was likely to have on the climate

> I guess that's kilograms of carbon but I can't relate to that. I don't know what that means. You know I could, well I do know what it means but, you know, it just looks like a number to me. (D1i[1])

As this quotation suggests, on their own, the monitors seem ill equipped to introduce carbon as a rationality of government. The same was true with kilowatt hours. Instead, the majority of interviewees had opted to use the pounds and pence metric built in to the monitors as this helped them to make sense of the real-time information being provided in familiar ways:

> A kilowatt-hour to most people is an abstract figure, isn't it, whereas pounds and pence, you know what you're spending. (D3i)

By designing the monitors to provide readings in pounds and pence, the manufacturers might be interpreted to have preempted this apparent failure of *carbon* as a governmental rationality, and to have taken advantage of an already pervasive financial rationale. Indeed, interviewees often used well-known financial metaphors when discussing how they interpreted the readings provided by the monitors:

> If you turn the kettle on that will come on and you can see the dial go up and then you can look down to see how many credits you've got left for that day … or if you're in the red or the black. (S4i)

Further, and as this quotation illustrates, the metrics the monitors used encouraged householders to compare their current energy consumption against their previous and potential future consumption. By contrast, other forms taken by carbon governmentality, such as in carbon footprinting techniques or CRAGs (e.g. Paterson and Stripple 2010), often encourage comparisons with people in other countries or with one's friends and neighbours.

In these respects, the manner in which the monitors operated as part of a broader programme of carbon governmentality can be seen, from the outset, not to exist in isolation, but as linked in and hybridized with preexisting governmental rationalities and programmes. In particular, they did so in ways that served to further an individualized and monetized understanding of climate change.

[1] To preserve anonymity, all interviewees have been given a unique identifier. The letter refers to the kind of monitor they were using (S = Solo, D = Duet, T = Trio). The number (1, 2, 3, 4) refers to the individual household, and the signifier i or ii refers to whether the quotation is drawn from the first (i) or second (ii) interview.

Another interesting aspect of this meshing together of different forms of governmentality is that the designers of these particular monitors intentionally made them resemble a car dashboard. For example, each monitor had a 'speedometer' dial to provide information on how much electricity was being used at any one moment in time, and a 'fuel tank' to show how much energy had been used so far during the day. Again, despite being part of a broad programme of carbon government, the monitors seem not to introduce a pure form of carbon rationality, but instead to connect this with other prevalent and well-established forms of practice and the kinds of rationality that govern them. Indeed, one might speculate that these particular monitors were designed to appeal most strongly to those individuals most 'captured' (Reckwitz 2002) by car-driving practices. In this respect, the form taken by this carbon governmentality should be seen not as a monolithic formation, but as something that spreads unevenly as it connects only partially with preexisting governmentalities, and as each of its potentially infinite number of hybrid forms is effective only at moulding the subjectivities of certain subjects.

These initial insights suggest that, on its own, carbon represents a weak governmental rationality that apparently fails to make sense to householders and thus to have any direct influence on their household practices. Further, they show not only that the RTDs were designed to take advantage of and to link in with preexisting governmental rationalities, but also that householders themselves actively meshed rationalities together to help them make sense of the feedback they received. In this regard, it is perhaps pertinent to question whether or not the use of the increasingly familiar technique of metering and monitoring aspects of everyday life (Shove and Pantzar 2005) may in fact serve to hide carbon within other more dominant mentalities of rule (e.g. financial accounting, obeying the speed limit etc.). If so, questions are raised as to whether or not alternative technologies of government might be more effective at making carbon dioxide both comprehensible and important in its own terms. Further still, the frequency with which householders chose to use the financial, rather than carbon or energy metric, illustrates how monitoring and metering domestic energy use can monetize and individualize carbon and climate change in such a way that they come to be embedded within dominant, neoliberal mentalities of rule rather than serving to challenge them (e.g. Strengers 2011).

In all, these observations raise interesting questions about the relationships between different forms of governmentality (carbon, energy, financial). To some extent all of these forms seem able, at least in some respects, to borrow one another's metrics and rationalities. This implies that, for at least some of the time and in certain spaces, 'carbon-calculating individuals' can live happily alongside or even inside 'energy citizens' and rational utility maximizers. Significant challenges seem likely to arise, however, at the points where these distinct identities begin to conflict and fragment.

Carbon-Calculating Individuals?

The second core theme to emerge from the interviews suggests that a narrow focus on the 'carbon conscious *individual*' in studies of carbon governmentality might be misplaced. Paterson and Stripple recognize that individual practice is shaped by 'large-scale historical/structural forces' (2010: 344) such as capital and state and, rather than focussing on either individuals or structures, they call for analyses that address how *both* function together to introduce carbon governmentality to practice. The empirical evidence from the Visible Energy Trial adds nuance to this view, however, by suggesting that many mediating influences exist between large-scale structures and individual practice. Specifically, the trial revealed that relationships between individual subjectivity and the form of carbon governmentality introduced by the monitors are very strongly shaped by preexisting social relationships that already serve to govern energy use in the household in various ways.

In almost all cases, the monitors were used by a single household member who was predominantly (although not always) the oldest male. In each case, however, this individual explained how they had tried to engage other householders in using the monitor. For example:

> My wife hasn't even taken the slightest bit of interest in it. I'll be honest with you, and it's quite disappointing really because I'm trying to educate her but it's just not working (laughs). (T2ii)

Nonetheless, despite this general lack of interest from other household members (a point which reinforces those made earlier about the general lack of strength of carbon as a governmental rationality within these households), with the exception of one single-person household, all interviewees were clear that even if they themselves wished to change their everyday practices to save money, energy or to cut carbon, they were limited in their ability to do so by those they shared their households with. In some, admittedly relatively rare, instances this had led to new forms of cooperation among household members as they held family meetings and discussions to work out where they were wasting energy and where it might be saved. For instance:

> You look at it and you go: 'Guys, just have a look at this a second because that's week one, that's week two and that's week three. Look at that there – someone's left the oven on', and we can drill into that day and I could tell you that somebody's turned the oven on to cook something and then gone off and played [the computer game] Football Manager for two hours. So we've been burning 2.2 kilowatts for two hours there. (D2i)

More commonly, and as the following quotations show, the new carbon governmentality introduced by the monitors had managed only to impact a single householder and s/he had subsequently attempted, typically unsuccessfully, to introduce it to others:

I couldn't go ahead and do a lot of changes without her agreeing. I mean she's been rather resistant to putting in the new kind of electric light bulbs and we're now sort of 9/10ths the way, but she still wants one or two in certain places, like next to the bed, where she still wants the old light bulb. It's illogical, but there it is (laughs). (S3ii)
If I'm here we tend to stay on top [of our energy use], but if I'm away it kind of like distracts. My wife kind of distracts from the goal. That's the problem. (T2i)

Throughout the interviews, it was therefore clear that the monitors had generated new forms of both cooperation and conflict among household members.

Related to this point, a few interviewees reported that the monitors had allowed them to observe more closely how much energy other householders were using and had even, in some instances, provided new 'evidence' in apparently long-running household disputes about 'wasting' energy:

I'm still quite a pushy Dad when it comes to 'you've left things on'. And if [my kids] say 'Oh yeah yeah, whatever', I will say 'Look look it shows you there how much we're using ... if you're leaving that light on, it does have an effect!' They can see it now. There's more evidence. (D4ii)

In this respect, and following Foucault (1977), the monitors may be seen to act as a kind of panopticon, introducing a new form of surveillance in which householders' energy use may be observed at any moment in time, even whilst they remain unaware of such observation. Here, however, where the panopticon works by getting individuals to internalize their own punishment, in the case of the RTD monitors no such 'punishment' exists unless it is enforced by someone else, in this case the 'pushy Dad', or if the individuals have first accepted that saving energy and carbon is a good thing. In this study, and as the next section elaborates further, such an ethic of energy and carbon saving did not appear to have developed among most householders. To the extent that it did exist, it demanded regular reminders from, and associated arguments with, the usually lone householder who had taken up the new carbon governmentality.

These observations show that whilst the monitors attempted to extend a form of carbon governmentality into households, typically this had only been effective with a single householder, who subsequently had to negotiate with other household members if this new governmental programme was to make any impact on their collective daily practice. This suggests, first, that carbon governmentality operates unevenly across different householders. Whilst some household members embrace the new carbon rationality and attempt to make practical changes, others may take up alternative rationalities, for example around saving money or energy, or may simply resist or ignore the monitors. Second, RTD monitors can simultaneously introduce a form of 'governmental power' in which individual householders internalize the new governmental rationalities and apply them to their own everyday conduct, and a form of 'disciplinary power' is imposed by one household member on others. This differentiated and complex relationship between the monitors and households suggests, therefore, that the crucial relationships to observe may not be those between carbon governmentality and individual subjectivities, but rather how carbon governmentality connects with existing relationships between householders and with other household rationalities, such as the desire to use old incandescent light bulbs for a warm, well-lit home or the desire to leave the oven on whilst playing computer games so as to live out a convenient and no-hassle lifestyle.

A narrow focus on how carbon governmentality modifies *individual* subjectivities is therefore unlikely to offer an accurate impression of the extent to which such a governmentality has taken hold. Instead, the interviews suggest that even when individual subjectivities have apparently been modified, the practical effectiveness of such changes is limited by the social contexts in which they are situated, the other people and subjectivities that exist within those contexts and the alternative forms of discipline and governmentality that swarm around within contemporary households (e.g. Donzelot 1979). In short, these observations suggest that a focus on 'My Space' (Paterson and Stripple 2010) will be incomplete as long as it continues to neglect 'Our Space'.

Resistance, Rejection and Retraction

Whilst the first two themes have explored the relative weakness of carbon as a governmental rationality and the ways this played out in negotiations among different household members, the third and final theme I highlight focuses on a number of more direct ways householders apparently resisted the RTDs and the ways they attempted to extend a form of carbon governmentality into the household (see Chapter 11 for further examples of resistance to climate governance). Crucially, whilst this resistance was often strong, in only very few cases did it result in wholesale rejection of the monitors. Rather, it appeared that the monitors' particular form of carbon governmentality was resisted or rejected in relation to some areas of daily practice, whilst simultaneously being taken up in others. Specifically, I highlight three forms of this resistance: first, the ways householders simply refused to apply a carbon rationality to some areas of their everyday lives; second, a more general rejection of the ways the monitors served to territorialize climate change at the level of the household and the associated responsibilization this entailed (Lövbrand and Stripple 2006); third, the fact that in almost all cases, the monitors' rationalities – whether financial, energy or carbon based – appeared to retract and wear off over time.

First, despite stating that they used their monitors to help them save energy and cut carbon in various ways, several interviewees highlighted particular areas of their everyday lives where the monitors were simply not considered welcome:

> I think it's really important to get the balance right between the sort of big brother and monitoring things and also just allowing people to enjoy their lives and their homes. I mean I

have some really lovely lamps in my lounge that I got from Venice and I refuse to be beaten down on this. I want to enjoy them and don't want to use them less or sit there looking at horrible light. Because you see all this stuff in the media that makes you feel guilty about things, but I think life is for living. (T1i)

In other interviews, different areas of everyday life were mentioned, from fish tanks, to computers, keyboards and bread makers. The key point, however, is that the monitors were simply ignored in areas of practice where they appeared to threaten cherished ethics and aesthetics of the home (cf. Aune 2007). Despite the recognition that this resistance was something that one might potentially feel guilty about, these areas of daily practice were simply considered off-limits to the form of carbon governmentality extended by the monitors.

Where these initial examples highlight a degree of discretion on the part of the interviewee as to which areas of everyday life should fall under the new carbon governmentality, in other examples interviewees stated that they felt they had no choice but to resist or impose limits to it. In these cases, householders emphasized they simply could not live without certain appliances:

It is all the nasty things, like the shower and the dishwasher and all the heating element type things that do chuck it up. And we are a trapped audience there; we can't not use those things. (D4ii)

Note, in this and the previous quotation, that even where the monitors were resisted or had limits imposed upon them, the interviewees still appear to have accepted their moralizing gaze, as previously normal practices and household appliances had now become 'nasty things' that contribute to climate change. Again, what was considered a necessity and what discretionary differed enormously between different households. The important insight, however, is that different households, each with their own distinct and well-established ethics, aesthetics, routines and moral economies (Silverstone, Hirsch and Morley 1992) resisted the monitors in different and quite specific areas even if they apparently embraced them in others. Whilst the preceding sections highlighted areas where the monitors appeared to have enabled a form of carbon governmentality to link in with and come to work alongside other forms of governmentality that exist within household practices, in these instances this link was not made, creating spaces of resistance.

The second example of this theme relates to a more general rejection, by several interviewees, of the monitors' silent but insistent commands to save energy or to reduce carbon dioxide emissions. Here, rather than resisting the new rationalities of government in relation to particular areas of practice, as in the previous example, interviewees argued that targeting them as individuals and as householders was misplaced as long as other, larger actors, such as governments or energy companies, were (perceived as) failing to take meaningful action on climate change themselves. Here, and as the following two quotations illustrate, interviewees emphasized how even if they were to make changes to their everyday practices to try to save energy and carbon, the difference they could make would be marginal compared to what might be achieved through broader structural measures such as electricity market reform or binding political action:

> Can you explain why my first 350-odd units are priced at 17p each, and all subsequent units are 12p? Now wouldn't it be better to say, actually your first 350-odd units are charged at 12p, and then we're gonna sting you. And I bet you somehow my consumption will find a way to drop-off from almost nothing to something lower. (T4ii)
>
> I don't think politicians will ever have, not throughout the whole world, will ever have enough power to change people's habits.... I mean when you think of what we're doing [about climate change], we're only tinkering at the edges really. (S3i)

These quotations demonstrate a significant amount of scepticism on the part of interviewees as to whether individual households are the most appropriate site through which to try and manage the global climate change problem, and both interviewees hint at alternative sites and actors – the energy market and conventional politics – through which more significant action might be more easily achieved. Other interviewees mentioned local authorities, housing associations, supermarkets and do-it-yourself stores as other potential sites but, whichever sites were mentioned, the key point was to reject a form of carbon governmentality that functions predominantly through householders and individuals' everyday practices in favour of one that works at a broader, perhaps more structural scale.

This exhibition of what Macnaghten, Myers and Wynne (1995) call 'relational agency' – that individuals perceive their own agency in relation to that of other actors – is not new (see also Eden 1993 on the concept of 'actionable responsibility'), but, to my knowledge, it has not previously been interpreted as a form of resistance to the territorialization of climate and environmental change at the household level (although see Hinchliffe 1996). This interpretation poses interesting questions about the potentially infinite cartographies of global climate governance and the ways they compete or cooperate with one another. Further, if, as these quotations imply, 'energy citizens' might be happier to accept infrastructural changes and policy and market reform than to have to implement countless small-scale behavioural changes within their own everyday lives, it suggests that there may currently be limits to the extent to which climate change can indeed be governed 'beyond' the state.

Third, and finally, in the second round of interviews, conducted after the monitors had been in interviewees' households for at least twelve months, almost all interviewees stated that after an initial 'honeymoon period' (D4ii) that lasted for two to three months in which they regularly checked the monitor, they had subsequently stopped reading it so often. In parallel with this, interviewees argued that whilst they had taken a number of measures to try and save money, cut their energy use or reduce their carbon emissions when they first received the monitors, after the honeymoon period had ended they had decided that they had done all that they needed or wanted to do and felt that no further action was necessary on their part to go still further. Interviewees spoke of having 'plateaued out' (T1ii) or being on a 'level keel' (D3ii) and expressed satisfaction that they had controlled their energy use to this level, and also a sense that they needn't take any further action:

> Short of understanding that something is using more electricity than normal, we don't actively, I don't actively try to limit any more. (D3ii)

This suggests that after an initial period in which the impacts of the monitors' new forms of governmental rationality is relatively strong, this quite quickly retracts or wears off as other everyday life concerns again come to dominate. Whilst a simple interpretation of these insights might suggest that this retraction represents the gradual yet final rejection of the monitors' carbon governmentality, interviewees also suggested that even if they had stopped regularly checking the monitors or striving to reduce their emissions, their level of awareness about their carbon emissions and energy use had been altered permanently. In this respect, whilst the monitors' carbon governmentality might not have been accepted wholesale, and may not necessarily have had significant impacts on normal daily practice, it would be wrong to suggest it had had no impact at all. The key, albeit very difficult challenge, therefore, is to try and identify the subtle traces that the monitors' efforts to introduce new rationalities of government into household practices leave behind them and the extent to which these have impacts by linking together with other, more enduring and more powerful forms of governmentality in everyday domestic life.

Conclusions

The preceding analysis is drawn from repeat interviews with just twelve households. As such, it would be unwise to generalize too far, or to take these insights as representative of the full diversity of experiences that might emerge when householders interact with RTD monitors and different forms of carbon governmentality. Nonetheless, these findings do offer valuable insights into how a particular expression of carbon governmentality fares when it is introduced to everyday situations.

To summarize, this chapter has argued that, as part of a broader programme of carbon governmentality, RTD monitors can and should be understood as a key technical device that introduces a range of new governmental rationalities into households in the attempt to manage domestic energy use. Specifically, taking Paterson and Stripple's (2010) analysis as a point of departure, it has sought to extend analyses of carbon governmentality beyond pure and abstract programmes of government and to examine the messy actualities that emerge as these interlace with the existing concerns of everyday life. The core findings to emerge from this analysis are, first, that for householders, carbon dioxide appears to be a weak rationality of government that makes little sense and seemingly has little relevance to routine practice. Anticipating this, the manufacturers of the RTD monitors used in this study designed them so that other, more familiar governmental rationalities, such as pounds and pence, could be used instead, and indeed this was the rationality of choice for most interviewees. What is critical, therefore, is the extent to which these rationalities either work together or contradict one another. Whilst a full exploration of this issue is beyond the scope of this chapter, the suggestion here is that, at least when they are introduced by RTD monitors, carbon rationalities are quickly encompassed within and marginalized by other, more dominant and pervasive mentalities of rule that exist within the households of advanced neoliberal economies.

Second, the form of carbon governmentality spread by the monitors appeared to work with some householders but not with others. Although not always, this was typically the oldest male in the household who tended then to make efforts to engage and 'educate' other household members and, in so doing, met with various forms of both cooperation and conflict. Future analyses should therefore look not at the direct relationship between individuals and forms of carbon governmentality, but rather at how this is mediated through other, preexisting forms of social relation.

Third, as well as falling unevenly onto different household members, the new rationalities of government introduced by the RTD monitors also fell unevenly on different areas of practice. Some householders simply refused to compromise in certain lifestyle areas where others felt they had no choice but to reject the monitors' moralizing gaze in relation to particular everyday 'necessities'. In turn, this led some to reject the individualized responsibilization for addressing climate change that the monitors seemed to be advancing, calling for alternative cartographies of government in its stead. Further, and finally, after an initial honeymoon period, the impacts of the monitors appeared to wear off. The monitors were consulted less often and no further changes to everyday practice were attempted even whilst interviewees did acknowledge that their awareness of their energy use and its carbon implications had been permanently altered. These findings demand further work to understand the many and various ways carbon governmentality links together with and functions through different individuals, practices, sites, objects and even temporalities.

In all, these findings serve to highlight the unevenness of and the tensions at play in the exercise of governmental power. Further, they suggest that whilst understanding programmes of carbon governmentality remains valuable, it is perhaps more important to trace the particular contours it will follow, the forms of resistance it will encounter and the ways it links together with other forms of governmentality in everyday life. This highlights a need for more attention to be paid to the edges or the limits of different programmes of government and to the ways they repel one another or fuse together. This suggests that the origins of the conduct of carbon conduct's influence may lie in the hybrid forms that emerge from its interplay with the many preexisting, distinct and alternative modalities of power in everyday life. What is needed, therefore, is a more detailed examination of how forms of carbon governmentality function by building upon and making use of other, parallel forms of power – whether sovereignty, discipline or biopower – and of how these recombinations and intersections shape both the successes and failings of programmes of carbon government. In parallel, and to conclude, it suggests that the subjectivities of 'carbon calculating individuals' that are moulded through these always messy and complex arrangements will result in mongrel identities that, although they may contain traces of carbon governmentality, are rarely wholly or directly shaped by it.

References

AECOM Limited. (2011). *Energy Demand Research Project: Final Analysis*, St Albans: AECOM Limited.

Agrawal, A. (2005). *Environmentality: Technologies of Government and the Making of Subjects*, Durham, NC and London: Duke University Press.

Aune, M. (2007). Energy Comes Home. *Energy Policy* **35**, 5457–65.

Burgess, J. and M. Nye (2008). Rematerialising Energy Use through Transparent Monitoring Systems. *Energy Policy* **36**, 4454–9.

Burgess, J., T. Bedford, K. Hobson, G. Davies, and Harrison, C. (2003). (Un)sustainable Consumption. In *Negotiating Environmental Change: New Perspectives from Social Science*, ed. F. Berkhout, M. Leach, and I. Scoones. Cheltenham: Edward Elgar, 261–92.

Darby, S. (2006). *The Effectiveness of Feedback on Energy Consumption: A Review for DEFRA of the Literature on Metering, Billing and Direct Displays*. University of Oxford: Environmental Change Institute.

Darier, E. (1996). Environmental Governmentality: The Case of Canada's Green Plan. *Environmental Politics* **5**(4), 585–606.

Darier, E., ed. (1999). *Discourses of the Environment*. Oxford: Blackwell.

Dean, M. (1999). *Governmentality: Power and Rule in Modern Society*. London: SAGE Publications.

DECC. (2009a). *Smarter Grids: The Opportunity*. London: Department of Energy and Climate Change.

DECC. (2009b). *The UK Low Carbon Transition Plan: National Strategy for Climate and Energy*. London: Department of Energy and Climate Change.

Devine-Wright, P. (2007). Energy Citizenship: Psychological Aspects of Evolution in Sustainable Energy Technologies. In *Framing the Present, Shaping the Future: Contemporary Governance of Sustainable Technologies*, ed. J. Murphy. London: Earthscan, 63–86.

Donzelot, J. (1979). *The Policing of Families*. Baltimore, MD: Johns Hopkins University Press.

Eden, S. E. (1993). Individual Environmental Responsibility and Its Role in Public Environmentalism. *Environment and Planning A* **25**, 1743–58.

Foucault, M. (1977). *Discipline and Punish: The Birth of the Prison*. London: Penguin.

Foucault, M. (1980). *Power/Knowledge: Selected Interviews and Other Writings 1972–1977 (edited by Colin Gordon)*. New York: Harvester Wheatsheaf.

Hargreaves, T. (2010). *The Visible Energy Trial: Insights from Qualitative Interviews*. University of East Anglia: Tyndall Centre for Climate Change Research Working Paper No. 141.

Hargreaves, T., M. Nye and J. Burgess (2010). Making Energy Visible: A Qualitative Field Study of How Householders Interact with Feedback from Smart Energy Monitors. *Energy Policy* **38**, 6111–19.

Hargreaves, T., M. Nye and J. Burgess (2013). Keeping Energy Visible? How Householders Interact with Feedback from Smart Energy Monitors in the Longer Term. *Energy Policy* **52**, 126–34.

Hinchliffe, S. (1996). Helping the Earth Begins at Home: The Social Construction of Socio-Environmental Responsibilities. *Global Environmental Change* **6**(1), 53–62.

Lövbrand, E. and J. Stripple (2006). The Climate as Political Space: On the Territorialisation of the Global Carbon Cycle. *Review of International Studies* **32**, 217–35.

Lovell, H., H. Bulkeley and S. Owens (2009). Converging Agendas? Energy and Climate Change Policies in the UK. *Environment and Planning C* **27**, 90–109.

Macnaghten, P., G. Myers and B. Wynne (1995). *Public Rhetorics and Environmental Sustainability: Ambivalence and Effects*. Lancaster University: Centre for the Study of Environmental Change.

Miller, P. and N. Rose (2008). *Governing the Present: Administering Economic, Social and Personal Life*. Cambridge: Polity Press.

National Audit Office. (2011). *Preparations for the Roll-out of Smart Meters*. London: The Stationery Office.

O'Malley, P., L. Weir and C. Shearing (1997). Governmentality, Criticism, Politics. *Economy and Society* **26**(4), 501–17.

Oels, A. (2005). Rendering Climate Change Governable: From Biopower to Advanced Liberal Government? *Journal of Environmental Policy & Planning* **7**(3), 185–207.

Ofgem. (2010). *Smart Metering Implementation Programme: Statement of Design Requirements (Ref: 94b/10)*. London: Office of Gas and Electricity Markets.

Paterson, M. and J. Stripple (2010). My Space: Governing Individuals' Carbon Emissions. *Environment and Planning D* **28**, 341–62.

Reckwitz, A. (2002). Toward a Theory of Social Practices: A Development in Culturalist Theorizing. *European Journal of Social Theory* **5**(2), 243–63.

Rutherford, S. (2007). Green Governmentality: Insights and Opportunities in the Study of Nature's Rule. *Progress in Human Geography* **31**(3), 291–307.

Rutland, T. and A. Aylett (2008). The Work of Policy: Actor Networks, Governmentality and Local Action on Climate Change in Portland, Oregon. *Environment and Planning D* **26**, 627–46.

Shove, E. (2010). Beyond the ABC: Climate Change Policy and Theories of Social Change. *Environment and Planning A* **42**, 1273–85.

Shove, E. and M. Pantzar (2005). *Metering Everyday Life. Paper to the Society for the Advancement of Socio-Economics (SASE) 17th Annual Meeting*, Budapest, 30 June 2005.

Silverstone, R., E. Hirsch and D. Morley (1992). Information and Communication Technologies and the Moral Economy of the Household. In *Consuming Technologies: Media and Information in Domestic Spaces*, ed. R. Silverstone and E. Hirsch. London: Routledge, 9–17.

Slocum, R. (2004). Consumer Citizens and the Cities for Climate Protection Programme. *Environment and Planning A* **36**, 763–82.

Strengers, Y. (2011). Negotiating Everyday Life: The Role of Energy and Water Consumption Feedback. *Journal of Consumer Culture* **11**(3), 319–38.

Van Vliet, B., H. Chappells and E. Shove, ed. (2005). *Infrastructures of Consumption: Environmental Innovation in the Utility Industries*. London: Earthscan.

Walker, G. and N. Cass (2007). Carbon Reduction, 'the Public' and Renewable Energy: Engaging with Socio-Technical Configurations. *Area* **39**(4), 458–69.

Wilhite, H. and R. Ling (1995). Measured Energy Savings from a More Informative Energy Bill. *Energy and Buildings* **22**, 145–55.

8

Translation Loops and Shifting Rationalities of Transnational Bioenergy Governance

Jarmo Kortelainen and Moritz Albrecht

Introduction

In this chapter we introduce the concept of translation loops, which aims to delineate how governance principles are designed, transported and adapted to various destinations. We understand governance as a process which generates, transports and implements various norms, incentives and other means seeking to influence people's ways of acting and thinking. Our approach stems from a relational understanding of space (e.g. Massey 2005; Murdoch 2006), which helps us to see governance as a process consisting of and performed by circulations between multiple places, actors and devices. All the interrelated entities impact the governance process but their roles vary and while some deal more with the design of governance, others transfer regulation to multiple directions and some are seen as destinations of governance where the intended outcomes should take place. We call these spatially differentiated 'stages' translation loops, which involve distinct rounds of negotiations and calculations and possess varying rationalities (see Figure 8.1 in the next section). We argue that this is a helpful way to understand broad governance arrangements and their ability to affect over long distances and in variable contexts.

We support this argument by employing the translation loop concept in a preliminary analysis of the renewable and bioenergy policies of the European Union (EU) and Finland as part of EU climate change governance. The increased utilization of renewable resources for energy production is an important aspect of EU policy and the target is to increase the production of renewable energy significantly during the next ten years. Each member country has received distinct targets for its share of renewable energy and each operationalizes the overall EU policy in its own way. We employ Finland, a country with a strong emphasis on wood-based bioenergy, as an example.

We approach EU bioenergy governance (BEG) as a process of governance generation carried out through five translation loops possessing different spatial domains and rationalities. The entire EU BEG is steered by the idea of marketization, which means that the implementation of the policy relies on the establishment of economic markets for bioenergy. This is clearly visible in EU regulation and national policies which aim to generate markets for bioenergy by providing guidelines, support and subsidies. The loops form a successive and interlinked chain of negotiations and calculations where the renewable energy targets and rules are defined in EU policy design processes (first loop) and then operationalized to such policy instruments as subsidies and taxation benefits, for instance, in each member state to enable the emergence of new production and markets (second loop). At the regional scale (third loop), locally specific development and strategy work combine bioenergy issues with regional policy and try to enhance local bioenergy production. Then the bioenergy production and markets themselves (fourth loop) materialize the governance and its targets in forests, heating plants and other production units. Furthermore, there is a critical debate over the results of bioenergy policies in Finland (fifth loop), as quite strong criticism has emerged emphasizing the uncertainties and negative effects of growing bioenergy production.

We analyse the implementation of BEG and its challenges in dealing with complex and geographically variable contexts as a governmental mode of power. We see it especially useful to approach BEG and its translation loops through the concepts of 'governmental rationalities' and 'regimes of practices' because they help in understanding how a transnational governance system has to work through multiple institutional contexts to make its presence felt in the intended regions and places. We pay specific attention to the uncertainties and challenges that may derive from market-based governance arrangements when put into practice. In Finland, the growing utilization of tree stumps and roots for energy provides an example. The growing wood energy market has started to change forest management and landscapes in various ways and has led to the removal of tree stumps for energy production. Markets have gone 'underground' because previously worthless stumps and roots have become valuable items in bioenergy production. The removal and storage of spruce stumps and roots has become a normal part of logging operations. However, a critical debate over the effects of their removal has emerged which challenges, among other things, the original rationale of BEG to decrease carbon emissions. Using this example we seek to demonstrate our translation loop approach as well as highlight the need for critical forums to raise problems of marketized governance. We base our study on textual materials consisting of EU administrative documents as well as national and regional government documents in Finland, academic literature and newspaper articles on stumps and bioenergy.

Governmental Rationalities and Market Generation

Governance is not only an effort of government but is also formed by actors from various social spheres and numerous material and cognitive devices which make its shape hybrid and contextually variable. The spatial reach of governance varies and the prefix 'transnational' emphasizes arrangements which aim to influence over extensive distances and to go beyond national boundaries. When focussing on these kinds of transnational activities we are dealing with power topologies which, according to John Allen, 'come into play when the reach of actors enables them to make their presence felt in more or less powerful ways that cut across proximity and distance' (2011: 284). Power is not something that an actor or group has (or has not) but it is a dispersed and relational phenomenon generated through practices. A topological understanding of power emphasizes relationships between actors which enable some actors to reach and be present in distant places. To be influential and have their presence felt in distant destinations it is not sufficient that governance designers generate standards, norms, incentives or similar elements but actors also have to be enlisted and devices developed to transport and link governance to distant and diverse contexts. Foucault's concept of governmentality (e.g. 1991a,b) provides a framework for studying the art of governing at a distance (see also Miller and Rose 2008), and therefore provides a helpful tool for this study. However, the term *art of governing at a distance* must be approached with caution when using a topological view of power. While Allen points out the provisional character of its outcomes, he highlights the processes between the 'here and there' within power topologies, an aspect specifically taken up by this study and its focus on the translation loops of transnational governance (2004: 24).

Studies on governmentality enable us to evaluate heterogeneous governance processes influenced by multiple institutional settings and their varying rationalities. As Merlingen states, rationalities are 'discursive formations, intimately linked to structures of power that produce effects of truth with regard to specific fields of governance' (2003: 366). However, rationalities are by no means neutral knowledge, but rather knowledge claims developed by their producing regimes of practice (Lemke 2001: 192). Furthermore, as Dean points out, regimes of practice are embedded in material and institutional settings but are also constituted by the minds of the involved actors (2010: 32). They reproduce and translate certain rationalities through their means of knowledge production and distribution, which are based on what they aim to govern. In the following, regimes of practice are understood to consist of specific networks of actors involving particular goals and practices to achieve their aims. Thus, regimes of practice are not all-encompassing constructs residing at a certain scale or site. While subject to change themselves (and their guiding rationalities), regimes of practice also transform governmental rationalities moving through different sites and contexts to satisfy their ends. According to Gordon, rationalities of government aim to enable the involved actors to comprehend the realm to be governed and involve such questions as what, for whom, by whom and how to govern specifically defined features (1991: 3).

In governance processes, actors highlight problematizations related to their stated reality, which require governmental action and reproduce regimes of practice to address these problematizations (e.g. Merlingen 2003; Miller and Rose 2008; Dean

2010). Regarding transnational governance such as EU bio/renewable energy policy, we must assume that problematizations (e.g. combating climate change) promoted by the initiating regimes of practice face multiple rationalities throughout the transnational process, and become subject to change themselves. As Prince and Dufty point out, governmental rationalities are subject to translation and change when they move across a geographical space (2009: 1752). Hence, EU governmental rationalities of combating climate change affect governance processes and their practices and are themselves transformed when they travel across various institutional and spatial domains.

Governance generation is imagined here as a relational and circular process comprised of multiple regimes of practice and their rationality to generate and translate properties which design, transport, implement and assess governance in different contexts. Governing principles and standards are therefore designed chiefly within the initiating regimes of practice and rooted in the rationalities of such regimes. They are transported to various sites and regimes of practice where they are fine-tuned and operationalized to be applicable; furthermore, they are conveyed to even more numerous sites where they cause changes in material practices. In addition to material practices, this process also entails epistemological features, and, echoing Dean, we can ask 'how does thought seek to transform' governmental practices (2010: 42). The process generating governance consists of additional long-distance transportation of knowledge and material elements, and of additional local processes of circulation where negotiations and calculations take place. Topologically interpreted, any form of transnational governance exists only if its presence is felt and it impacts numerous and distant places (Allen 2011). This means that governance has to be transferred over great distances by means of multiple actors and devices including rules, norms, ideas or directives. Furthermore, governance is attached to different contexts by forcing or attracting actors and entities to join the governance network. This requires negotiations and calculations in which abstract governance principles are interpreted, contested and redefined. It is possible to imagine the context-dependent practices as loops – rounds of negotiations and calculations necessary to operationalize and materialize as well as further develop the abstract tools of governing. We call them 'translation loops' because they consist of forums in which actors are enlisted and devices developed to design, operationalize, materialize and contest the norms, incentives, standards or principles. In translation, common definitions are negotiated, roles and identities settled, calculations carried out and objectives set; in other words, original rationalities are shifted and translated to match the requirements and problematizations of the respective regimes of practice. As Huxley points out, such transformations include multiple, possibly conflicting rationalities, an aspect vividly displayed by the debate on stump removals discussed later in this chapter (2007: 187).

If we take a closer look at the EU BEG network through this 'translation loop lens', we can recognize several such loops where the translation occurs. It is a government

(EU and nation-states) initiated and operationalized process and its spatiality, although transnational, is shaped by the scalar and territorial character of governmental and intergovernmental arrangements. However, it is important to note that BEG is far from a neatly hierarchical and territorially nested governmental system because the loops themselves connect governance to numerous other actors and their worlds have their own spatial reach. In addition to governmental actors it involves, for example, a scientific community, business actors, consumers and civil organizations and the combination of actors and their scale of networks varies in different loops. Figure 8.1 seeks to illustrate the loop system of EU BEG.

Although the loops are presented as successive rounds, it is important to acknowledge the simultaneity of work in various sites. In addition to the successive circulation, there are direct links between the sites in the form of negotiations, knowledge production and exchange as well as mobility of actors. Although each translation process within each loop is unique, it is never a closed system but involves and is related to all the other loops as well as numerous other actors. The combination of relations makes each loop different. This means that all loops in the system participate in the operations of all the other loops. This is especially evident in the design process (first loop), which must be informed in one way or another by all the other loops to competently design regulatory tools capable of travelling and adapting to various contexts. Thus, the picture of translation loops (Figure 8.1) also includes direct connections between all the loops.

As discussed earlier, BEG is designed and specifically operationalized to work through the economic markets. Economic markets are not a closed and self-sustaining sphere maintained by the 'invisible hand' of market forces. They are, in contrast,

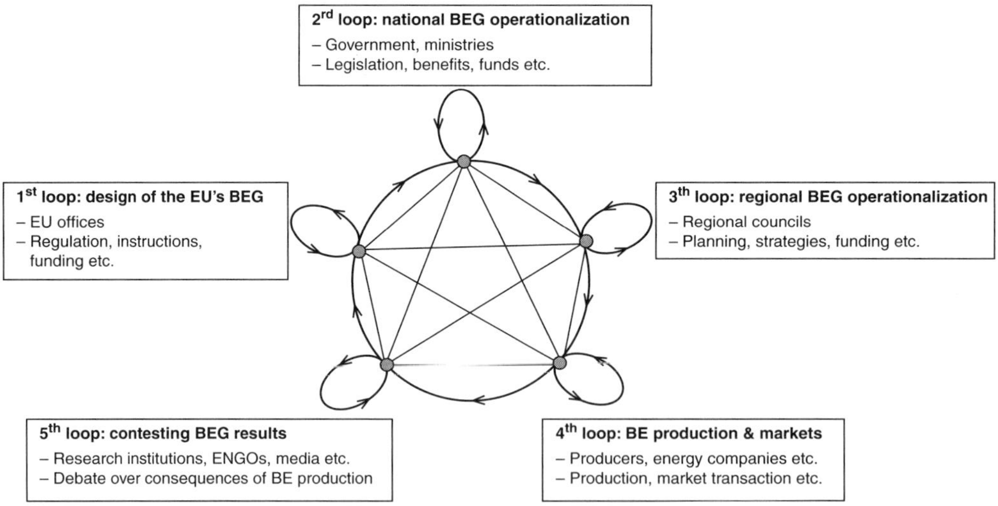

Figure 8.1 Governance generation and loops of translation in the Finnish bioenergy governance system.

established, interrupted and maintained by countless actors and activities. The marketization of BEG governance can be seen as an economic experiment in which the EU and governments try, through conscious design as well as trial and error, to materialize markets which would achieve the climate change targets (see Callon 2009).

Bioenergy markets are formed by creating various types of benefits which encourage economic actors to behave in a certain way. The concept of governmentality provides us with a perspective of how to approach the system of loops where a governance arrangement including an environmental target and rationality is shifted to various market systems determined by rationalities increasingly based on economic aspects and calculations. In addition to directly ordering people to think or act in a distinct way, their conduct can also be influenced by various less prescriptive economic rules, norms or incentives which guide their behaviour but still make them feel like they are acting autonomously (e.g. Miller and Rose 2008). Economic markets are considered a 'core loci' of such freedom. They are performed as autonomous spheres where independent economic beings behave according to their apparent free will and economic calculations. In the following, with the help of the translation loops concept, we briefly present how the EU BEG is generated, transported, materialized and contested.

Design and Operationalization of EU BEG

The *first translation loop* in the EU BEG consists of negotiations, planning and decision making which have sought to design the targets and more general standards for European Union energy policy. The development of this bioenergy and renewable energy policy has been a long and slow process. The first steps were taken after the oil crisis of the 1970s and several attempts have since been made to harmonize it. However, for a long time renewable energy was not seen to be so important as to require deeper coordination by the EU. This started to change in the late 1990s when the liberalization of the energy market occurred and the climate change issue entered the political arenas. These developments led to the Directive on Renewable Energy (RED) in 2009, which was the first regulatory attempt to create all-encompassing and binding targets for EU renewable energy policy (Hildingsson, Stripple and Jordan 2012).

The EU BEG is guided by the commission's directorate-general for energy and its respective sub-directorates and related agencies (see European Commission 2012). Sub-directorate C1, Renewables and CCS Policy, with its main office in Brussels, can be considered the nerve centre of this loop. The initial motivation for increasing renewable energy production comes from global climate change agreements and their requirements to decrease carbon dioxide emissions. In addition, the EU hopes to increase the independence of its internal energy supply and achieve increased economic development through energy production, especially in rural areas. The EU aims to increase the use of renewable energy to 20 per cent of total energy consumption by the year 2020 (RED 2009). This 'de-carbonization' of the EU energy sector (European Commission 2011: 4) is guided by the RED, which seeks to act in concert with several related EU financial support mechanisms and programmes. Financial support for RED implementation can, for instance, be granted from European Regional Development and Cohesion Funds, to name just one source (ECOFYS 2011). Together with its predecessor, the Biomass Action Plan by the European Commission (2005), RED is an important device in the EU BEG because it sets standardized targets and offers guidelines for the use of the aforementioned funds. It is common to all, it travels to all member countries unchanged and carries with it certain definitions, rules or guidelines which determine the ways each member state can implement its renewable energy policies.

The RED is full of definitions and guidelines which aim at generating economic markets for bioenergy. In this way the EU is, through its regulation and directives, creating certain kinds of economic markets to achieve the goals set for renewable energy production. Aside from this main policy aim, increased energy autonomy, economic development and employment possibilities are also promoted as positive effects (European Commission 2005; RED 2009; European Commission 2011). All in all, the whole idea of increasing the share of renewable energy to mitigate climate change is based on a rather simple calculation and assumption that renewable energy production produces less CO_2 than production from fossil sources.

The directive moves forward and enters the *second loop* of translation in each member state. It consists of negotiations and calculations seeking to operationalize the EU targets to national measures and regulation. It should be pointed out that the RED, despite demanding national renewable energy action plans (NREAP), insists that member states develop solutions based on a common EU energy market and establish transfer systems. However, most NREAPs employ strategies based on their own national resources to reach RED targets, despite policy instruments which foster such market cooperation between member states (European Commission 2011). The national targets set by the EU vary between 10 per cent and 49 per cent because of the countries' different opportunities to utilize renewable resources. In Finland, for instance, the share of renewable energy was 28 per cent in 2009 and the official goal set by the EU for 2020 is 38 per cent (RED 2009).

Regulatory tools favouring domestic renewable sources are not a new thing in Finland; on the contrary, they have been developed and used for decades. Kivimaa and Mickwitz (2011) have divided the development of Finnish bioenergy policy into three phases. The first one started from the energy crisis of 1970s and its target was to secure self-sufficiency in energy supply and employment by promoting the use of wood for energy. The second phase, support for wood- and industry-based bioenergy, started in the early 1990s and its goals also included environmental aspects including CO_2 emissions. The third phase started at the turn of the century and is tightly linked with the climate change policies. The strategic decisions related to the EU RED and its targets represent the most recent element in this phase.

The Finnish government has chosen to reach the target by generating a more extensive and profitable market for wood energy. The energy division of the Ministry of Employment and Trade coordinates bioenergy policy in Finland. Finland's national action plan for promoting energy from renewable sources (Energy Department 2011) presents, for instance, the following measures which aim to generate and facilitate wood energy markets:

- Feed-in tariffs for wood energy, which guarantee a fixed minimum price for produced energy.
- Energy subsidies for harvesting of small-sized wood.
- Introducing an electronic forest resources database.
- Developing price statistics systems.
- Developing measuring methods for energy wood.
- Support for research and development in bioenergy production.

Generally, feed-in tariffs are the most prominent incentive utilized by member states when they attempt to perform bioenergy markets in accordance with RED aims; Finland is among the last member states to implement them (Energy Department 2011; ECOFYS 2011). Further financial incentives to increase the use of wood energy consist of investment support and tax reductions for energy/heat production facilities, farms and private households. Furthermore, energy wood harvesting from young forest stands and first thinnings are eligible for subsidies (Energy Department 2011).

In Finland, the *third translation loop* takes place in regional planning and strategic work. The country is divided into eighteen regional councils responsible for development programmes and regional planning in their territories. Each has produced a programme for promoting bioenergy production. Regional councils relate bioenergy and its production to regional policy programmes, regional economies and employment. We use North Karelia – the easternmost region in Finland with about one hundred sixty-five thousand inhabitants – as an example. This peripheral region has been highly dependent on forestry during its history and has suffered from job losses and out-migration for decades. Recently, regional actors have started to place high expectations on bioenergy as a new growth sector, and in 2007 the North Karelian Regional Council published a bioenergy programme providing guidelines for developing the sector (Pohjois-Karjalan maakuntaliitto 2007). The programme was published following the invitation by the Commissions Biomass Action Plan (European Commission 2005) to prepare national biomass action plans. Thus, it predates the EU RED and does not directly reflect the directive and its targets. Contrary to the RED, though the biomass action plan's aim to reduce carbon emissions was secondary to the primary aim of energy security, it played a significant role (European Commission 2005; RED 2009).

The regional example is important in our approach for two reasons. On one hand, the regional programme work demonstrates the multi-sited and dispersed nature of BEG design because it takes place simultaneously in the EU centres and national governments, as well as regional planning offices. On the other hand, the regional programme example shows the properties of the operational environment and rationalities into which the targets and measures caused by the directive have had to be attached. In other words, the regimes of practice have generated specific rationalities which affect the ways governance is taking place.

The North Karelia Bioenergy Program (Pohjois-Karjalan maakuntaliitto 2007) sought to develop preconditions for the growth of what it calls the 'bioenergy cluster', which consisted of energy production (heat and electricity), bioenergy technology and research and development. It set a goal to develop sustainable energy production, strengthen know-how and enhance the export of bioenergy-related know-how and technology. The report maintained that sustainable energy production required, among other things, intensification of wood harvesting and stump removals on a larger scale.

The publication used the standard terminology of contemporary regional development and spoke, for example, about innovation, learning, networking and benchmarking. It placed an emphasis on the environmental consequences of bioenergy production, pointing out such negative effects as problems in nutrient balance, ecological issues and landscape disturbances, and called for an assessment of environmental consequences. However, the programme's key message was to enhance the development of the bioenergy cluster as a regional growth industry generating economic growth, giving employment opportunities and intensifying the use of natural resources.

Both the national policies and regional programmes entail regimes of practice developed over a longer period of time and that form the sociopolitical environments where the second and third loops of translation take place. While the new governance problematizations will likely transform such old regimes to some degree, new eventually contrary or overlapping regimes of practice are created at the same sites. In other words, these operationalization loops possess rationalities that most likely differ from the rationalities of the regimes initiating BEG design in EU agencies. The rationalities of both national and regional regimes are shaped by their emphasis on economic and industrial development. As a consequence, sustainability and climate change – major arguments of the EU BEG – become secondary issues within the operationalization processes nationally and regionally. These political-economic rationalities in the second and third loops are present in regard to the lack of transnational approaches in the NREAPs (e.g. European Commission 2011). Increasing 'trans-European' bioenergy trade could drain bioenergy investments by the national/regional economies and thus national approaches are favoured to achieve national RED targets.

Materializing BEG

The *fourth translation loop* consists of actions and calculations carried out by economic actors. This translation is important because it possesses a 'rationality jump' from a political process to economic practices and calculations in the markets. Using this jump metaphor we want to highlight the rationality gap between the political process of governance design and operationalization, on one hand, and the economic markets materializing it, on the other. The BEG has to cross the gap and survive a fundamental change of rationalities because the public good of political actors shifts to the private profit of market agents as the main criteria for calculations and motivation of agency. The great challenge is to make BEG jump over the gap, that is to guarantee that the actors operating according to the market rationale produce the initial targets and intended results of the EU governors.

The EU BEG is based on the assumption that renewable energy usage emits less carbon than fossil fuels. Thus, its goal is to enhance and expand markets for renewable energy by ordering specific targets for each member state which support the emergence of markets in their own way. The processes of BEG marketization form new markets, include new human and nonhuman entities and define a new role for them as economic actors or goods. In Finland, one of the consequences is that tree stumps have become goods in the growing markets of bioenergy as Figure 8.2 shows.

Stumps and roots had long been a problem for the forest industry. They represented a potential resource but their profitable utilization was difficult. For decades there were numerous plans and experiments for their utilization but all these attempts proved fruitless (Hakkila 2004; Walmsley and Godbold 2010). In practice, stumps and roots were worthless in economic terms and the boundary of the wood markets

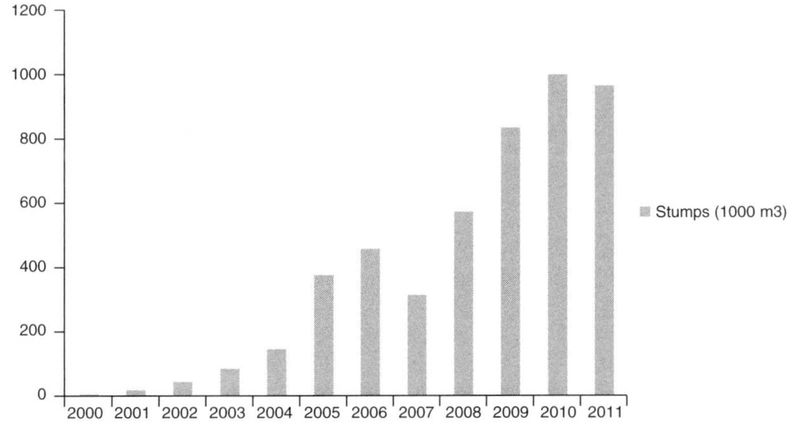

Figure 8.2 Consumption of stumps and roots for energy production in Finland 2000–10 (*Source*: METLA 2011).

was located above the ground. However, the benefits and financial support originating from the BEG systems changed the situation and the markets moved underground.

The removal of stumps and roots is not directly subsidized by the state, but their utilization has been strongly affected by the emergence of the bioenergy markets. On the harvesting side, most of the subsidies go to thinning and harvesting of smallwood, while on the production side, wood energy plants get feed-in tariffs, operational aid and investment support (Häkämies 2011). The emerging wood energy markets – performed by the benefits described earlier – have increasingly incorporated different kinds of wood materials and created new markets in which stumps and roots have become valuable items. Hence, the extension of wood markets underground owes much to indirect financial incentives targeting wood chip production for energy, small combined heat power plants fired by wood fuels and the general support schemes for energy production from renewable energy (Energy Department 2011).

Marketization of stumps has generated an entire market assemblage – the 'bioenergy-cluster' – consisting of technological innovations and systems, new modes of entrepreneurship and novel logistic arrangements. While transforming traditional regimes of practice in Finnish forestry, this also created new regimes spanning the national and regional level. In Finland, only spruce stumps are utilized in energy production. They are removed, and after drying for two to three years, usually at the logging site, the stumps are chipped and burned in plants producing heat or electricity. The growing energy use of stumps has created markets for specific machinery and transport systems needed in this process. New forest management instructions and guidelines have also been developed to guide harvesting of stumps and roots as well as other wood energy sources. Finland is also the world's first country to integrate the removal of stumps into sustainable forest certification criteria (see PEFC 2009; FSC 2011). These as well as guidelines by the Forest Development Centre Tapio restrict and guide removal practices.

The utilization of stumps and roots for bioenergy production avoids the dilemma involving the use of first thinnings and young stands for that production. Subject to bioenergy subsidy, the latter two are also unsubsidized resources for the pulp and paper industry. Thus, their utilization collides with the profitability of certain pulpwood, which is largely produced by the same actors, the large Finnish forest corporations. Within this context, subsidies are partially regarded as market distortion as they reduce the profitability of timber. While this is not the case for stumps or roots, it demonstrates the shift to more profit-seeking rationalities in the implementation processes.

Contesting the Results

When a governance system is planned to work through economic markets it has to deal with externalities; that is, all the effects of economic activity that do not incur costs or provide benefits to economic actors. They are external to the economic markets because they do not exist or affect the cost-benefit calculations of those actors (see Callon 1999). Sometimes a critical translation loop may emerge contesting and assessing the results – the externalities as well – and their 'compliance' with the original aims and targets of the governance system. The contestation loop can consist of more organized forums, as well as open and public discussions and forums, and at its best provides input to and steers the governance design process.

We can see such a round of negotiations and calculations concerning the BEG as forming the *fifth translation loop*. A lively public discussion concerning the influences of bioenergy production has emerged in Finland, in relation to transnational conflicts too, and the following critiques have appeared in the agenda:

- *the climate change criticism:* all operations do not reduce carbon emissions.
- *'loss of biodiversity' criticism:* intensification of resource use reduces biodiversity.
- *'loss of nutrients' criticism:* more efficient utilization of wood removes nutrients from the forest.
- *economic criticism:* bioenergy production is not economically profitable.
- *aesthetic criticism:* energy wood production creates negative changes in the landscape.

These topics have also appeared throughout the academic debate on these issues (see Stupak et al. 2007; Eräjää et al. 2010; Walmsley and Godblom 2010). The most interesting discussion has concerned stumps and greenhouse gases. There have been competing calculations of how the energy use of roots and stumps affects carbon emissions and two camps of scientists have presented quite contrary arguments and calculations of the climate change benefits of removing and burning stumps and roots. The researchers of the Finnish Environment Institute have argued that their use as energy causes carbon releases similar to burning coal. The critical argument goes as follows: stumps and roots are important carbon sinks which reserve carbon in the ground for decades. Burning releases carbon into the atmosphere and is thus in opposition to climate change mitigation targets for reducing CO_2 discharges in the short run (Repo, Tuomi and Liski 2011). Jari Liski, a researcher at the Finnish Environment Institute, argued in a newspaper interview that 'the removal of stumps would have to continue for about 20 years before the emissions per produced amount of energy fell below the emissions of oil and gas.... If we want to reduce emissions within 20–50 years, switching from fossil fuels to burning stumps is not justifiable' (Saavalainen 2010). This research finding provoked a lively public debate and caused some to question the climate change benefits of stump burning.

On the opposite side stood academics from the Finnish Forest Research Institute who claim that calculations should cover a much longer time period and consider the broader view of forests as a growing carbon reserve. Their argument is based on a comparison between the burning of wood and coal, and their conclusion is that the amount of CO_2 emissions is lower in stump burning. Two professors at the Finnish

Forest Research Institute put it in this way: 'According to certain studies stumps decompose much faster in the forest than the calculation of the Finnish Environment Institute assumes. If we base the calculations on these studies, wood would be a relatively carbon-neutral fuel within a 30-years time frame' (Asikainen and Ilvesniemi 2011).

The discussion of the carbon emissions of stump burning is a vital issue for the EU BEG because the entire policy is based on the assumption that favouring renewable energy production would reduce greenhouse gas emissions. Without judging which argument has greater validity, we can learn a number of lessons from this debate. Firstly, it shows the uncertainties that the marketization of BEG faces when it occurs in various and complex socioecological milieus, for instance, the Finnish woodlands. When the materialization of BEG is given to the markets there is a possibility that negative externalities will emerge, elements that do not appear in the cost-benefit calculations of economic actors but which might work against the principal aims of the BEG. These consequences are also externalities to governmental actors, both at the national as well as regional scales because their rationalities are basically related to economic and regional policies. Secondly, the debate demonstrates the importance of the critical forums which enable the detection and evaluation of such externalities. Consequently, it is important for a governance system to acknowledge these debates and create feedback links to design and operationalize governance. The externalities have to be internalized into the BEG system and bioenergy markets through the development of new regulatory tools in agencies of the EU and member states. Thirdly, the controversy indicates an opposite 'rationality jump' back to politics with the environment as the source of problematization and argument.

Conclusions

In this chapter we have introduced an approach which imagines governance as a chain of translation loops; rounds of negotiations and practices aiming to design, transport and adapt BEG to various contexts and, furthermore, to challenge its results. This has been supported by a preliminary analysis of the EU BEG employing a governmentality perspective focussing on the rationalities and regimes of practice involved in the process. In the loop imagery, BEG circulates from the translation loop of EU policy design to the loops of national and regional governments, and, finally, to the loops of materialization, that is the businesses and transactions in the wood market. We maintain that the translation loop concept provides us with helpful imagery to understand how power topologies work and how transnational governance can be effective and sustainable across huge distances (see Allen 2011).

The rationalities to combat climate change within EU policy shift when the governing tools travel from the transnational nodes towards the sites of materialization. Therefore, sites and regimes of practice closer to the materialization of bioenergy production are assumed to reproduce rationalities based on economic calculations of bioenergy development and production. These processes include what we call 'rationality jumps', major shifts of rationalities in the move towards the sites or loops of materialization. While the material aim – the increase in bioenergy production – remains the same within these processes, the problematizations related to the varying rationalities differ, enabling such 'jumps'. Aside from the property of evoking tensions and contesting practices between the various loops and their regimes of practice, 'rationality jumps' also include enabling properties, as in the case of tree stump removal. A challenging 'rationality jump' occurs when rationalities of climate change policies shift to the market rationalities of economic calculations. Using the tree stump issue we discussed the challenges that externalities of the marketized BEG and the 'rationality jumps' between translation loops create for the BEG systems.

Our analysis also showed that an opposite 'rationality jump' occurred when a more spontaneous critical loop emerged challenging the results of policy. The way bioenergy production is problematized in the *fifth translation loop* resembles more the initial rationalities concerning climate change and carbon emissions. This loop is more heterogeneous, but the arguments that set the loop in motion and maintain it in the centre of discussion are based on environmental rationalities and regimes of practice. Thus, a second, reverse 'rationality jump' takes place when the critical regimes of practice in the fifth loop share a common ground with the initiating motives of the EU. While these jumps relate to each other, and are critical aspects in the process of BEG, the second jump is more extensive in its spatial reach and governmental impact. In addition, because of the multiplicity and heterogeneity of regimes of practice in the fifth loop, it practically combines regimes at the transnational level with regimes of the second and third loops, and this second rationality jump contains a multiplying, translational effect on the rationalities based on those regimes with varying problematizations.

The example of stump removals in this chapter displays the ways such a mismatch between the rationalities of materialization and assessment can result in conflicts about materialization practices. The socioecological contexts of the sites of materialization are partially excluded from the rationalities of economy and markets which steer implementation. Still, the emerging conflicts and contests, framed by such a 'rationality jump', can entail a positive capacity for the development of BEG processes. It reintroduces governance because new problematizations are highlighted and a rethinking of perceived negative materialization externalities is fostered. The inclusion of the fifth loop into the EU BEG governance system is a demanding challenge because it should somehow be institutionalized to be effective but at the same time it should be kept open to facilitate the influx of different rationalities and heterogeneous information streams. In this way negative externalities can be detected, discussed and internalized into the BEG system and the markets materializing its targets.

Acknowledgements

This study has been funded by The Academy of Finland (Project: 14878) and has been supported by the strategic funding of the University of Eastern Finland (Project: 931429).

References

Allen, J. (2004). The Whereabouts of Power: Politics, Government and Space. *Geografiska Annaler B* **86**(1), 19–32.

Allen, J. (2011). Topological Twists: Power's Shifting Geographies. *Dialogues in Human Geography* **1**(3), 283–98.

Asikainen, A. and H. Ilvesniemi (2011). Kanto on kivihiiltä vähäpäästöisempi polttoaine. *Helsingin Sanomat*, 23 March 2011.

Callon, M. (1999). Actor-Network Theory – The Market Test. In *Actor Network Theory and After*, ed. J. Law and J. Hassard. Oxford: Blackwell, 181–95.

Callon, M. (2009). Civilizing Markets: Carbon Trading between *in vitro* and *in vivo* Experiments. *Accounting, Organizations and Society* **34**(3–4), 535–48.

Dean, M. (2010). *Governmentality, Power and Rule in Modern Society*, second edition. London: SAGE.

ECOFYS (2011). *Financing Renewable Energy in the European Energy Market*. ECOFYS.

Energy Department (2011). *Finland's National Action Plan for Promoting Energy from Renewable Sources Pursuant to Directive 2009/28/EC, revised version*. Helsinki: Ministry of Employment and Economy of Finland.

Eräjää, S., P. Halme, J. S. Kotiaho, A. Markkanen and T. Toivanen (2010). The Volume and Composition of Dead Wood on Traditional and Forest Fuel Harvested Clear-Cuts. *Silva Fennica* **44**(2), 203–11.

European Commission (2005). *Communication from the Commission: Biomass Action Plan*, COM (2005) 628 final. Brussels: Commission of the European Communities.

European Commission (2011). *Review of European and National Financing of Renewable Energy in Accordance with Article 23(7) of Directive 2009/8/EC*, SEC (2011) 131 final. Brussels: European Commission.

European Commission (2012). *European Commission – Energy: Mission Statement of DG Energy*. Brussels: European Commission. [http://ec.europa.eu/dgs/energy/mission_en.htm]. Accessed 24 January 2012.

Foucault, M. (1991a). Politics and the Study of Discourse. In *The Foucault Effect: Studies in Governmentality: With Two Lectures by and an Interview with Michel Foucault*, ed. G. Burchell, C. Gordon and P. Miller. Chicago, IL: University of Chicago Press, 53–72.

Foucault, M. (1991b). Governmentality. In *The Foucault Effect: Studies in Governmentality: With Two Lectures by and an Interview with Michel Foucault*, ed. G. Burchell, C. Gordon and P. Miller. Chicago, IL: University of Chicago Press, 87–104.

FSC (Forest Stewardship Council) (2011). *FSC Standard for Finland*. Helsinki: Finnish FSC Association.

Gordon, C. (1991). Governmental Rationality: An Introduction. In *The Foucault Effect: Studies in Governmentality: With Two Lectures by and an Interview with Michel Foucault*, eds. G. Burchell, C. Gordon and P. Miller. Chicago, IL: University of Chicago Press, 87–104.

Hakkila, P. (2004). *Puuenergian teknologiaohjelma 1999–2003. Metsähakkeen tuotantoteknologia loppuraportti*. Teknologiaohjelmaraportti 5/2004. Helsinki: Tekes.

Hildingsson, R., J. Stripple and A. Jordan (2012). Governing Renewable Energy in the EU: Confronting a Governance Dilemma. *European Political Science* **11**, 18–30.

Huxley, M. (2007). Geographies of Governmentality. In *Space, Knowledge and Power: Foucault and Geography*, ed. J. W. Crampton, and S. Elden, Farnham: Ashgate, 185–204.

Häkämies, J. (2011). *Valtion mahdollisuudet uusiutuvan energian hankkeiden edistämisessä Kymenlaaksossa*. [http://www.gasum.fi/yritysinfo/Energiaseminaarit/Kouvola2011/Sivut/JyriHakamies.aspx]. Accessed 19 January 2012.

Kivimaa, P. and P. Mickwitz (2011). Public Policy as a Part of Transforming Energy Systems: Framing Bioenergy in Finnish Energy Policy. *Journal of Cleaner Production* **19**, 1812–21.

Lemke, T. (2001). 'The birth of bio-politics': Michel Foucault's Lecture at the Collège de France on Neo-Liberal Governmentality. *Economy and Society* **30**(2), 190–207.

Massey, D. (2005). *For Space*. London: SAGE.

Merlingen, M. (2003). Governmentality: Towards a Foucauldian Framework for the Study of IGOs. *Cooperation and Conflict: Journal of the Nordic International Studies Association* **38**(4), 361–84.

METLA (Finnish Forest Research Institute) (2011). *Metsätilastollinen Vuosikirja; Finnish Statistical Yearbook of Forestry*. Vantaa: Forest Research Institute.

Miller, P. and N. Rose (2008). *Governing the Presence: Administering Economic, Social and Personal Life*. Cambridge: Polity Press.

Murdoch, J. (2006). *Post-Structuralist Geography: A Guide to Relational Space*. London: SAGE.

PEFC (Programme for the Endorsement of Forest Certification) (2009). *Criteria for Group Certification; Level of Forest Centre or a Forest Management Association*. Helsinki: PEFC Finland.

Prince, R. and R. Dufty (2009). Assembling the Space Economy: Governmentality and Economic Geography. *Geography Compass* **3**(5), 1744–56.

Pohjois-Karjalan maakuntaliitto (2007). *Pohjois-Karjalan bioenergiaohjelma 2015*. Julkaisu 105. Joensuu: Pohjois-Karjalan maakuntaliitto.

RED (2009). *Directive 2009/28/EC of the European Parliament and the Council of 23 April 2009 on the promotion of the use of energy from renewable sources and amending and subsequently repealing Directives 2001/77/EC and 2003/30/EC*.

Repo, A., M. Tuomi and J. Liski (2011). Indirect Carbon Dioxide Emissions from Producing Bioenergy from Forest Harvest Residues. *GCB Bioenergy* **3**, 107–15.

Saavalainen, H. (2010). Kantojen poltosta syntyy kivihiileen verrattava päästö. *Helsingin Sanomat*, 16 September 2010.

Stupak, I., A. Asikainen, M. Jonsell *et al.* (2007). Sustainable Utilisation of Forest Biomass for Energy – Possibilities and Problems: Policy, Legislation, Certification, and Recommendation and Guidelines in the Nordic, Baltic, and Other European Countries. *Biomass & Bioenergy* **31**, 666–84.

Walmsley, J. D. and D. L. Godbold (2010). Stump Harvesting for Bioenergy – A Review of the Environmental Impacts. *Forestry* **83**(1), 17–38.

9

Governing Mobile Species in a Climate-Changed World

Juliet J. Fall

Introduction

In the spring of 2011, Sir Richard Branson created a furore: he announced his intention to introduce endangered lemurs from Madagascar to his island retreat in the Caribbean. One of his advisors allegedly informed him that they would have a much better life there than where they currently live and would thrive (Lara Mostert quoted in Harrison 2011). Environmentalists and scientists argued that this was in contravention to the IUCN Position Statement on Translocation of Living Organisms (IUCN 1987), suggesting that the environmental assessment mentioned by Branson had only examined the suitability of the island as lemur habitat, and had not taken into account the possible effect of the omnivorous lemurs on the native fauna and flora. The ensuing debate involved casting globally mobile lemurs as cuddly threatened victims versus alien invasive monsters. This eventually led Branson to reconsider his plan, initially preferring to keep lemurs in cages instead of releasing them.

Beyond the anecdote of publicity-seeking billionaires rushing in with poorly designed ideas for conserving charismatic species, globally mobile species raise a number of important questions in a context of global climate change, questions related to human and nonhuman agency about what moves where, and to shifting scales of environmental governance. In this chapter, I specifically explore how calculative practices and scientific discourses on biodiversity frame how mobile species are governed within global environmental policies. These new calculative practices have rethought 'nature' as 'biodiversity' – taking accounting paradigms to new objects – and individual species have been categorized as more or less desirable in particular places. Because this labelling is controversial – and paradoxical – when considering the issue of invasive species on a global scale, particularly within a context of climate change where it can no longer be assumed that specific biogeographical conditions are immutable in specific places – this example helps us to broaden and deepen the analysis of governmentality that has fruitfully been used to discuss the governing of the global climate (Lövbrand and Stripple 2011; Chapter 1).

While there has been a strong focus in the literature on governmentality on issues relating to climate, applying a similar framework to biodiversity is more unusual. Yet such a framework helps us to focus specifically on the invention of the 'something new' that is made in a particular place, and at a particular time, and that is presented as requiring global governance: in effect analysing how something is problematized (Dean 1999: 28). It makes us look in depth at how the dispersed knowledge practices, techniques and calculative methods have rendered nature manageable by creating new experts and expertise, crafting controversial management and extermination programmes for objects that by their very nature transcend the scale of the state. Across the social sciences, the burgeoning field of biosecurity studies, which examines mobile threats to human, animal and plant life, has been informed and driven by a number of theoretical currents, including a sustained interest in governmentality and biopolitics (Braun 2006; Lakoff and Collier 2008; Collier 2009), as well as the interrogation of spatial processes of categorization and boundary making (Hinchliffe et al. 2012; Fall 2013). This rebranding of the centuries-old battle with anything potentially threatening to life has largely been concerned with keeping out certain things from certain places while allowing others to circulate (Hinchliffe et al. 2012). This takes place within a particular global geographical imagination of an increasingly networked planet that enhances transfers 'from "zoonotic pools" in the Global South and East towards the North and West, a reversal of the pre-colonial tendency for disease to follow empire' (Hinchliffe et al. 2012: 2). Furthermore, because this is happening in conditions of rapid climate change and land use change, new and uncertain conditions of possibility have been created globally for mobile species and pathogens.

This example of emerging biopolitics – in which the living is governed, mostly through containment, surveillance and isolation in new ways – is thus a productive example through which to imagine the challenge of global environmental governance in tension between traditional topographical ways of thinking space and scale (with the roles of municipalities / sub-state jurisdictions / regions / states and so on forming the main focus, as in much international relations literature and political science) and topological ways of thinking through networks (Allen 2011b; Paasi 2011). Like many other examples described in the emerging literature on global environmental networks, these networks are simultaneously 'global and local, state and nonstate' (Betsill and Bulkeley 2006), requiring new thinking that connects the interactions between supranational and subnational state and nonstate actors. This mirrors the Foucauldian insistence on heterogeneous and dispersed governing practices, mapping out a micro-physics of power that unfolds in seemingly disparate places (see Chapter 1). As the diversity of relevant actors has drafted, modified, subverted or adapted legislative instruments to develop new governance tools, the usual ways of constructing the global through additions of the local are obviously challenged, yet continue to be referred to as unproblematic. John Allen, in his long-standing project of rethinking spaces of power and governance, suggests cautiously that the Deleuzian term of *assemblage* offers potential for thinking through this sort of complexity:

> Insofar as relationships of whatever kind are assembled across space and over time, the fact that there may be a mix of space/times embedded in the practices of the diverse actors that give shape to regional political or, for that matter, environmental assemblages should alert us to the fact that we have moved beyond the mappable coordinates and distances of Euclid. A world in which power and authority is easily locatable and capable of extension over fixed distances with more or less uncomplicated reach over a given territory sits awkwardly alongside a geography of co-existence and entanglement where proximity and presence are themselves not straightforward givens. (2011b: 155)

The purpose of this chapter is to add a mobile dimension to such topological thinking: for here it is more than just linking together distant places and things in topologically creative ways, while avoiding what Allen calls the pitfalls of 'thin description' (2011b: 156) that would be little more than a jumble of disparate but apparently connected elements. Instead, the very things that are connected are themselves in motion, making the assemblage itself – always already open-ended in any case – in flux. Underpinning the idea of Allen's assemblage was a surprisingly static conception of topographical space, with things clearly rooted somewhere: 'The way that things hold together in assemblages is down to their relatedness, so to speak, rather than simply their topographical setting and location. In that sense, geometric scale and the idea that actors move up and down them, or "jump" them even [Chapter 8], is somewhat misplaced, as is the conventional assumption that power and authority is something that can merely be extended outwards through networks across a flat surface' (Allen 2011b: 156). The scale jumpers here are more vital, more mobile than expected, and they shift and challenge our topographical and topological ways of thinking about space and environmental governance. Allen's later work with Hinchliffe and others (Hinchliffe et al. 2012) on biopolitics acknowledged this, as I develop in more detail later in this chapter. This follows a section on the emergence of invasive species as a global problem. This section lays out how climate change creates new conditions of possibility for globally mobile species, in a context of heightened fears of both environmental and sociopolitical changes. I explore how this links up to the creation of new governable domains through new ways of thinking of nature as nationally listed biodiversity within specific calculative and accounting practices. The chapter ends with a discussion of one example of national and local strategies in Switzerland, illustrating how these are much more ad hoc both in their emergence and application than might be expected.

Moving Species, Changing World

As global trade and travel have increased exponentially, various agencies have made increasingly global attempts to govern the circulation of living matter through specific biosecurity policies (relevant conventions include EPPO, OIE, CBD Clearing

House Mechanism, Bern Convention). The increased global circulation of certain species has therefore become a cause for concern, and is in particular framed as a global problem: the survival of native species versus the exponential success of new arrivals. Countries which ratified the Convention on Biological Diversity are legally required to set up national strategies to govern and control the circulation of nonnative plants and animals defined as invasive. This is presented as a global swarming of species labelled variously as invasive, exotic, nonnative, nonindigenous or alien, with the discussion of such terms the subject of heated debate. This is deemed the second global cause of species extinction and biodiversity loss – the first cause being habitat loss – and the first cause of species extinction on islands (Burgiel and Muir 2010). Plants and animals are seen as out of place and out of control beyond their native habitats, categorized as dangerous and singled out for destruction. Increasingly, and beyond the wordplay on the term *(s)warming*, this issue is linked explicitly to climate change, because

> combined, the complexity of the interaction of these two global drivers – climate change and invasive species – increases dramatically, and evidence is rapidly growing on how climate change is compounding the already devastating effects of invasive species. Climate change impacts, such as warming temperatures and changes in CO2 concentrations, are likely to increase opportunities for invasive species because of their adaptability to disturbance and to a broader range of biogeographic conditions and environmental controls (Burgiel and Muir 2010: 4).

At the same time as the movement of some species is associated with increased threat, the ability of desirable species to adapt when faced with shifting ecological conditions is seen as fundamental for their survival. A paper in *Nature* a few years ago suggested that: 'we predict, on the basis of mid-range climate-warming scenarios for 2050, that 15–37% of species in our sample of regions and taxa will be "committed to extinction"' (Thomas et al. 2004: 145) because 'the consistent overall conclusions across analyses establish that anthropogenic climate warming at least ranks alongside other recognized threats to global biodiversity. Contrary to previous projections, it is likely to be the greatest threat in many if not most regions' (Thomas et al. 2003: 147). The ability of species to move to new climatically suitable areas is assumed to be hampered by habitat fragmentation and destruction, and their possible lack of competitiveness with mobile invasive species is also mentioned as problematic. Just as invasive species moving around the world in an uncontrolled manner challenges attempts to fix and assign nature to particular places, climate change directly requires that movement be encouraged. To some extent, this need for movement has already been recognized in the design of protected areas and national and regional biodiversity strategies, with new emphasis on corridors, continent-wide networks and stepping stones, but these can have the unwanted effect of making invasive species more likely to spread, as well.

Climate change thus creates a new context for environmental governance, traditionally based on the idea of setting aside areas for nature conservation and, to varying extents, letting nature take its course there – in a sense 'sitting still' to allow it to be governed, as Conca has argued in another field (2005). But this is more than a shift in scale. Scaling up environmental policies around invasive species to a global level is not unproblematic in any case: plants defined as 'globally invasive' have inevitably moved from somewhere, and any global attempts to control them have to take into account this question of geographical origin, and the very liveliness and adaptability of the species in question. The simple addition of local scenarios (adding up national Black Lists of species for instance into a global Black List) does not make any sense. Instead, it is this very disorder that is the problem: unlike pollution or greenhouse gases that are a problem regardless of their location, these species are only a problem when they are literally in the wrong place (Gabrys 2009 makes a similar point about waste). Global, in this context, is therefore an effect, not a condition, and is uneven rather than uniform. The global here is made through specific assemblages of things, species, people, science, policy, technical instruments and practices: forming tangled webs of different length, density and duration, experienced differently in different places, the consequence of specific connections and encounters (Braun 2006). Any examination of moving species must therefore take into account the particular techniques, data, artefacts and so on that are deployed not only to constitute the problem but also to subsequently – often simultaneously – enact particular governance strategies to fix the natural in specific places all the while global conditions are changing.

Furthermore, because these tales tell of swarming, invading, foreign, and out-of-control natures, opportunistically playing on other social fears (Fall and Matthey 2008; 2011; Rémy and Beck 2008), this often becomes a highly charged and emotional debate, creating new challenges for adapting governance structures to local contexts. At the same time, as in the case of Branson's lemurs facing habitat destruction in their native land, there are now calls for assisted relocation of individual threatened species that would need to move to survive in the face of changing climate, because modified landscapes do not provide species with suitable conditions to migrate (Butler 2012), or because such species are geographically isolated on islands or mountain tops.

Constituting a Global Biodiversity Polity through Calculative Practices

The starting point that allows this way of thinking about species mobility is the idea of biodiversity, a term and way of framing the issue of environmental change that has durably modified not only how we think about nature and the environment, but also who is identified as responsible for framing and solving related problems (Takacs 1996; Mauz and Granjou 2010). The crucial role of conservation biology, and disciplines related to biology more widely, is well known and documented. The idea of biodiversity centres on an accounting paradigm that involves thinking of nature as individualized species and specific assemblages, paradoxically reflecting both a carefully evolved order and a capacity for change. Yet despite this focus on change, it is the question of order and permanence of nature that appears to be particularly prevalent in the popular imagination, and that receives the most attention as it is translated into governance policies. Representations of nature as the Garden of Eden, with Nature viewed as a single entity reflecting a divine and perfect order (Macnaghten and Urry 2000), continues to orient many policy debates. It leads to policies aimed at maintaining permanence, as the English terms *conservation*, *preservation* or *protection* reflect. Yet this focus on stability is recent: when the first acclimatization societies were founded in France and Britain in the middle of the nineteenth century, moving species around the globe had fundamentally positive connotations. Scientists embraced the noble aims of improving the supposedly defective colonial landscapes and rendering the metropolis exotic and cosmopolitan (Osborne 2000; Smout 2003). Today, however, managing and controlling the environment involves not the addition but the subtraction of selected species. As has been noted in a number of other cases, this framing of an environmental problem is intimately connected to both the technical 'fix' proposed to resolve it and the governance structures designed to manage it.

Changing climates complicate the ecological story, for while many species will need to move to survive, these changes in ecological assemblages are simultaneously cast as the very problem requiring fixing. The fixing very much relies on the calculative practices implied by the idea of biodiversity: listing, counting, categorizing species (as suggested by Lövbrand and Stripple 2011 for carbon accounting). If in the case of climate change such practices rely on constituting flows of carbon as objects to be governed, then species management relies on constituting species as equally governable unique objects. The risks and potential impacts of what is constituted as a unique and urgent threat imply designing new rationalities and techniques through which 'emergency programmes' of government can be constituted. Campaigns are organized to eradicate, control and monitor chosen species in particular places, and measures are put in place to ensure that new species are not introduced into specific environments constituted as coherently bounded. This policing of movement, and tensions between practices of adaptation and mitigation in the face of change, require rethinking the particular topologies of mobile species. As Hinchliffe and colleagues have written:

> Biopolitics draws attention to shifts in the governing of life. The register and object of reference for biopolitics was, initially at least, not the people but a territorially less precise population, a risk pool literally inscribed through number, tabulation and calculation. Through surveillance and accumulation of data, population allowed for discernible patterns, for calculable probabilities, such that life could be regulated through its eventual qualities. This is what Foucault meant when referring to biopolitics as taking control of life through its regularisation. And yet, as becomes clear in Foucault's later work, populations are embedded in an expanding space of circulation with the result that the promotion of life is always in tension with its regulation. (Hinchliffe et al. 2012: 10)

Yet in this case, and somewhat ironically, the privileged scale of governance, as formally enshrined by the Convention on Biological Diversity, is always that of the state, somewhat surprisingly for a problem intrinsically framed as global and transnational. Thus emergent, and often improvised, forms of techno-governmentality that involve extermination and the assumption of movement and change as suspect are established and legitimized through the framing of this biodiversity polity, all the while movement and boundary crossing is valued as a saving strategy for other charismatic species threatened by shifting environmental conditions brought on by climate change.

These spaces of circulation are framed and bounded through metaphors or metonymies of contamination and invasion, created bounded spaces that are cast as under constant threat of breaching, and this despite the boundary-challenging background biophysical context of global climate change. There has, however, been much agonized debate about the specific terms used to frame and name these species (Warren 2007; 2008; Head and Atchison 2008; Richardson, Pysek, Simberloff et al. 2008). Critiques of the terms used in this debate have been twofold, largely following disciplinary traditions: natural scientists worry that emotive categories (alien, invasive, native and so on) are scientifically inaccurate and counterproductive, and require refining and maintaining to be more useful. Some have suggested that words like *invasion* and *alien* are scientifically counterproductive because subconscious associations with preconceived terms, particularly emotive ones, can lead to divergent interpretations and a confusion of concepts, clouding conceptualization of the processes they are meant to describe, leading to 'widely divergent perceptions of the criteria for "invasive" species' (Colautti and MacIsaac 2004: 135), making generalization difficult or impossible. This, they argue, has led to deep divisions between invasion ecologists (Davis and Thompson 2001; Davis, Thompson and Grime 2001). Many natural scientists perceive vocative terms – such as *swarming*, *military invasion* or *natural enemies* – as simple communication tools that allow people to build on their experience when they extend familiar relationships to unfamiliar contexts. In the natural sciences, where much is inferred rather than directly observed, metaphors can make the difference between comprehension and confusion, helping to get a message across (Chew and Laubichler 2003). However, their repetitive use reifies categories, assigning them a normative force, their materiality reinforced by scientists' repeated, compelling use. They intersect with other sets of metaphors, creating 'a common ground between discourses … in a complex interplay of science, media, and policy' (Larson 2005: 245). Others, mainly but not exclusively social scientists, therefore, have vigorously critiqued the fundamentally political nature and assumptions of such categories and metaphors (Larson 2005; Coates 2007). The focus has been on the intimacy between social metaphors and claims about exotic nature that play on feelings of insecurity (Beck 1999; Clark 2002; Chew and Laubichler 2003; O'Brien 2006; see Fall 2011, 2013; Fall and Matthey 2011 for more details of these debates). This has been a nasty conversation, with tempers flaring and accusations of racism and xenophobia clouding exchanges.

A New Governable Domain: Making Species into Objects of Governance

Parties to the Convention on Biological Diversity are required to set up Black Lists and Watch Lists of species. Article 8 specifically requires signatory countries to list existing invasive species, then prevent new introductions and control or eradicate established ones. There is, of course, no legally binding global Black List, because inevitably each species originally comes from somewhere, although such lists exist largely for communication purposes (Figure 9.1).

In some cases, one singular species can be both globally threatened – and therefore on a Red List in one country – and designated as an invasive species – and therefore on a national Black List in another. The issue is fundamentally about scales (and times), as species move across the globe, rather than a clear-cut story of global goodies and baddies, as overlapping place-based identities are in constant tension, as the following example illustrates:

> Very often you have species that need protection in their native range, and at the same time are invasive elsewhere, I don't know, just the example of the rabbit that is very critically, very endangered in Spain, its native range, and it has a huge impact in terms of biodiversity, because it was the basis of the trophic chain in Spain. But at the same time, it's one of the worst invasives in Australia, and in other countries where it creates a lot of huge economic impacts (T16).[1]

On the global scale, the IUCN/SSC Invasive Species Specialist Group, taking over from the dissolved Global Invasive Species Programme (GISP), whose core funding from The Nature Conservancy and to a lesser extent IUCN led to its closure in March 2011,[2] is the only existing global body concerned with invasive species. It produces a list of 'One Hundred of the World's Worst Invasive Alien Species' that details species from the major taxonomic groups. This not however the global equivalent of national Black Lists:

> For black list we intend very different issues, so it's very unclear what the mission is, and that is the problem – like the definition you use in Switzerland is different from the one we use in Italy or the one used in Austria and Germany. So, black list is not a clearly defined term. In general, we consider black lists more for trade regulations. If you think of a black list as a list of species that are potentially harmful, so you regulate input into country, then it is quite a del-

[1] Some quotes are translated by me from the original French, one interview carried out in English. Interviews were carried out by Juliet Fall or Irène Hirt or both, during a research project funded by the Fondation Boninchi, from January 2010 to June 2011. Laurent Matthey, Irène Hirt and Marion Ernwein were involved at various crucial stages, for which I am immensely grateful. Anonymity was granted, and transcriptions are numbered T1 to T16.

[2] The GISP Partnership is now officially in a dormant phase, 'pending clarification on future funding, and any remaining activities will be undertaken by CABI; an intergovernmental organisation, a founding partner of GISP and a global leader in the prevention and management of invasive species', as explained in an e-mail by its former executive director, document available here: http://www.bgci.org/resources/news/0794/.

Figure 9.1 Title page from Lowe, S., Browne, M., Boudjelas, S., and De Poorter M. (2000) *100 of the World's Worst Invasive Alien Species: A Selection from the Global Invasive Species Database* (U.S.G.S. photo by G. Rodda).

> icate issue of course, because we all live in a free trade regime, so you have to be very careful when you regulate that trade of species.... So in general, these lists of species that are regulated internationally are for health problems or for agricultural problems. There is no possibility to have a global list of regulated invasive species just for their environmental impact. (T16)

Thus while the '100 Worst' list drafted by GISP has succeeded in raising awareness of the issue on a global level, it is clearly a hybrid: a tool that continuously negotiates its status between policy and successful awareness raising. The success of this global list led to the creation of others, at other scales, such as the 'Worst European invasive species', drafted by GISP as a mix of species to reflect what is perceived as a broad problem, and to present it as such. In the same vein, the European Environment Agency has drafted a Black List of alien species that have been shown through risk assessment to pose risks to the environment, economy or human well-being (EEA 2010). The range of scales and the diverse actors involved in such listing exercises is increasingly complex, as each list has a different legal standing, ranging from totally informal to legally binding. In the next section, I explore one example of a national strategy, using this to illustrate the remarkable spatial fragmentation and the challenge for coordinated governance in a changing world.

An Example of a National Strategy in a Context of Global Governance

Around 350 so-called exotic species can now be found in Switzerland, from a known total of 2,943 species known to be established in the country (Moser 2002). Those having appeared in the country after the year 1500 are listed as 'exotic', that is from a date set rather arbitrarily at the time of the discovery of the New World, a date that has much less biological pertinence in Europe than in other contexts where the moment of contact with species from afar was more easily dated, and indeed sometimes catastrophic. While the vast majority of exotic species cause absolutely no environmental problems, a minority do become invasive. Recent climate change has added an additional layer of complexity to this tale, as plants formally requiring specific care from gardeners to survive harsh winters are thriving in warmer conditions. These include, for example, palms planted as ornaments along picturesque lakeshores specifically to add a hint of exoticism to the landscapes in the so-called Italian-speaking Swiss Riviera. These plants are now finding the local chestnut forests a suitable habitat, creating an unexpected new problem that has particular resonance in an area where chestnut forests are culturally coded as a landscape icon.

Despite the grouping of a minority of problematic exotic species under the single category of 'invasive species', their actual existence within a single category is unclear in practice. Ecological specificities and cultural differences between how each species is engaged makes the single category less pertinent than might be expected, as is discussed later on in this chapter. In Switzerland, the creation of a collective of invasive plants ironically heavily relied on the presence of one specific plant. Ragweed (*Ambrosia artemisiifolia*) can create severe respiratory reactions in some people, and yet it has been celebrated as a useful champion by scientists who have strategically used the increased political interest in it to speak about invasive species in general. This raised the political profile of all other invasive plants by focussing on the threat to human health.[3]

> We are lucky to have, well, lucky and unlucky … but yes we are also lucky to have Ragweed to help and understand this issue (T1).

[3] Giant Hogweed (*Heracleum mantegazzianum*) can also cause skin blistering on contact, and this is also mentioned in many circles, but mobilization around this particular threat has been less crucial in making the issue of invasive plants public. This reference to human health in the case of these two plants has had the unforeseen effect of making some people think that *all* plants on the Black List are dangerous to people – as we discovered during our focus groups.

The international legal requirement to control species is made concrete in the latest version of the Ordonnance fédérale sur la dissémination des organismes dans l'environnement (ODE) that came into effect on 1 October 2008. This targets the protection of humans, animals, the environment and the preservation of biological diversity. Its most recent revision reinforces the precautionary principle in the use of organisms, a reinforcement of surveillance and a wider injunction to inform buyers about species offered for sale. This built upon the list drafted by the secretariat of the Swiss Commission for Wild Plant Conservation (CPS), that the Federal Office of the Environment asked this semi-autonomous body to produce.

At the very least, all the species listed on the official Swiss Black List and Watch List are subjected to a precautionary principle. The ODE also forbids the direct use in the environment of the listed species, seen to not only have environmental impacts but also consequences for the economy and on human health. This further recommends eradicating the species on these lists, but without making this a legal requirement. However, what the term *the environment* legally covers is not as straightforward as might be assumed: private lands are excluded, for instance. A number of other legislative tools mention invasive species, in particular the Loi sur la protection de l'environnement (art. 1 et 29a) that also recommends a precautionary approach. The La loi sur la protection de la nature et du paysage also governs the release of foreign animal and plant species, and Article 23 in particular stipulates that permission should be sought before species, subspecies and breeds of foreign animals and plants are released, although quite how this should be done in practice remains unclear. This definition of *foreign* includes, in this case, species foreign to 'certain regions', which presumably might include regions within the country. One specific plant (Ragweed) has led to further legislative instruments.

However, the extent to which these various legal instruments can actually be implemented is highly debatable, and not only because of the number of potential actors and spaces involved in a country with extreme devolution to local scales. A brief excursion to a Swiss garden centre or florist, for instance, will show that several of the species on the Black List remain on sale and continue to be planted in gardens – sumacs, buddleias, golden rods and so on – all attractive plants brought to Europe for their ornamental qualities. This is partly because existing laws only apply to public land, not private gardens, but also because the responsibility for the implementation of this governance framework lies with local communities (*communes*), with no formal federal or cantonal coordination, and at best only ad hoc discussions between countries (*cantons*). The ultimate responsibility rests with individual communes, whose capacity to mobilize resources – and show interest – is very varied. When we contacted the forty-five individual communities within the Canton of Geneva, only twelve replied, and of these only three told us they had a specific policy in place, including the large City of Geneva. Additionally, of course, controlling motivated mobile beings is difficult in practice, and – somewhat as with climate change – a key policy objective is mitigation and damage limitation, not eradication. One interview partner expressed the official aim as being not the eradication of these species, but the elimination of the varied negative effects of each (T4), and in specific places.

Concluding Remarks: Governing Complex Spaces

In this chapter, I have argued that the identification of invasive species as a problem requiring global governance in a time of global change relies on the maintenance of the category of global 'invasive species' as internally coherent and global in reach, showing how this is constantly reworked in various places, appealing to the specific biographies of key charismatic species. At the same time, it is clear that establishing a 'global' category for plants that inevitably come from somewhere, and 'global black lists' of universal value, is highly problematic. I have noted how the apparent coherence of the category of invasive species is continually challenged and resisted by the very vitality and biographies of the individual species collected under this label, specifically as the environmental conditions they live and spread in change, leading to curious interpretations of policies that lead citizens to assume that all exotic, or all invasive, species are a threat to human health. At the same time, I have tried to give some feel for the complexity linking up the calculative practices of listing and categorizing with the political imperatives of finding suitable governance mechanisms at appropriate scales.

Throughout, I have shown how the political need to maintain this category of species deemed 'invasive' relies on an understanding of ecological scale not as something that is produced, articulated, and used to interpret the outcomes of ecological change and spatiotemporal difference in socialized landscapes, but rather rests upon an unfortunate ontological commitment to (political) scales as something reified. Because the protection of biodiversity has been drafted as the responsibility of states, and the nativeness of specific species is defined in reference to the political territories of these, the subsequent inevitability of states as the appropriate scale of environmental governance of mobile species comes to be taken for granted. This idea of nature as biodiversity – inevitably and inescapably planned and governed through and by states, rather than, say, as a collective good for all humanity – renders such confusions inevitable. Thus the problem is inescapably reduced to one of finding new national and subnational control measures, new biosecurity procedures and new international and national governance mechanisms to limit mobility with little understanding that it is specifically the choice of scale that made the political object in the first place.

Changing climates, and the inevitability of the need for change and movement for the survival of many species, makes it even more urgent to move beyond the confusion between what we mean when we talk about mobile species and scales. For while presenting 'invasive species' as a collective has raised its profile on the world stage, conjuring up images of catastrophic global collapse on par with what happened on oceanic islands, it is also clear that it has hindered clear alternative geographical imaginations of socio-ecological change that might be more open to the vitality and creativity of these motivated global travellers.

References

Allen, J. (2011a). Topological Twists: Power's Shifting Geographies, *Dialogues in Human Geography* November, **1**, 283–98.

Allen, J. (2011b). Powerful Assemblages? *Area* **43**(2), 154–7.

Beck, U. (1999). *World Risk Society*. Cambridge: Polity Press.

Betsill, M and H. Bulkeley (2006). Cities and the Multi-Level Governance of Global Climate Change. *Global Governance* **12**, 141–59.

Braun, B. (2006). Environmental Issues: Global Natures in the Space of Assemblage. *Progress in Human Geography* **30**(5), 644–54.

Burgiel, W. S. and A. A. Muir (2010). *Invasive Species, Climate Change and Ecosystem-Based Adaptation: Addressing Multiple Drivers of Global Change*. Washington, DC: Global Invasive Species Programme (GISP).

Butler, K. (2012). *Should We Move Creatures Threatened by Climate Change?* Econumdrums. Mother Jones. [http://motherjones.com/environment/2012/01/species-relocation-climate-change].

Chew, M. K. and M. D. Laubichler, (2003). Natural Enemies – Metaphor or Misconception? *Science* **301**(5629), 52–3.

Clark, N. (2002). The Demon-Seed: Bioinvasion as the Unsettling of Environmental Cosmopolitanism. *Theory Culture Society* **19**(1–2), 101–25.

Coates, P. (2007). *American Perceptions of Immigrant and Invasive Species: Strangers on the Land*. Berkeley: University of California Press.

Colautti, R. J. and H. J. MacIssac (2004). A Neutral Terminology to Define 'Invasive' Species. *Diversity and Distributions* **10**, 135–41.

Collier, S. J. (2009) Topologies of Power: Foucault's Analysis of Government Power beyond Governmentality. *Theory, Culture and Society* **26**(6), 78–108.

Conca, K. (2005). *Governing Water: Contentious Transnational Politics and Global Institution Building*. Cambridge, MA: MIT Press.

Davis, M. A. and K. Thompson (2001). Invasion Terminology: Should Ecologists Define Their Terms Differently than Others? No, Not if We Want to Be of Any Help. *Bulletin of the Ecological Society of America* **206**.

Davis, M. A., K. Thompson and J. P. Grime (2001 (Jan–Mar)). Charles S. Elton and the Dissociation of Invasion Ecology from the Rest of Ecology. *Diversity and Distributions* **7**(1/2), 97–102.

Dean, M. (1999). *Governmentality: Power and Rule in Modern Society*. London; Thousand Oaks, CA: Sage Publications.

EEA (European Environment Agency) (2010). *Towards an Early Warning and Information System for Invasive Alien Species (IAS) in Europe*. Technical Report No.5. Luxembourg: Office for Official Publications of the European Union.

Fall, J. J. (2011). Invasions étranges, invasions étrangères, ou quand cygnes et écureuils bouleversent les frontières. I. Dubied, A.; Gerber, D.; Fall, J. J. Fall. 2011, ed. *Aux frontières de l'Animal : mises en scènes et reflexivités*. Coll. Travaux de sciences socials, Genève/Paris: Droz.

Fall, J. J. (2013). Biosecurity and Ecology: Beyond the Nativist Febate. In *Biosecurity: The Socio-Politics of Invasive Species and Infectious Diseases*, ed. K. Barker, A. Dobson and S. Taylor. Abingdon: Routledge. http://www.routledge.com/books/details/9780415534772/.

Fall, J. J. and L. Matthey (2008). Aliens: Plantes mobiles et experts en mutation. EspacesTemps.net, Mensuelles. In the air, 25 June 2008 http://www.espacestemps.net/en/articles/aliens-en/.

Fall, J. J. and L. Matthey. (2011). *De plantes dignes et d'invasions barbares: Les sociétés au miroir du vegetal*. VertigO – la revue électronique en sciences de l'environnement. *Débats et Perspectives, online 27th septembre 2011, URL :* http://vertigo.revues.org/11046.

Gabrys, J. (2009). Sink: The Dirt of Systems. *Environment and Planning D: Society and Space* **27**(4), 666–81.

Gurevitch, J. and D. K. Padilla (2004). Are Invasive Species a Major Cause of Extinctions? *Science* **19**(9), 470–4.

Harrison, D. (2011). *Sir Richard Branson's 'Eco-Island' Plans Hit by Row over Non-Native Lemurs*, online: [http://www.telegraph.co.uk/news/worldnews/centralamericaandthecaribbean/britishvirginislands/8455570/Sir-Richard-Bransons-eco-island-plans-hit-by-row-over-non-native-lemurs.html] Accessed 16 April 2011.

Head, L. and J. Atchison (2008). Cultural Ecology: Emerging Human-Plant Geographies. *Progress in Human Geography*, 1–10. DOI: 10.1177/0309132508094075.

Hinchliffe, S., J. Allen, S., Lavau, N. Bingham and S. Carter (2012). Biosecurity and the Topologies of Infected Life: From Borderlines to Borderlands. *Transaction of the Institute of British Geographers*. Article first published online: 10 August 2012.

International Ragweed Society. (2010). *"Who are we?"* [http://www.internationalragweedsociety.org/01WhoAreWe/index01WhoAreWe.html] Accessed November 2010.

IUCN (1987). *IUCN Position Statement on Translocation of Living Organisms: Introductions, Reintroductions and Re-stocking*, Species Survival Commission in collaboration with the Commission on Ecology, and the Commission on Environmental Policy, Law and Administration. Approved by the 22nd Meeting of the IUCN Council, Gland, Switzerland, 4 September 1987.

Kowarik, I. (2003). Human Agency in Biological Invasions: Secondary Releases Foster Naturalisation and Population Expansion of Alien Plant Species. *Biological Invasions*, **5** (4) 293–312.

Kowarik, I. (2011). Novel Urban Ecosystems, Biodiversity, and Conservation. *Environmental Pollution* **159**, 1974–83.

Lakoff, A. and S. J. Collier (2008). *Biosecurity Interventions: Global Health and Security in Question*. New York, Columbia University Press.

Larson, B. M. H., B. Nerlich and P. Wallis, (2005). Metaphors and Biorisks: The War on Infectious Diseases and Invasive Species. *Science Communication* **26**(243), 243–68.

Lowe S., M. Browne, S. Boudjelas and M. De Poorter (2000). *100 of the World's Worst Invasive Alien Species: A Selection from the Global Invasive Species Database*. Published by The Invasive Species Specialist Group (ISSG) a specialist group of the Species Survival. Auckland, New Zealand.

Lövbrand, E. and J. Stripple (2011). Making Climate Change Governable: Accounting for Carbon as Sinks, Credits and Personal Budgets. *Critical Policy Studies* **5**(2), 187–200.

Mabey, R. (2010). *Weeds: How Vagabond Plants Gatecrashed Civilisation and Changed the Way We Think about Nature*. London: Profile Books.

Macnaghten, P. and Urry, J. (2000). Bodies of Nature: Introduction. *Body & Society* **6**(1) 1–11.

Mauer, T. M. J., M. J. Russo and M. Evans, (2001). Element Stewardship Abstract for Centaurea Maculosa (Spotted Knapweed), *The Nature Conservancy. Arlington: Virginia*. [http://tncweeds.ucdavis.edu/esadocs/documnts/centmac.rtf.] Accessed November 2010.

Mauz, I. and C. Granjou, (2010). La construction de la biodiversité comme problème politique et scientifique, premiers résultats d'une enquête en cours. *Sciences Eaux et Territoires*. **3**, 10–13.

Moore, A. (2008). Rethinking Scale as a Geographical Category: From Analysis to Practice. *Progress in Human Geography* **32**, 203–25.

Moser, D. M., A. Gygax, B. Bäumler, N. Wyler and R. Palese (2002). Rote Liste der gefährdeten Farn- und Blütenpflanzen der Schweiz. BUWAL-Reihe 'Vollzug Umwelt'.

O'Brien, W. (2006) Exotic Invasions, Nativism and Ecological Restoration: On the Persistence of a Contentious Debate. *Ethics, Place and Environment* **9**(1), 63–77.

Osborne, M. A. (2000). *Acclimatizing the World: A History of the Paradigmatic Colonial Science*. Osiris, 2nd series, **15** Nature and Empire: Science and the Colonial Enterprise) 135–51.

Paasi, A. (2011). Geography, Space and the Re-Emergence of Topological Thinking. *Dialogues in Human Geography* November, **1**, 299–303.

Rémy, E. and C. Beck (2008). Allochtone, autochtone, invasif: catégorisations animales et perception d'autrui. *Politix* **21**(82), 193–209.

Richardson, D. M., P. Pysek, D. Simberloff, M. Rejmanek, and A. D. Mader, (2008). Biological Invasions – The Widening Debate: A Response to Charles Warren. *Progress in Human Geography* **32**, 295–8.

Smout, T. C. (2003). The Alien Species in 20th-Century Britain: Constructing a New Vermin. *Landscape Research* **28**(1), 11–20.

Takacs, D. (1996). *The Idea of Biodiversity: Philosophies of Paradise*. Baltimore, MD: Johns Hopkins University Press.

Thomas, C. D. A. Cameron, R. E. Alison, M. Bakkenes, L. J. Beaumont, Y. C. Collingham, B. F. N. Erasmus, de S. M. Ferreira, A. Grainger, L. Hannah, L. Hughes, B. Huntley, A. S van Jaarsveld, G. F. Midgley, L. Miles, M. A. Ortega-Huerta, A. P. Townsend; O. L. Phillips, and S. E. Williams (2004). Extinction Risk from Climate Change. Letters to Nature. *Nature* **427**, 145–8.

Warren, C. R. (2007). Perspectives on the 'Alien' Versus 'Native' Species Debate: A Critique of Concepts, Language and Practice. *Progress in Human Geography* **31**(4), 427–46.

Warren, C. R. (2008). Alien Concepts: A Response to Richardson et al. *Progress in Human Geography* **32**(2), 299–300.

Willis, K. J. and H. J. B Birks (2006). What is Natural? The Need for Long-Term Perspective in Biodiversity Conservation. *Science* **314** (5803) 1261.

Withgott, J. (2004). Are Invasive Species Born Bad? *Science* **305** (5687), 1100b–1101.

10

Measuring Forest Carbon

Heather Lovell

Introduction

This chapter assesses insights about climate governance from two theories of the book – governmentality and actor-network theory (ANT) using a specific case, namely international efforts to measure the carbon stored in forests. Accurate measurement of forest carbon has become a key element of attempts to integrate forests more fully into international climate policy. There is widespread agreement about the value of including forests in a post-Kyoto United Nations (UN) treaty, as tropical deforestation and forest degradation constitutes an estimated 20 per cent of global greenhouse gas emissions (IPCC 2007). The basic idea of the current UN Framework Convention on Climate Change (UNFCCC) policy initiative on forests, termed 'Reducing Emissions from Deforestation and Forest Degradation' (widely referred to as 'REDD+'[1]), is a simple one: to financially reward developing countries who protect their forests. However, the translation of this overarching policy objective into a workable set of policy guidelines, standards and practices has proved much more difficult, and work is still ongoing to decide upon key issues.

REDD+ is an area of policy making, common to much climate change policy, where scientists are centrally involved: policy makers seek their expertise, and scientific research and new techniques shape the scope and substance of REDD+ policy. This chapter explores the role of scientists working on forest carbon measurement issues, in particular remote sensing scientists, and the ways they have framed the REDD+ policy debate. It draws on empirical findings from interviews, participant observation

The author would like to thank the UK Nuffield Foundation for generously funding the research that led to this chapter under its New Career Development Fellowship Scheme. Also thanks to Professor Donald MacKenzie, my mentor for the fellowship, for providing valuable comments on an earlier draft; and to my remote sensing colleague in the School of GeoSciences at Edinburgh University, Dr Iain Woodhouse, for his time and support in explaining remote sensing, as well as Professor Martin Herold and Dr Brice Mora at the University of Wageningen for their collaboration and valuable input into the GOFC-GOLD Sourcebook survey.

[1] The term 'REDD' has over time evolved to 'REDD+', with the '+' signifying activities related to sustainable management and enhancement of forest carbon stocks (the text after the semicolon in the Bali Action Plan, see page 5).

at conferences and meetings and a survey of science authors of a widely used technical handbook for measuring forest carbon[2] in order to demonstrate how and why the measurement of forest carbon has come to dominate policy discourse. The chapter argues that such issues of complex climate governance increasingly demand a mix of theoretical perspectives to further our understanding of them. In broad terms, political science theories tend to overlook the material substance of policy (although there are of course exceptions, see for example Bennett 2010). Science and technology studies perspectives such as ANT, which do valuably focus on materiality, typically underplay the role of power and politics in processes of change.

Of particular interest in the case of forest carbon – a contemporary example of climate governance – is the activity of nonstate organizations (NGOs, science institutions) in defining forest carbon measurement standards, going beyond the methodological guidance that the Intergovernmental Panel on Climate Change (IPCC)[3] set out in its 2003 Good Practice Guidance for Land Use, Land Use Change and Forestry (Penman et al. 2003). The UNFCCC has agreed that the 2003 IPCC Good Practice Guidance (GPG) should form the basis for measuring forest carbon under REDD+ (see UNFCCC 2007, Decision 2/CP.13(6)). However, the IPCC GPG alone does not provide sufficient detail to operationalize REDD+ and make it work. A huge amount of activity has therefore emerged concentrating on the precise practices and technologies of forest carbon monitoring, resting on IPCC guidelines but fleshing them out through provision of best practice examples, guidelines, standards, protocols and manuals. This 'IPCC plus' area of activity forms the primary empirical focus of the chapter: Who is doing this work and why? How have scientists been involved? How are new practices and technologies shaping the evolution of REDD+ policy, and vice versa? Answers to these questions take us away from traditional state-orientated theories of the policy process because a range of nonstate actors – science institutions, private companies, nongovernmental organizations (NGOs) – have played a leading role in shaping policy and practice. Further, it is not just people who have influence in the complex policy domain of forest carbon governance, but also the forests themselves and the rapidly evolving satellite monitoring technologies and models. Interdisciplinary theories that take into account these myriad aspects of international forest carbon policy and practice are required.

[2] An online survey of authors (#44) of a leading handbook for forest carbon measurement (the 'Sourcebook', published by the international organization Global Observation of Forest and Land Cover Dynamics (GOFC-GOLD) (65 per cent response rate), was conducted in late 2011 in conjunction with colleagues at Edinburgh University (Woodhouse), and the University of Wageningen (Mora, Herold). Other empirical research includes: participant observation and work shadowing of a remote sensing scientist at the University of Edinburgh for two months (summer 2010); semi-structured interviews (#15) (2010–11) with remote sensing and other forest scientists in academia, NGOs, the UN and the private sector; attendance at a number of REDD and forestry seminars and conferences (e.g. Miombo Conference, 2009, Commonwealth International Forestry Conference, July 2010; UK Government Earth Observation Workshop, June 2010). This primary empirical research has been complemented by a policy literature review of existing standards, protocols and guidelines for forest carbon MRV, and of international climate policy relating to forests and REDD+.

[3] An expert science body established by the United Nations in 1988 to provide advice on climate change to governments.

The structure of the chapter is as follows. First, the international forest carbon policy context is briefly explained – highlighting recent decisions as well as providing some historical perspective to contemporary debates. Second, in the main section of the chapter, governmentality and ANT are used as lenses to examine the most interesting empirical findings. Governmentality helps explain the popularity of the concept of 'Measurement, Reporting and Verification' (MRV) in forest carbon policy debates, and allows for a critical analysis of the role of the state as a knowledgeable overseer or 'shepherd' guiding the diverse mix of organizations engaged in enacting REDD+ MRV. Actor-network theory (ANT) is used to assess the implications of forests being living, rather unstable things, not easily enrolled into networks centred on the policy goal of accurate, long-term measurement. By bringing together these two theoretical perspectives, the chapter interweaves ideas about the role of forests as unpredictable actors in the policy process with the more conventional policy science approach of governmentality, which brings to the fore valuable political questions to do with the how and why of forest carbon governance. At its simplest, using these two theories in conjunction fills the gaps in conceptual analysis that arise from using only one such approach. But, as the empirical analysis demonstrates, what ANT adds more fundamentally to conventional political science approaches to climate governance is a deeper understanding of the capacity of nonhuman things such as forests to 'bite back' (Pinch and Swedberg 2008: 2), that is to influence and even reframe the course of policy. It points to a richer conceptualization of climate governance that does not separate the power and agency of humans and nonhumans in making and remaking policy.

A Brief History of Forests in International Climate Change Policy

The core objective of REDD+ is to financially compensate developing countries (termed 'Non-Annex One' countries, under the UNFCCC) that are able to prevent deforestation and degradation of their carbon-rich tropical forests. The 2007 Bali Action Plan at the UNFCCC Conference of the Parties (COP) 13 marked the formal agreement to reinvigorate the role of forests within the UNFCCC, stating that a comprehensive approach to mitigating climate change should include: 'Policy approaches and positive incentives on issues relating to reducing emissions from deforestation and forest degradation in developing countries; and the role of conservation, sustainable management of forests and enhancement of forest carbon stocks in developing countries' (Decision 1/CP.13, para 1b iii UNFCCC 2007).

Much could be said about the history of forests within the UNFCCC (see Fogel 2005; Bäckstrand and Lövbrand 2006; Stephan 2012 for excellent, detailed discussions). In short, forests have had a rather controversial role, largely stemming from unease on the part of developed countries regarding the implications of including tropical forests in an international agreement. Under the Kyoto Protocol's Clean

Development Mechanism some forestry projects were allowed, but in practice only a handful of projects came to fruition (because forest projects were quickly discovered to be less profitable, more complicated and risky than other types of Clean Development Mechanism project; see Fogel 2005). It was hoped, therefore, that with the initial REDD agreement at the UNFCCC COP-13 meeting in Bali a new era of forest protection under the UNFCCC had been born. Since the 2007 Bali COP more detail has followed at each Conference of Parties and interim UNFCCC meetings (notably meetings of the Subsidiary Body on Scientific and Technical Advice (SBSTA)), including about how to measure carbon in forests. But progress has been slow – held up in part by problems with the overall climate negotiations – leading to the rather curious current situation characterized by extensive, detailed debate and activity about how to do REDD+ (i.e. how to operationalize it), yet without a final REDD+ policy agreement in place.

Leaving that anomaly aside for the moment, one issue it is important to appreciate with regard to REDD+ forest carbon measurement is that Annex One developed countries have been measuring and reporting on carbon in their forests (and other carbon 'pools') for some years, as signatories to the Kyoto Protocol and UNFCCC. The UNFCCC (under Article 4) requires annual reports on greenhouse gas sources and sinks, with further more detailed carbon pool measurements required for countries signed up to the Kyoto Protocol (see Watson 2009: 8 for a helpful summary of these complex rules), meaning that developed countries have built up considerable experience and expertise on forest carbon measurement. In contrast, non-Annex One developing countries – the focus of REDD+ – have in the main not been reporting to the UNFCCC on their national inventories of emissions and sinks (including forests). The UNFCCC has not required them to do so on an annual basis, and most have not: a study of ninety-nine tropical forest countries found that less than 20 per cent of them have submitted a complete greenhouse gas inventory to the UNFCCC, and only three out of the ninety-nine countries have 'very good' capacity for monitoring forest carbon (Herold 2009).

So it is evident that knowledge and experience of forest carbon measurement in non-Annex One countries is patchy at best. This 'capacity gap' is further compounded by anticipated differences in the degree of accuracy required to measure the carbon stock of forests at a national scale (as for Annex One developed countries) compared with measuring carbon in specific forest projects (as is under discussion as part of REDD+, for non-Annex One developing countries) and, crucially, in a situation where financial compensation is involved. Much policy discussion about REDD+ has therefore centred on the issue of achieving so-called compliance grade measurement, reporting and verification (MRV), namely more accurate monitoring sufficient to reassure investors. Scientists have played an important role in this policy discussion, including in convincing and reassuring policy makers that achieving REDD+ policy (however it may be finalized) is technically possible, as the introduction to the popular

'Little REDD+ Book', produced by the Global Canopy Programme (a forest research organization) neatly summarizes: 'Agreement on REDD is within reach. The spread of new technologies such as satellite monitoring is overcoming some long-standing technical barriers. Collaboration by scientists, economists and policy makers at the UNFCCC, IPCC and other forums, is helping to clarify outstanding methodological issues' (Global Canopy Programme 2009: 8).

As the quote suggests, various organizations – intergovernmental and nongovernmental organizations, the private sector, science/research centres – have been busy developing methodologies and guidance for countries in preparation for REDD+. But even before the fresh policy impetus of REDD+, forest carbon measurement guidance was issued by the Intergovernmental Panel on Climate Change (IPCC) in response to requests from Annex One countries (who were measuring forest carbon for their annual reports, as previously outlined). In 2003 the IPCC Good Practice Guidance for Land Use, Land Use Change and Forestry was published (Penman et al. 2003), which then subsequently merged with agricultural guidance to become Volume 4 – Agriculture, Forestry and Other Land Use Guidelines (AFOLU) – of the IPCC for National Greenhouse Gas Inventories Guidance, published in 2006 (Eggleston et al. 2006)). As mentioned, it is currently agreed that the 2003 IPCC Good Practice Guidance should form the basis for measuring forest carbon under REDD+ (see UNFCCC 2007, Decision 2/CP.13(6)). However, the IPCC guidelines alone do not provide sufficient detail to operationalize REDD+ and make it work, as evidence from research of forest and remote sensing scientists (discussed later in this chapter) reveals. A report for UNEP neatly summarizes the state of play:

> As the leading intergovernmental scientific body for the assessment of climate change, the IPCC is in the foremost position to form the basis of any future accounting guidance under an international climate change convention. If a mechanism for reduced emissions from degradation and deforestation (REDD) emerges … *it is likely to build on the principles and good practice guidance already established in IPCC guides*. (Watson 2009: 26, emphasis added)

The following section explores how a mix of state and nonstate organizations have focused on developing and standardizing forest carbon measurement methodologies; united by a shared view of the central role of measurement in operationalizing forest carbon policy.

Governmentality – Understanding the Consensus around Forest Carbon 'Measurement, Reporting and Verification'

In common with other multilevel environmental governance debates (Bulkeley and Betsill 2003), the discussions around REDD+ are complex – policy makers and others shaping the policy process are dealing with myriad issues of forest science and forest policy operating at a variety of spatial scales. As is typical within new arenas of policy making, language and practices are particularly important because in numerous subtle ways they shape choices about rules and norms, thereby framing the policy debate and setting it in a particular direction (Dryzek 1997; Owens and Cowell 2002). In other words, how a policy problem like REDD+ is first discussed and operationalized affects how it is conceived, and thus the choice and acceptability of subsequent solutions (Rein and Schon 1993; Laws and Rein 2003). The distinctiveness and applicability of governmentality as a theoretical approach here lies in its bringing together of discourse with practices and technologies. Further, it is a theory that explicitly embraces a range of different types of policy actor: both state and nonstate, and thus it is highly applicable to the hybrid mix of organizations involved in forest carbon policy (see Table 10.1). A strength of governmentality in analysing issues of contemporary climate governance such as forest carbon is its broad framing of government as 'any more or less calculated and rational activity, undertaken by a multiplicity of authorities and agencies, employing a variety of techniques and forms of knowledge' (Dean 1999: 11). As discussed in the introductory chapter to this volume as well as in Chapter 4, the term *governmentality* has multiple meanings: first, as a description of the type of government in a specific historical era; second, as a process of transforming institutions of the state; and third as a more general analytical concept that helps distinguish between different types of governmentality or 'regimes of practice' often simultaneously in operation (Dean 1999; Oels 2005). In this chapter, *governmentality* is used in the third sense to analyse the 'visibilities, knowledge, techniques and practices, and identities' (Dean 1999: 23) of contemporary forest carbon governance, centred on accurate measurement.

The Discourse of Forest Carbon MRV

The dominant REDD+ governmentality is about 'Measurement, Reporting and Verification' or 'MRV' (see Lövbrand and Stripple 2010, 2011 for related discussions about climate MRV). The discursive framing or storyline of this MRV governmentality is that in order for developing countries' forests to be successfully incorporated into the UNFCCC under REDD+, robust so-called compliance grade systems of MRV must be developed and systematically implemented at a national level. These national systems need to measure changes in forest cover (i.e. deforestation, reforestation – the quantity of forest) as well as the quantity of carbon per hectare – the carbon stock or concentration (see Copenhagen Accord: Decision 4/CP.15, UNFCCC 2009a). The MRV storyline has constructed global tropical forests as wild, unmanageable and unquantified: untamed forests that need to be brought to order and under control through systematic measurement by nation-states (Lövbrand and Stripple 2010). Forest carbon MRV is about monitoring, data gathering and surveillance of forests.

The dominance of the framing of forest carbon policy around MRV is key for understanding the vast amount of work devoted to devising a workable, robust, global system of measurement for carbon in forests, which has in turn shaped international climate negotiations. Taken together, this discursive framing and its associated set of practices constitute a governmentality of MRV. Acknowledging this allows much of what is taking place in forest and climate governance to be better understood, with the MRV governmentality acting to 'shape the realm of the possible', as Lövbrand and Stripple convincingly argue in their analysis of space-based monitoring of climate change: 'To understand the current shape of climate politics and possible climate policy futures, we must pay more attention to how practices like MRV have enabled the climate as an administrative domain and how they shape the realm of the possible' (Lövbrand and Stripple 2010: 21).

But despite the daunting size of the task – the unwieldy, huge 'administrative domain' of tropical forests – there is an optimism to the MRV storyline: a belief in the UNFCCC position that one can 'know' the forest and translate it into policy through a systematic process of measuring, quantifying, creating data infrastructure, verifying and so on. It is an optimistic 'measure and manage' discourse, as an extract from an interview transcript with Dr James Baker, director of the Clinton Climate Initiative's Global Carbon Measurement Program illustrates: 'Why is measurement important? The short answer is: You can't value what you can't measure.... Only by measuring forest carbon reliably can countries access the carbon markets and other sources of investment capital that would make forest preservation economically viable. *Accurate information truly underpins everything communities, governments, and charities are doing to halt tropical deforestation*'(2012, emphasis added).

A surprisingly diverse range of organizations participates in the MRV storyline (see Table 10.1), from community-based forest groups to national foresters, scientists and financial organizations.

With the high degree of consensus gathered around the MRV storyline (i.e. with virtually no one saying 'don't bother to accurately measure carbon in forests'), attention has been heavily concentrated on how to measure forest carbon, not why. How has this situation emerged? Why has an MRV governmentality come to dominate? Part of the answer to this question relates to existing forestry and climate expert knowledges and practices that have shaped the direction of forest carbon policy. Several spheres of expertise, which predate the emergence of REDD+ – timber management, forest ecology, global land use change studies, global carbon budget science, remote sensing and community forest management – have all been important influences on the direction of REDD+ policy in different ways. Whilst there is not scope within this chapter to examine all of these bodies of knowledge and expertise, I focus on arguably the most influential one – that of remote sensing – which has come to be particularly important, providing the technological capability, via a network of sophisticated satellite sensors, to measure forest carbon with high frequency at a global scale – a large-scale, high-tech surveillance system that promises high-quality 'eco-knowledge' (Luke 1999). Foucault's ideas about 'biopower' are instructive here. *Biopower* is a term Foucault

Table 10.1 *Examples of use of the Measurement, Reporting and Verification (MRV) storyline by diverse organizations – state and nonstate*

Organization	Position on Forest Carbon MRV	Reference
Centre for International Forestry Research (CIFOR)	'Full national participation in a global REDD+ system requires a far better MRV system than currently exists, and there is a huge capacity gap' (p. xiv). 'Better MRV systems are needed to develop performance-based payment systems. One concern is that high transaction costs (e.g. for forest carbon monitoring) prohibit the inclusion of local communities.... Recent work on community monitoring demonstrates that their costs can be substantially lower compared to professional surveys, and the accuracy is relatively good' (p. xiv).	Angelsen 2009 (ed.) Realising REDD+: National strategy and policy options. http://www.cifor.org/publications/pdf_files/Books/BAngelsen0902.pdf.
The Meridian Institute	'A GHG-based instrument that rewards REDD on the basis of quantified emission reductions and/or removal enhancements requires agreement on standards for MRV' (p. ix).	REDD Options Assessment Report (OAR) – 2009, http://www.redd-oar.org/links/REDD-OAR_en.pdf
The World Bank (Forest Carbon Partnership Facility)	'The UNFCCC has made it clear that the eligibility of countries for REDD-plus financing within global compliance markets or non-market compliance mechanisms will be contingent on their having compliance-grade MRV and accounting of emissions and removals' (Baker et al. 2010: 35). 'Addressing methodological issues such as Reference level and measurement, reporting and verification (MRV) is a key entry requirement for REDD+ programs' (FCPF 2011 annual report, p. 26).	http://www.forestcarbonpartnership.org/fcp/sites/forestcarbonpartnership.org/files/Documents/PDF/Oct2011/FCPF_Carbon_AR_FINAL_10_3.pdf

UN REDD Programme	'Monitoring systems that allow for credible measurement, reporting and verification of REDD+ activities are among the most critical elements for the successful implementation of any REDD+ mechanism'.	http://www.un-redd.org/UNREDDProgramme/InternationalSupport/MeasurementReportingandVerification/tabid/1050/language/en-US/Default.aspx
Global Canopy Programme	'As part of a future UN climate treaty, REDD+ could pay forest-owning nations who protect their forests. In return, these nations will have to measure, report and verify (MRV) the effectiveness of their actions, both in terms of the carbon retained in their forests and the involvement of indigenous peoples and local communities in the process. In addition, on-the-ground implementation of REDD+ will require robust but cost-effective monitoring to ensure that funds are being well spent. Communities should have a central role in developing and implementing these MRV and Monitoring systems'.	http://www.globalcanopy.org/projects/community-mrv
Group on Earth Observations (GEO) Global Forest Observations Initiative (GFOI)	'Forest monitoring is receiving particular attention because of the potential inclusion of forests in future international climate change agreements. Robust measurement, reporting and verification (MRV) systems are a fundamental forerunner to any such inclusion' (p. 2). 'To operate efficiently and sustainably, national forest monitoring systems require a continuous, timely and affordable supply of observations' (p. 1).	Geo Forest Carbon Tracking Report on the Concept Phase for developing a Global Forest Monitoring Network (2010) and GEO Global Forest Observations Initiative (GFOI) Implementation Plan (2011)

(2008) introduced to denote a type of governmentality first arising in the eighteenth century that 'seeks to foster and use the forces and capabilities of living individuals that make up a population' (Oels 2005: 190). With biopower, the population, rather than individuals or territories, is the object of government. Although Foucault was referring to human populations, it is possible to apply his ideas about biopower to other nonhuman populations, such as forests. In related ideas about biopolitics, it is aspects of the power and authority to control life that move to centre stage: biopolitics is about the 'struggle for life' – the infiltration of political issues into the biological sphere (humans, nature, genomes, ecosystems, biological science) with the objective of creating a surplus of power from 'lifeforms'. Whilst the literature on biopower and biopolitics is extensive (Foucault 2008; Lemke, Moore and Casper 2010), applying its essential elements to forest carbon, one can view forests as a type of 'carbon population', with the work of the UNFCCC, IPCC and nonstate actors on REDD+ as part of a political power struggle to measure, understand and therefore control this ecosystem.

Further, a key element of theories of biopower and biopolitics is the prioritization of science knowledge: in order to successfully govern populations one must understand how they function, both as individuals and as a whole. Foucault used the metaphor of the state as a shepherd tending to his flock. Rutland and Aylett, in their paper on climate governance in the city of Portland, explore this metaphor further in discussion of the role of knowledge in effective governing: 'The shepherd, Foucault argues, comes to control his flock because he has *clear and precise knowledge of each of its members*. He uses this knowledge to attend to and care for each of them in a way that allows him to guide them as he wishes. Thus, *the creation and use of specific forms of knowledge becomes central to the exercise of governing*' (2008: 630, emphasis added). The key point Rutland and Aylett emphasize is the 'clear and precise' knowledge the shepherd has 'of each of its members'. In other words, populations are successfully governed if the individuals, and their relations with others, are understood. There are interesting parallels here with Rose-Redwood's (2006) account of house numbering in the United States as a way of bringing these households into the realm of the state; with Rose-Redwood arguing that 'geo-power' (i.e. geographical statistics and information) is a prerequisite of biopower (see also Oels 2005). Translating ideas about biopower to REDD+ policy, the subject to be governed is both the individual tree and the forest, and hence the desire to 'know' each individual tree – to collect data from and understand each tree – must be balanced against the obvious practical challenge of doing this at a global scale, for all tropical forests. Further, biopolitics is about power over a population (Agamben 1998), optimizing the 'life' or vitality of a population (Lemke et al. 2010), as well as establishing its defining characteristics (Foucault 2008). In these varied but closely related understandings of biopolitics and biopower, the role of science measurement and data is central to understanding and controlling forest populations and their carbon content, resulting in categorization of forests as either 'carbon productive' (uniform, high density, historical inventory, cloud free) or 'carbon deviant' (patchy, mixed species, hollow trees, hilly terrain), based on ease of carbon measurement by remote sensing methods. Remote sensing dominates REDD+ MRV policy discourse and practice because it is the only set of technologies to provide a means of measuring and monitoring the whole forest population at a global scale. However, remote sensing is less convincing at providing tree-level data and information on changes below the forest canopy (forest degradation and conservation), meaning that other more local-scale MRV practices (forest ecology; forest community monitoring) are also jostling for attention within the MRV governmentality.

For the remainder of this chapter, I first explore in more detail the relationship of remote sensing to the MRV governmentality, and, second, discuss the role of forest carbon standards in attempting to bring together different types of MRV practice (site-based and space-based data; expert and lay knowledges).

The Techniques and Practices of Forest Carbon MRV: Part I – Remote Sensing Science

For the remote sensing science community, MRV is a set of practices done from space using a range of satellite sensors to capture relevant forest data at a global scale. Remote sensing is presented as the only set of technologies that matches the scale of the task, that is global, accurate, wall-to-wall coverage of tropical forests, for example:

> satellite data … play an essential role in delivering the necessary time-series global, annual, high resolution wall to wall forest information products which developed and developing countries will need as part of their MRV systems (Baker et al. 2010: 42); and
>
> The scale of sampling must match the scale of the subject to be measured, in this case the carbon stocks of tropical forests. (UNFCCC 2009b: 28, para 93)

Remote sensing science presents forest carbon MRV as necessitating an ambitious, high-cost and high-tech approach – involving the development of a whole new global data infrastructure and monitoring system, with parallels frequently drawn to the establishment of the World Meteorological Organisation (WMO) in terms of the amount of data required and the size and credibility of the international organization needed to manage it:

> Inter-governmental organizations, in particular the UN system, have a key role in operationalising a data supply infrastructure. The experience of weather-related satellite data and the role of the WMO in providing the infrastructure is an important and successful example (UN-REDD Programme 2008: 5, para 7, emphasis added); and
>
> Compared to the global networks which are the basis for meteorological and oceanographic monitoring, the international coordination and coherence of terrestrial observation and reporting for forests is much less advanced. (GEO Forest Carbon Tracking 2010: 2)

The obvious appeal of remote sensing to policy makers is the potential for a technically reliable global system of MRV, which, although costly to implement, would then in theory deliver a long-term robust and tangible solution to the problem of gaining accurate data on forest carbon. As noted, the science of remote sensing provides an appealing body of knowledge and expertise to REDD+ policy makers about the global forest carbon population, seemingly without need to first measure, monitor and understand the individual trees.

The remainder of this section concentrates on one international organization – the Global Observation of Forest and Land Cover Dynamics (GOFC-GOLD) – and the good practice guide or 'Sourcebook' its REDD Working Group has developed, driven by UNFCCC REDD policy. GOFC-GOLD is a UN intergovernmental organization (a panel of the larger Global Terrestrial Observing System (GTOS)) with a strong earth observation and remote sensing remit. Its overall objective is to 'improve the quality and availability of observations of forests and land cover at regional and global scales and to produce useful, timely and validated information products from these data for a wide variety of users' (GOFC-GOLD 2012). The GOFC-GOLD REDD+ Sourcebook (hereafter referred to as 'the Sourcebook') provides a detailed field guide or 'how to' book – a set of 'agreed upon rules' (Timmermans and Epstein 2010: 71), with its stated objective of 'complementing' IPCC guidelines, 'the Sourcebook aims to provide additional explanation, clarification, and methodologies to support REDD early actions and readiness mechanisms' (GOFC-GOLD 2009: 2).

The discussion in this section is based on empirical findings of a survey conducted of GOFC-GOLD authors (#44; response rate 65 per cent) in late 2011, as well as a related set of interviews (#15). Survey participants were asked several questions, including how they came to be involved in the GOFC-GOLD Sourcebook, how they account for its influence and their views about the possibility of scientists developing forest carbon measurement standards that meet policy makers' expectations. The organization GOFC-GOLD and its Sourcebook were selected for investigation after an initial two month period of enquiry in mid-2010 on forest carbon MRV. The Sourcebook has become one of the most well-referenced science-based guidelines for REDD (UNFCCC 2009b; Baker et al. 2010; GEO Forest Carbon Tracking 2010), as well as being cited widely in preliminary interviews and at conferences as a 'state of the art' science text. Further, the Sourcebook has a strong focus on new, emerging technology (it is updated each year in time for the annual intergovernmental negotiations under the UNFCCC), with a particular emphasis on remote sensing, which enabled further investigation into how this science knowledge and expertise has been applied and translated into policy.

Here I briefly review survey findings about the role of knowledge and the extent to which the Sourcebook has influenced the policy process through the provision of expert knowledge. The main findings are summarized in Table 10.2.

Table 10.2 *GOFC-GOLD author survey findings and analysis*

Finding	Illustrative GOFC-GOLD author survey responses
Methodological feasibility of REDD+ – the Sourcebook has been important in giving policy makers confidence that REDD+ is technically possible	‘The Sourcebook passed the following key message to policy makers: monitoring and reporting emissions from deforestation and forest degradation in tropical countries is technically feasible with acceptable accuracy for a REDD mechanism’ [respondent 4, Q7];
	‘When policy makers were concerned on the feasibility of REDD, the GOFC GOLD sourcebook sent a positive message, which I think was influential on REDD negotiations’ [respondent 13, Q7].
	‘The Sourcebook helped in solving the questions over the current technical capacity to measure forest carbon stock changes in developing counties which were some of the main concerns at the beginning of the REDD+ negotiations’ [respondent 26, Q7].
Relationship with IPCC – The focus of the Sourcebook has been on translating IPCC guidelines, providing detailed country case studies and examples to show what achieving REDD+ might mean in practice.	‘The Sourcebook makes more user-friendly methods described in the IPCC guidelines providing additional information to guide stakeholders in implementing those methods ensuring that the principles on which IPCC methods have been built (i.e. transparency, completeness, consistency, comparability and accuracy) are respected in monitoring forest carbon stocks and carbon stocks changes’ [respondent 14, Q7].
	‘[the] Sourcebook is like a guidebook developed on the basis of IPCC Guidelines with case studies and examples for those who would like to learn and implement about the MRV and REDD plus issues in climate change and carbon stock of forest ecosystems’ [respondent 21, Q7].
Scepticism about policy makers’ knowledge and comprehension of forest carbon measurement - e.g regarding specific areas of knowledge to do with classifying different types of forest carbon measurement, but also more broadly in terms of a lack of understanding of how science is done, the science process.	‘One point which is essential is to separate out is the issue of assessing reduced deforestation (including conservation) and issues of assessing [forest] degradation (including forest enhancement and sustainable management of forest), since the first 2 are primarily based on area change and the last 3 primarily on stock change.... *Most policy makers hardly understand the difference*’ [respondent#1, Q8].
	‘The way forest carbon is measured will influence the results – no two methods will yield exactly identical results and so if international policy makers are waiting for the scientific community to settle on one agreed upon approach for measurement then I am not optimistic’ [respondent 11, Q8].

One might expect scientists who study forests to be quite cautious overall about the ease and potential of getting to 'compliance-grade' MRV of forests, and results from the GOFC-GOLD author survey to some extent back this up. As the selection of survey findings presented in Table 10.2 illustrates, there is doubt amongst some in the science community that forests can ever be subject to a workable system of MRV and made governable within international climate policy, as one survey respondent comments: 'The GOFC-GOLD Sourcebook has definitely influenced the negotiators and bureaucrats in REDD – it has told them that "yes, we can" measure forest change globally, but I'm not sure I agree on that. In that sense the book has been a bit misleading' (GOFC-GOLD survey, Respondent #6, 2011). But it is a mixed picture, with other remote sensing scientists much more certain about the potential of remote sensing to provide an accurate, global measure of forest carbon. It is clear that the practices that populate the forest carbon governmentality of MRV are not yet fixed, and indeed if remote sensing is the forerunner of these practices then it signals that the ambitious discourse of MRV might be somewhat ahead of the technological capabilities of MRV.

The Techniques and Practices of Forest Carbon MRV: Part II – Standards

One important way to discern the key technologies and practices of government in any particular policy sector is to assess the standards (the rules, principles and 'how to' guides) used to operationalize government. It is through standards and guidelines that existing forest carbon expertise and knowledge is translated into REDD+ MRV policy and practice. Timmermans and Epstein, in their detailed review of the sociology of standards, define the standardization as: 'a process of constructing uniformities across time and space, through the generation of agreed-upon rules' (2010: 71). Standards are valuable in areas of policy flux because they represent something relatively durable and thereby provide a 'marker' or identity for organizations. The formalization of knowledge into standards and guidelines is a way of enacting and formalizing power. From a governmentality perspective, standards are a kind of technology or technique that facilitates the 'conduct of conduct': controlling behaviour and practices across time and space and making them uniform, 'ubiquitous but underestimated phenomena that help regulate and calibrate social life by rendering the modern world equivalent across cultures, time and geography' (Timmermans and Epstein 2010: 70). In the case of forest carbon MRV, standards have become a key avenue for translating knowledge and expertise into the policy realm, both as a route to enacting that expertise, and also in gaining legitimacy and credibility for the organizations involved. Table 10.3 highlights a small selection of the many standards, protocols and 'how to' guides that populate the field of forest carbon MRV.

As Table 10.3 demonstrates, once again we see active in this policy area a diverse range of organizations: government and nonstate; science and non-science. As noted,

Table 10.3 *Examples of forest carbon MRV standards*

Name of Organization	Type of Organization	Forest carbon MRV standard	Comment	Key references
Centre for International Forestry Research (CIFOR)	Research-orientated NGO	Forest Carbon Database (2010)	The aim of the CIFOR Forest Carbon Database is to provide open access information for scientists and other practitioners, to avoid duplication, allow comparison and so forth of forest carbon data; based on IPCC guidelines (see page 11 of User Guide)	http://www.cifor.org/publications/pdf_files/Books/BKurnianto1001.pdf; http://carbonstock.cifor.cgiar.org/
Intergovernmental Panel on Climate Change (IPCC)	Intergovernmental UN science agency; under auspices of UNFCCC	2003 IPCC Good Practice Guidance (GPG) for Land Use, Land Use Change and Forestry (LULUCF)	Providing international principles and guidance for measuring and reporting on carbon stock changes and emissions. IPCC was invited to do this GPG under the Marrakesh Accords	IPCC GPG for LULUCF (2003), see: http://www.ipcc-nggip.iges.or.jp/public/gpglulucf/gpglulucf_contents.html IPCC (2006) for ALOFU
Clinton Climate Initiative (CCI)	NGO	Carbon Measurement Collaborative	'CCI helps countries comply with international measurement and reporting verification (MRV) standards, building a credible database on which to advance international agreements on deforestation. CCI also helps to ensure that critical, national-scale satellite information on forests is delivered regularly and routinely to countries that depend on it' [quote from CCI website, 14 January 2011]	http://www.clintonfoundation.org/what-we-do/clinton-climate-initiative/forestry/measuring-carbon
World Bank Forest Carbon Partnership Facility (FCPF)	Intergovernmental, UN	Guidelines and Templates		http://www.forestcarbonpartnership.org/
Group on Earth Observations (GEO) Global Forest Observation Initiative	Intergovernmental, voluntary partnership of governments and international organizations	In preparation (see Implementation Plan), 2014+ will be developing methods and protocols		Group on Earth Observations (GEO) Global Forest Observation Initiative Implementation Plan (2011)

(continued)

Table 10.3 *Continued*

Name of Organization	Type of Organization	Forest carbon MRV standard	Comment	Key references
GOFC GOLD	International state-based organization	GOFC GOLD Sourcebook – updated at each COP, provides additional information on methods and procedures for MRV	[see main text for empirical discussion]	
Climate Community and Biodiversity Alliance (CCBA) standards	Partnership of international NGOs, inc. CARE, Rainforest Alliance, Conservation International	Carbon standard for voluntary market land-based projects that encompasses not just carbon reductions but other positive project benefits (poverty alleviation etc.)	The goal of CCB standards is to 'catalyze a robust carbon market for multiple-benefit forest carbon projects' (CCB standards, second edition, p. 4) Rests on IPCC guidelines for AFOLU (2006), 'or a more robust and detailed methodology' (p. 14, second edition CCB standard)	http://www.climate-standards.org/
Verified Carbon Standard (VCS)	Founded by carbon market organizations and NGOs	Leading voluntary carbon market standard	Often used in conjunction with CCB standards [as CCBA does not actually issue credits]. Has a separate AFOLU Expert Committee.	http://www.v-c-s.org/develop-project/agriculture-forestry-projects

for forest carbon MRV the 2003 IPCC Good Practice Guidance for Land Use, Land Use Change and Forests (LULUCF) is in effect the baseline standard. The IPCC GPG provides international principles and guidance for measuring and reporting on carbon stock changes and emissions for different land use categories and carbon pools, including forests. IPCC Monitoring Standards (Tier 1, 2, 3) set different levels of accuracy for forest carbon measurement: Tier 1 is the most basic, using standardized continent-scale data (averages for six vegetation types); Tier 2 has some country-level data, based on forest inventories; and Tier 3 is the most robust standard, based on individual sample plots. So whilst the standards in Table 10.3 rest on the IPCC GPG, there remain a host of other secondary activities not defined by the IPCC, that have not yet been standardized and that must be agreed upon to develop a workable system of MRV for REDD+, and it is in this fast-moving policy space – stemming out from IPCC GPG and UNFCCC policy decisions and beyond – that competing practices of the MRV governmentality are jostling.

Remote sensing science, as discussed, is the knowledge domain and set of technologies associated most closely with forest carbon MRV, but there are other competing practices, in particular ground-based local community forest monitoring (i.e. a set of MRV practices carried out by forest communities themselves based on their local knowledge and capabilities), with its approach and methodologies derived from the discipline of forest ecology. According to the UNFCCC, REDD+ MRV must include both types of MRV practices: ground-based and remote sensing; indeed it has been quite explicit that both are required to develop accurate systems of measurement, reporting and verification (see Decision 4/CP.15, 1d(i) UNFCCC 2009a). This UNFCCC stance helps explain the degree of consensus around MRV: to some extent there is not competition between different MRV practices (ground and space-based) because both are required. The challenge, and to date largely unresolved difficulty where the tensions do emerge, is therefore about how to integrate different types of MRV practices, each with different definitions about what to measure and how, but with a shared commitment to the overall objective of forest carbon MRV. The proliferation of standards on forest carbon MRV is in large part in response to the continuing uncertainty: the practices and technologies of the MRV governmentality are not yet settled.

One useful way to understand how these different types of forest carbon MRV expertise are jostling together is from a 'bottom up' perspective looking at the forests and trees themselves and their capacity to be measured. How might these practices consolidate and become stable within the MRV governmentality? The ANT concept of translation is helpful in providing a framework for analysing and explaining how forests (as living, rather unstable things) might become stabilized as part of networks based on the policy goal of accurate measurement. One of the main elements of ANT is the idea that nonhuman things (such as trees, forests) have agency, that is that they can affect and influence processes of change. Trees are living things and can catch fire, succumb to disease, be hollow (and therefore store much less carbon), hide beneath clouds (making satellite measurements difficult), be blown over, have more/less roots than supposed and so on. Hence even a cursory ANT-inspired focus on the agency of the forests suggests they are perhaps not amenable to being enrolled or translated into reliable 'actor networks', and that measuring carbon in forests to compliance grade standard for REDD+ is therefore unlikely to be a straightforward task. Indeed, a key theme of research in ANT concerns how durable materials help hold a network together, as Latour suggests: 'whenever we discover a stable social relation, it is the introduction of some non-humans that accounts for this relative durability' (1991: 111). Clearly forests – comprised of trees and multitudes of other living things – are not durable in the same way that, for example, buildings, power stations or rail tracks might be. Indeed, much recent forest science has highlighted the fragility of tropical forest ecosystems (see Malhi et al. 2008).

So an ANT perspective adds to the concept of governmentality by suggesting that part of the reason for the ongoing state of flux around MRV technologies and practices – the multitude of standards, the unclear role of different sets of expertise, the jostling of techniques – is to do with the forests themselves. ANT provides a useful reminder that stable relations between actors, entities and places are not yet formed for forest carbon measurement: in ANT terminology *translation* has not yet happened. *Translation* describes the process of building actor networks 'during which the identity of actors, the possibility of interaction and the margins of manoeuvre are negotiated and delimited' (Callon 1986: 203). Callon identifies four stages or 'moments' of translation: *problematisation* (where relevant actors and their identities are defined), *interessement* (where attempts to stabilize the actors are made), *enrolment* (negotiations about the relationships between actors are ongoing) and *mobilisation* (when the extent to which the actor network represents other similar situations, or wider populations, is tested; its generalizability and capacity to 'travel'). It is difficult to judge precisely where forest carbon MRV sits in this four-stage model of translation. Indeed, Callon himself cautions in his case study of scallop farming that these distinct phases of translation are not in reality neatly separable, explaining:

> Translation is a process before it is a result. That is why we have spoken of moments which in reality are never as distinct as they are in this paper. Each of them marks a progression in the negotiations which result in the designation of the legitimate spokesmen who, in this case study, say what the scallops want and need, and are not disavowed. (1986: 224)

The role of standards is important in providing some sort of stability and in making more durable the identity of relevant actors, and therefore perhaps fits best with the moment of *interessement*. But standards are also about making things the same across space – about geographical uniformity, and are therefore also a tool of mobilization. Following Murdoch's (1998) insightful analysis of translation in policy, forest MRV could alternatively be characterized as a network of 'variation and flux' where links between entities in the networks are only provisional, what Murdoch terms 'spaces of negotiation'.

Summary and Conclusions

This chapter has introduced and discussed one example of contemporary climate governance, namely managing and measuring the carbon in tropical forests. The thrust of REDD+ policy is about making forests governable, bringing them into the realm of the state through collecting data, standardizing and quantifying them. The specific appeal of remote sensing can be better understood by conceptualizing forest carbon MRV as a type of governmentality; a governmentality enabled through obtaining forest carbon data and improving knowledge of the forests. The case study provides a nice illustration of both the challenges and opportunities in achieving successful global climate governance in an area of policy characterized by fast-moving science. The opportunity is to embrace new research findings in order to make significant and rapid reductions in global greenhouse gas emissions via avoided deforestation and forest degradation. The challenge is to knowledgeably evaluate the latest forest carbon science and methods to support policy and forest communities through guidance and standards, thereby making policy objectives achievable. One key mechanism through which this translation of science knowledge appears to be happening is through the development of standards, guidelines and protocols. A focus on the role of standards is a useful entry point to understanding climate governance, involving a complex mix of science, technologies, policy and politics. Standards necessarily have to leave out data and ideas that do not fit: the objective of standardization is creating simplified categories and rules, and in the process discarding things, ideas and knowledge. So whilst the multiple standards emerging in forest carbon MRV provide valuable stability, capturing the uncertainty and dissent amongst the multiple expert communities on forests is also important if a long-term, workable solution to forest carbon MRV is to be developed.

Theories relevant to climate governance shed light on particular aspects of forest carbon MRV. For governmentality its strength lies in explaining the emergence of MRV discourse and its practices, showing how they are interrelated. The role of knowledge is central to this relationship between discourse and practices in governmentality; as Rose-Redwood explains, 'most studies of governmentality share a common concern for critically examining the role that knowledge production has played in the formation of modern governmental practices' (2006: 469). In fast-moving policy sectors such as forest carbon, the role of knowledge becomes even more important – it is a challenge for governance to keep pace with science and technological developments. The danger in the forest carbon case is whether the UNFCCC knows forests well enough to 'shepherd' the forest carbon 'population'. In a situation of continuing uncertainty about the technologies and practices of forest carbon MRV, remote sensing has obvious appeal for policy makers keen to identify an authoritative, reliable and global-scale set of forest carbon monitoring technologies. But those involved in REDD+ policy must take care to gauge and solicit a wide range of science expertise, otherwise there is a danger that one particular area of expertise (remote sensing) comes to dominate policy, because it fits with dominant rationality of 'measurement, reporting and validation'. Governmentality perspectives about the role of knowledge, and how and in what circumstances knowledge becomes powerful, are key to understanding how REDD+ policy has evolved.

An actor-network approach is a useful complement to governmentality because it reminds us to appreciate that the 'stuff' of climate policy, whether it be forests, the atmosphere or wall cavity insulation, is not necessarily pliable or reliable, and therefore perhaps not so amenable to being governed in the way traditionally conceived of by many political science theories, including governmentality (see Chapters 2 and 7; as well as Foucault 1991; Oels 2005; Rutland and Aylett 2008; Lovell and Liverman 2011). So for ANT its strength is in explaining some of the inconsistencies and difficulties inherent to measuring forest carbon, with a focus on the trees and forests themselves. Without close attention to the substance of what is measured and monitored, there is a danger that policy objectives may be overly ambitious. Nonhuman actors have the capability to disrupt governmentalities, particularly those such as forest carbon MRV – centred on notions of surveillance and control – by resisting monitoring, measuring and other policy objectives which demand durability and stability. Forest ecosystems are inherently dynamic, diverse and often fragile. ANT, particularly ideas about translation, therefore points to ways that governance may be disrupted and fall apart, bringing a valuable sense of instability and flux to governmentality and other political science theories of climate governance.

References

Agamben, G. (1998). *Homo Sacer: Sovereign Power and Bare Life*. Stanford, CA: Stanford University Press.

Angelsen, A. (ed.) (2009). *Realising REDD+: National Strategy and Policy Options*. Bogor, Indonesia: Center for International Forestry Research (CIFOR).

Bäckstrand, K. and E. Lövbrand (2006). Planting Trees to Mitigate Climate Change: Contested Discourses of Ecological Modernization, Green Governmentality and Civic Environmentalism. *Global Environmental Politics* **6**, 50–75.

Baker, J. D., K. Andrasko, M. Bartlell, R. deFries, M. Herold, D. Murdiyarso, D. Pandey, D. Singh and C J. Souza (2010). *MRV for REDD-plus for Classes of Countries: Requirements, Readiness and Capacity Gaps*. Washington: World Bank, Forest Carbon Partnership Facility.

Bennett, J. (2010). *Vibrant Matter: A Political Ecology of Things*. Durham, NC: Duke University Press.

Bulkeley, H. and M. Betsill (2003). *Cities and Climate Change: Urban Sustainability and Global Environmental Governance*. London and New York: Routledge.

Callon, M. (1986). Some Elements in a Sociology of Translation: Domestication of the Scallops and Fishermen of St. Brieuc Bay. In *Power, Action, Belief*, ed. J. Law. London: Routledge and Kegan Paul, 196–233.

Clinton Climate Initiative (2012). *Preserving Forests: A Conversation with Dr. D. James Baker*, from http://www.clintonfoundation.org/what-we-do/clinton-climate-initiative/i/preserving-forests-a-conversation-with-dr-d-james-baker. Accessed 23 January 2012.

Dean, M. (1999). *Governmentality: Power and Rule in Modern Society*. London: Sage Publications Ltd.

Dryzek, J. (1997). *The Politics of the Earth: Environmental Discourses*. Oxford: Oxford University Press.

Eggleston, H. S., L. Buendia, K. Miwa, T. Ngara and K. Tanabe (2006). *IPCC Guidelines for National Greenhouse Gas Inventories – Vol. 4 Agriculture, Forestry and Other Land Use*. Kanagawa, Japan: IPCC, National Greenhouse Gas Inventories Programme, Institute for Global Environmental Strategies.

Fogel, C. (2005). Biotic Carbon Sequestration and the Kyoto Protocol: The Construction of Global Knowledge by the Intergovernmental Panel on Climate Change. *International Environmental Agreements: Politics, Law and Economics* **5**, 191–210.

Foucault, M. (1991). Governmentality. In *The Foucault Effect: Studies in Governmentality*, ed. G. Burchell, C. Gordon and P. Miller. London: Harvester Wheatsheaf, 87–104.

Foucault, M. (2008). *The Birth of Biopolitics: Lectures at the College de France, 1978–1979*. London: Palgrave Macmillan.

GEO Forest Carbon Tracking (2010). *Report on the Concept Phase for Developing a Global Forest Monitoring Network*. Geneva: Group on Earth Observations (GEO) Forest Carbon Tracking Task.

Global Canopy Programme (2009). *The Little REDD+ Book – An Updated Guide to Governmental and Non-Governmental Proposals for Reducing Emissions from Deforestation and Degradation*. Oxford: Global Canopy Programme.

GOFC-GOLD (2009). *A Sourcebook of Methods and Procedures for Monitoring and Reporting Anthropogenic Greenhouse Gas Emissions and Removals Caused by Deforestation, Gains and Losses of Carbon Stocks in Forests, Remaining Forests and Reforestation. Report version COP15–1*. Global Observation of Forest and Land Cover Dynamics (GOFC-GOLD).

GOFC-GOLD (2012). *GOFC-GOLD: Global Observation of Forest and Land Cover Dynamics – Home Page*. from http://www.fao.org/gtos/gofc-gold/. Accessed 30 July 2012.

Herold, M. (2009). *An Assessment of National Forest Monitoring Capabilities in Tropical Non-Annex I Countries: Recommendations for Capacity Building*. London and Oslo: The Prince's Rainforests Project and the Government of Norway.

IPCC (2007). *Climate Change 2007: The Physical Science Basis – Summary for Policymakers*. Paris: Intergovernmental Panel on Climate Change (IPCC).

Latour, B. (1991). Technology is Society made Durable. In *A Sociology of Monsters: Essays on Power, Technology and Domination*, ed. J. Law. London and New York: Routledge, 103–31.

Laws, D. and M. Rein (2003). Reframing Practice. In *Deliberative Policy Analysis: Understanding Governance in the Network Society*, ed. M. A. Hajer and H. Wagenaar. Cambridge: Cambridge University Press, 172–206.

Lemke, T., L. Moore and M. Casper, (2010). *Biopolitics: An Advanced Introduction*. New York: New York University Press.

Lovell, H. and D. M. Liverman (2011). Understanding Carbon Offset Technologies. *New Political Economy* **15**, 255–73.

Luke, T. (1999). Environmentality as Green Governmentality. In *Discourses of the Environment*, ed. E. Darier. Oxford: Blackwell Publishers, 121–51.

Lövbrand, E. and J. Stripple (2010). *Governing the Climate from Space: Monitoring, Reporting and Verification as Ordering Practice. 2010 International Sociological Association Annual Convention*, New Orleans.

Lövbrand, E. and J. Stripple (2011). Making Climate Change Governable: Accounting for Carbon as Sinks, Credits and Personal Budgets. *Critical Policy Studies* **5**, 187–200.

Malhi, Y., J. T. Roberts, R. A. Betts, T. J. Killeen, W. Li and C. A. Nobre (2008). Climate Change, Deforestation, and the Fate of the Amazon. *Science* **319**, 169–72.

Murdoch, J. (1998). The Spaces of Actor-Network Theory. *Geoforum* **29**, 357–74.

Oels, A. (2005). Rendering Climate Change Governable: From Biopower to Advanced Liberal Government. *Journal of Environmental Policy and Planning* **7**, 185–207.

Owens, S. E. and R. Cowell (2002). *Land and Limits: Interpreting Sustainability in the Planning Process*. London: Routledge.

Penman, J., M. Gytarsky, T. Hiraishi, T. Krug, D. Kruger, R. Pipatti, L. Buendia, K. Miwa, T. Ngara, K. Tanabe and F. Wagner (2003). *IPCC Good Practice Guidance for Land Use, Land-Use Change and Forestry*. Institute for Global Environmental Strategies, Kanagawa, Japan: National Greenhouse Gas Inventories Programme.

Pinch, T. and R. Swedberg, ed. (2008). *Living in a Material World: Economic Sociology Meets Science and Technology Studies*. Cambridge, MA: MIT Press.

Rein, M. and D. Schon (1993). Reframing Policy Discourse. In *The Argumentative Turn in Policy Analysis and Planning*, ed. F. Fischer and J. Forester. London: UCL Press Ltd, 145–66.

Rose-Redwood, R. S. (2006). Governmentality, Geography, and the Geo-Coded World. *Progress in Human Geography* **30**, 469–86.

Rutland, T. and A. Aylett (2008). The Work of Policy: Actor Networks, Governmentality, and Local Action on Climate Change in Portland, Oregon. *Environment and Planning D* **26**, 627–46.

Stephan, B. (2012). Bringing Discourse to the Market. *Environmental Politics* **21**, 621–39.

Timmermans, S. and S. Epstein (2010). A World of Standards but not a Standard World: Toward a Sociology of Standards and Standardization. *Annual Review of Sociology* **36**, 69–89.

UN-REDD Programme (2008). *Role of Satellite Remote Sensing in REDD: Issues Paper*. Rome: FAO, UNDP, UNEP.

UNFCCC (2007). *Report of the Conference of the Parties on its thirteenth session, held in Bali from 3 to 15 December 2007*. United Nations Framework Convention on Climate Change (UNFCCC).

UNFCCC (2009a). *Copenhagen Accord: Decision 4/CP.15 Methodological guidance for activities relating to reducing emissions from deforestation and forest degradation and the role of conservation, sustainable management of forests and enhancement of forest carbon stocks in developing countries*. Copenhagen: UNFCCC.

UNFCCC (2009b). *Technical Paper FCCC/TP/2009/1 – Cost of implementing methodologies and monitoring systems relating to estimates of emissions from deforestation and forest degradation, the assessment of carbon stocks and greenhouse gas emissions from changes in forest cover, and the enhancement of forest carbon stocks.*. UN Framework Convention on Climate Change (UN FCCC).

Watson, C. (2009). *Forest Carbon Accounting: Overview and Principles*. Addis Ababa, Ethiopia: UNDP-UNEP CDM Capacity Development Project for Eastern and Southern Africa.

11

Climate Security as Governmentality: From Precaution to Preparedness

Angela Oels

Introduction

From the early 2000s onwards, climate change has increasingly been articulated as a security issue. The UN Security Council had climate change on its agenda in 2007 and 2011. The UN secretary-general produced a report to the UN General Assembly in 2009 entitled *Climate Change and Its Possible Security Implications*. What is the meaning of this security turn in climate discourse? To what extent does it differ from previous discourse – enough to call it a discursive shift? Is the new wording related to new practices of securing in the face of climate change or does the debate remain at the level of rhetoric? Scholars have employed a range of theoretical approaches to tackle these questions: discourse analysis, the Copenhagen School and neo-Marxist poststructuralism. The overall finding of these approaches is that the security rhetoric does not seem to be linked to any new governmental action on the ground.

Researchers have utilized the framework of the Copenhagen School to argue that despite the scale of the dramatic securitizing moves in the field of climate change, this has not yet enabled a political state of exception – or in fact any significant action to combat climate change (Trombetta 2008). From a neo-Marxist poststructuralist point of view, Swyngedouw (2010) has argued that what he calls the 'climate apocalypse' is just the opium of the people, distracting them from the fact that the structural causes of climate change remain untackled. From a discourse analytical perspective, Detraz and Betsill (2009) have developed a more nuanced argument. They show that what appears to be a discursive shift is better understood as a continuity of an environmental security discourse which defines security in line with human security. Detraz and Betsill welcome such continuity because they argue that the environmental security discourse is conducive to enabling effective action to address climate change, while what they call environmental conflict discourse is not. However, this effective action could not be traced at the time of completion for their research. So overall, existing research remains sceptical about the relevance of security discourses for the politics of climate change. If we trust the existing research, climate security may sound a loud alarm, but its impact on policy making has thus far been negligible.

In contrast to the existing research outlined, this chapter presents the findings of Foucauldian studies of climate security which argue that there is evidence of a change in international climate policy (Methmann and Rothe 2012; Oels 2012; Oels 2013). The security turn in climate politics is neither a continuity nor is it without effect. I show that there has been a shift from a configuration that I call 'tolerable levels of climate change', characteristic of the 1990s and early 2000s, to 'climate apocalypse', which is the most recent configuration. While 'tolerable levels of climate change' is based on an apparatus of risk management that seeks to keep greenhouse gas concentrations in the atmosphere at 'safe' levels, climate apocalypse insists on the fundamental unpredictability of climate change and proposes resilience in the face of sudden disruptions.

I introduce the concept of governmentality as a framework for the analysis of 'security'. I suggest that Foucault's work enables us to distinguish between at least four modes of securing, each linked to a technology of power: sovereign power, discipline, liberal biopower and advanced liberal government. Each of these four technologies of power draws on characteristic techniques – for example, sovereign power draws on the law to ban undesirable actions, discipline employs surveillance to induce desired behaviour and liberal biopower draws on statistics to target governmental interventions to at-risk groups. Foucault, however, emphasized that these technologies of power continue to coexist. The latest technologies of power like liberal biopower and advanced liberal government draw on elements of sovereign power and discipline and reconfigure them to serve their purposes. A topological governmentality analysis (Collier 2009) must therefore study the coexistence and redeployments of elements of all four technologies of power in current problematizations of climate security.

I begin the chapter by introducing the work of other researchers on climate security in order to highlight the limitations of the theoretical frameworks chosen and the arguments made. Next, I introduce Foucault's concept of governmentality as the framework for analysis. In the main part of the chapter, I present the Foucauldian argument that climate security marks a discursive shift from a conceptualization of climate change as controllable to the idea of 'climate apocalypse'. I present evidence for new practices of resilience which are facilitated in the name of preparedness for 'climate apocalypse'. Section five returns to the work of other authors and discusses their approaches in light of the Foucauldian-inspired studies. By doing so, I seek to highlight the added value of the governmentality approach to climate security. The conclusion summarizes what can be gained from studying security as governmentality – on a theoretical level and for the empirical case of climate security.

Other Approaches to Climate Security

Scholars have made a number of attempts in the literature to analyse the political implications of producing climate change as a security issue. The main research interest is to ask first, whether it is possible to find evidence of a discursive shift and second, to assess if this shift has enabled or is likely to enable tougher climate policy. *On one hand*, those following the Copenhagen School (Buzan, Waever and de Wilde 1998) have asked whether the securitization of climate change has been successful (i.e. accepted by a relevant audience) and if it is therefore capable of enabling extraordinary policy measures (Stripple 2002; Brauch 2009). In his application of the Copenhagen framework written more than a year before the failure of the Copenhagen summit, Brauch claims that the articulation of climate change as a security issue is used 'to legitimate extraordinary and costly measures that require a progressive increase in energy efficiency and a decarbonisation of the energy system by increasing renewable energy sources' and to justify the 'allocation of significant public funds' (2009: 71). While demands for such action have certainly been made in the name of climate security, there is no evidence whatsoever of this action on the ground. Despite this, a successful securitization of climate change could also be used to enable repressive measures to uphold global order (for example responsibility to protect interventions after a climate-induced disaster) rather than to curb climate change. However, thus far no applications of the Copenhagen framework follow this line of investigation.

On the other hand, academics following the post-politics discussion have applied their framework to the issue of climate change. Eric Swyngedouw argues that the way climate change has been articulated as 'climate apocalypse' in public discourse is marked by populism that evacuates 'the political' from climate change debates. The climate apocalypse is constructed as 'a universal humanitarian threat' that turns 'us' into 'universal victims', glossing over social differences and conflicts of interests (2010: 221). The threat of climate change is constructed as an aberration from an otherwise unproblematic capitalist system: 'CO_2 stands here as the classic example of a fetishized and externalized foe that requires dealing with if sustainable climate futures are to be attained' (2010: 222). From Swyngedouw's perspective, the primary function of the securitization of climate change is 'to make sure that nothing really changes' (2010: 222). To hide the fact that what is being secured is the capitalist system, '[a]n extraordinary techno-managerial apparatus is under way … with a view to producing a socio-ecological fix to make sure nothing really changes' (2010: 222). An example of such a fix is the replacement of fossil fuels with nuclear energy. The Copenhagen School and the post-politics perspectives criticize the securitization of climate change as a mode of preempting political debate by moving an issue 'beyond politics' and therefore beyond questioning. While the Copenhagen School has advanced the hypothesis that drastic climate policy might be enabled as a result of the securitization of climate change, those following the post-politics debate claim that the climate apocalypse serves to make sure that nothing really changes.

From a discourse analytical perspective, the meaning of security can actually vary and different discourses of security are linked to different practices and policy implications. Nicole Detraz and Michele Betsill (2009) have analysed the articulation of climate change as a security issue in international policy venues from a discourse analytical perspective. Detraz and Betsill distinguish between two discourses: the *environmental conflict* discourse is said to concern itself with the national security of the state in the face of potential conflicts over scarce resources (2009: 305). The *environmental security* discourse, on the other hand, is said to focus on the human security of the population in the broadest sense of the term (2009: 306). In their findings, the authors claim that the politics of climate change have historically been informed by the *environmental security* discourse. They recognize that the problematization of conflicts as a result of climate change is a new development, and was most apparent at the 2007 UN Security Council session. However, Detraz and Betsill refuse to speak of a discursive shift because the problematization of conflicts as a result of climate change is said to be embedded in an *environmental security* discourse. As a result, they find that no policy changes can be traced. Detraz and Betsill argue that *national security* discourses are counterproductive in solving the problem of climate change, while *human security* discourses are conducive. This analysis demonstrates that different security discourses are linked to different policy implications. This distinction between different discourses significantly advances the academic debate on the securitization of climate change. However, while this analysis is a good starting point, I have at least two major problems with it. First, I question Detraz and Betsill's somewhat stereotypical assumption that the *environmental security* discourse is conducive to solving the problem of climate change while the *environmental conflict* discourse is judged as counterproductive. By contrast, I would like to demonstrate in this chapter that human security is also capable of legitimizing and mobilizing violence like international military interventions. Second, I disagree with Detraz and Betsill that no policy shifts can be traced. I argue that we must look for them in different places (for example, the field of professionals of (in)security) than those suggested by Detraz and Betsill. Moreover, some of these shifts are so recent that Detraz and Betsill could not map them.

While some have claimed that the securitization of climate change – if successful – might facilitate extraordinary measures to curb greenhouse gas emissions (Brauch 2009), I demonstrate that the enabled measures are neither exceptional nor geared towards halting global warming. It is exactly because climate change *will not* be prevented or slowed that security experts and politicians prepare themselves for upholding 'global order and world peace' in the face of climate-induced disruptions. I demonstrate in the following that the construction of climate change as *threat multiplier* has enabled *routine* measures of enhancing resilience to disruptions that could potentially result from the secondary impacts of climate change, such as disasters, migration and violent conflict.

Governmentality Studies: From a Categorical to a Topological Analysis

In his 1978 governmentality lectures, Foucault uses *security* as synonym for liberal biopower. According to this reading, security is a technology of power used to render certain subject-objects governable. Michel Foucault's governmentality lectures (2007) offer two inroads to the study of security. First, a categorical analysis allows us to distinguish between different forms of exercising power. Second, a topological analysis investigates how elements of the categorical ideal types are reconfigured and recombined in particular instances of rendering an object-subject governable. The following sections introduce each approach in turn.

Categorical Analysis: Modes of Power

The work of Michel Foucault (2007) has highlighted that government is not simply an activity undertaken by an individual actor called 'the' government. Instead government is better understood as:

> any more or less calculated and rational activity, undertaken by a multiplicity of authorities and agencies, employing a variety of techniques and forms of knowledge, that seeks to shape conduct by working through our desires, aspirations, interests and beliefs, for definite but shifting ends and with a diverse set of relatively unpredictable consequences, effects and outcomes. (Dean 2003: 11)

Government does not act upon 'given' subjects or objects 'out there'. Instead, it has to make the subject-objects of government knowable and thinkable before techniques for acting upon them can be conceived and established. These subject-objects are constituted in the act of their regulation, and the very categories of thinking about subject-objects are created in these acts of government (for the creation of the category 'lesbian' in practices of border control see Luibhéid 2002). This close link between knowledge and power, between thinking and the ability to act on something informed Foucault's governmentality lectures. In his 1977–8 lectures at the Collége de France entitled *Security, Territory, Population*, Foucault develops the concept of governmentality as an analytical framework for studying the configuration of power in the modern era. Leading scholars generally define the concept of governmentality 'as a political rationality that shapes the "conditions of possibility" for thinking and acting in a certain way' (Collier 2009: 96). A governmentality analysis demonstrates how problems are 'made thinkable and practicable [as] knowable and administrable domain[s]' (Rose, O'Malley and Valverde 2006: 86, in Collier 2009).

The value of Foucault's work on governmentality lies in the distinction he offered between at least three modes of exercising power in government. In Foucault's 1978 lectures, the term *governmentality* designates the historic era of biopower, namely 'the institutions, procedures, analyses and reflections, calculations, and tactics that allow the exercise of this … power that has the population as its target, political economy as its major form of knowledge, and apparatuses of security as its essential technical instrument' (Foucault 2007: 108). Foucault distinguishes between biopower and other 'diagrams' of power, namely sovereign power and discipline. Following Collier, I call these modes of exercising power 'technologies of power' (2009: 97). To my understanding, these are ideal types which cannot be observed as such in the real world. Foucault offered them to distinguish certain historic eras and to pin down 'what is general about a new class of governmental forms across a range of cases' (Collier 2009: 99). The identification of these historic eras with their characteristic form for the exercise of governmental power helps orientate the researcher as to which technologies of power might be most influential in contemporary government (Collier 2009: 97). Foucault has linked each historic era to characteristic knowledge systems and practices, both of which create (or fail to create) certain visibilities and certain subject-objects (Dean 2003). Sovereign power uses the law to rationalize the exercise of power and sanctions noncompliance with violence. Disciplinary power draws on surveillance and control in order to mobilize desired identities and behaviour patterns. Liberal biopower's key aim is to foster and optimize the life of the population to enhance its productivity. Liberal biopower uses statistics about the population to identify risk groups towards which costly interventions are targeted. Since Foucault's death in 1984, Niklas Rose has added advanced liberal government as a more recent form of risk management. Advanced liberal government often goes hand in hand with Michael Dillon's (2007) risk management through contingency. This is a governmental form of power that seeks to enhance life's capacity for adaptive emergence in the face of sudden and unpredictable shocks (Dean 2003; Dillon 2007). Table 11.1 offers an overview of these four 'technologies of power' in the form of ideal types abstracted by Foucault (and others) from certain historic eras.

Topological Analysis: Redeployments of Techniques of Different Governmentalities

Newer scholarship on Foucault's governmentality lectures proposes a '"topological" analysis of power that examines how existing techniques and technologies of power are re-deployed and recombined in diverse assemblies of biopolitical government' (Collier 2009: 79; on a similar note Adey and Anderson 2012). Foucault himself has said of the relationship between sovereign power, discipline and biopolitics that 'there is not a series of successive elements, the appearance of the new causing the earlier ones to disappear.... [W]hat above all changes is the dominant characteristic, or more exactly, the system of correlation between' them (2007: 8). This means that the focus of analysis should be on how the knowledge systems and practices characteristic of sovereignty and discipline are reconfigured and possibly redeployed in the era of liberal biopower. Sovereignty and discipline are 'governmentalized'; they are reconfigured and transformed by liberal biopower. In a liberal regime of biopower, which aims to foster the welfare of the population, it is tricky to exercise the *sovereign* right to kill. According to Foucault, killing can only be legitimate in liberal biopower when it is deemed necessary for the survival of the population or more specifically, a designated subgroup of the population. To legitimize the killing of others, they must first be

Table 11.1 *Reading 'securitization' with Foucault's concept of governmentality*

Technology of power	Sovereign power	Disciplinary power	Biopower I: Probability-based risk management	Biopower II: Risk management through contingency
Referent object of security	The nation-state	The individual body	The population as living entity (man-as-species)	International circulation of goods, services and people, especially emergent life
Mechanism of securing	Inclusion/exclusion	Normation	Normalization Distinguishing good and bad circulation, maximizing good circulation, keeping bad circulation at a tolerable level	Enhancing life's capacity for adaptive emergence, enhancing resilience and capacity to regenerate
Practices	-Rights and duties of government and subjects enshrined in law -Punishment for those that violate the law	-Normation: define an ideal as norm -stigmatize those deviating from the norm as abnormal -provide incentives to return to the norm	-Normalization: identify the statistical average as 'normal' -identify risk groups that deviate most from the norm -target measures/treatment on risk groups -state insurance	-build resilient communities -create markets: make risks transferable by speculating about futures in financial markets -private insurance -scenario planning -responsibilize subjects -preparedness
Subjects produced	Criminals	Abnormals Victims or villains	Risk groups	Adaptive life Responsible subjects

dehumanized in a racist sense. In liberal biopower 'the imperative to kill is acceptable only if it results not in a victory over political adversaries, but in the elimination of the biological threat to and the improvement of the species or race' (Foucault 2003: 256). The law that had such a prominent role in sovereign power is still in use in liberal biopower, but its function has changed away from punishment towards the betterment and education of life.

Under a regime of liberal biopower, security discourses have led to a proliferation of technologies of *discipline* into all areas of life (Lentzos and Rose 2009: 234). The liberal regime of biopower has reconceptualized discipline as a precondition and driving force of liberty in a 'free' market economy based on circulation (Lentzos and Rose 2009: 234). Lentzos and Rose identify important differences between traditional and contemporary forms of discipline: first, they argue that discipline today is not so much about securing an enclosed space as it is about securing circulation. Second, the authors argue that a 'plurality of agencies and forces' is involved in the act of securing, not just the state. Third, discipline gathers observations to identify patterns and regularities rather than fixed preconceived norms (2009: 234–5).

Advanced liberal government has facilitated the latest redeployment of elements of sovereign power, discipline and biopower. Advanced liberal government as a technology of power requires 'free subjects'. As Dean explains, 'in order to act freely, the subject must first be shaped, guided and moulded into one capable of responsibly exercising that freedom through systems of domination' (2003: 165). In that sense, '[t]his is a subject whose freedom is a condition of subjection' and vice versa (2003: 165). Advanced liberal government addresses the subjects of government as 'capable' and 'responsible' for their own risks (2003: 166). Security is achieved by stimulating insecurity in individuals, so that they will be mobilized to participate in their own securing (Lentzos and Rose 2009: 235). Communities are required to anticipate threats, take precautions and prepare for the unexpected (Lentzos and Rose 2009: 235). In these arrangements, the state is no longer seen as the ultimate guarantor of security and it is no longer taken for granted that the national territory is the 'natural' spatial reference point (Lentzos and Rose 2009: 233). Instead, advanced liberal government governs through 'community', targeting certain subgroups of the population in order to mobilize them to enter into partnership with state agencies, professionals and service providers, to become both objects and subjects of their government at the same time (Dean 2003: 170). A topological analysis has to examine specific problematizations of subject-objects, and which elements of sovereign power, discipline, biopower and advanced liberal government are correlated to render these subject-objects governable.

Collier has remarked that trying to fit descriptions of the workings of 'neoliberalism' into the framework provided by advanced liberal government has often obscured more than it has clarified (2009: 100). Instead, neoliberalism should be understood and studied as 'a form of thinking, a kind of reflection that aims to critique and remediate existing *mentalities* and practices of government that have become uncertain or problematic' (2009: 100). This implies not focussing too much on the 'conditions of possibility' of thinking and acting implied by certain texts, but instead engaging more with how interventions like those of neoliberal thinkers try to reconfigure and overcome the currently dominant 'conditions of possibility'. Their texts should be studied as responses to historically situated problems which they are trying to overcome with their 'specific activity of thought' (2009: 100). Rather than remain at the descriptive level of passive discourses, a focus on thought can trace and 'understand the processes of recombination and reproblematization through which contemporary government … is being refigured' (2009: 100). This also helps to mark how existing 'conditions of the possible' may be overcome by new strategies of 'thinking'. The following section illustrates what a topological analysis of climate security might look like and highlights the added value of adopting such an analytical framework.

Shifting Topologies of Climate Security: From Tolerable Levels of Climate Change to Climate Apocalypse

This section explores the topologies of power at work in climate security, drawing on existing Foucauldian studies of climate security (Methmann and Rothe 2012; Oels 2012, 2013). As the perception of climate change has shifted over the past twenty years, so too have the modes of 'securing' in the face of a changing climate. How have the practices of securing changed as a result of (or contributing to) the shifting problematizations of climate change? Based on the Foucauldian work on climate security, I argue that there has been a shift from a configuration that I term 'tolerable levels of climate change', characteristic of the 1990s and early 2000s, to 'climate apocalypse', the most recent configuration. While 'tolerable levels of climate change' is based on an apparatus of risk management that seeks to keep atmospheric greenhouse gas concentrations at 'safe' levels, 'climate apocalypse' insists on the fundamental unpredictability of climate change and proposes resilience in the face of sudden disruptions. The topology of power characteristic of each configuration is introduced in turn. The main focus of analysis will be to highlight how techniques of various technologies of power are redeployed by a regime of liberal biopower and/or advanced liberal government.

Tolerable Levels of Climate Change

In the early 1990s, climate change was defined as a problem of defining 'safe' levels of greenhouse gas concentrations in the atmosphere. The idea was that greenhouse gas emissions and global warming were not problems as such. The problem was if greenhouse gas concentrations in the atmosphere reached a critical threshold where 'dangerous' levels of climate change would occur. These were levels where ecosystems or the economy would fail to adapt and would therefore break down (UNFCCC 1995, Article 2). Article 2 of the United Nations Framework Convention on Climate Change (UNFCCC) signed in 1992 defines the aim of the convention as the 'stabilization of greenhouse gas concentrations in the atmosphere at a level that would prevent dangerous anthropogenic interference with the climate system' (UNFCCC 1995, Article 2). Comprehensive monitoring systems were set up to survey the earth system. Computer modelling was used to develop scenarios of future emission trends and their likely impact on the global climate system (Edwards 2010).

Overall, constructing the climate as in need of securing facilitates a large range of state interventions on its behalf (Luke 1999: 122). The idea of the climate regime in these early days was clearly that climate change was a knowable and therefore controllable phenomenon (Oels 2012). In the field of greenhouse gas emissions, there are clear biopolitical attempts aimed at 'organising circulation, eliminating its dangerous elements, making a division between good and bad circulation, and maximising the good circulation by diminishing the bad' (Foucault 2007: 18). We can see how a certain level of greenhouse gas emissions is considered unproblematic. Only the 'excess' emissions which threaten to destabilize the system are to be addressed and reduced to a 'tolerable' level, in line with economic cost-benefit analysis. Not until the 2009 Copenhagen climate summit did politicians declare that dangerous climate change could be prevented if the increase in global average temperature remained below two degrees Celsius compared to preindustrial levels. The Copenhagen Accord explicitly makes reference to the Intergovernmental Panel on Climate Change (IPCC) as an authority on whose work this target was supposedly based. However, the IPCC has never officially defined any 'safe' levels of climate change as huge uncertainties remain in climate science.

This biopolitical regime of keeping climate change at 'safe' levels is based on a proliferation of disciplinary elements. Under such a regime, the entire planet is subject to large-scale technological surveillance systems. The 'eco-knowledge' (Luke 1999) gained from these systems is used to subject the planet to techno-scientific management. Luke has pointedly argued that '[e]nvironments are spaces under police supervision, expert management or technocratic control; hence, by taking environmentalistic agendas into the heart of state policy, one finds the ultimate meaning of the police state fulfilled' (1999: 149). In the case of climate policy, the IPCC has played a key role in defining baselines and facilitating norms for the measurement and reporting of data for national reports on emissions and greenhouse gas sinks (Beck 2009).

Other phenomena also monitored by disciplinary systems of surveillance and modelled in global computer models were the possible impacts of climate change in industrialized and developing countries. Vulnerability mapping was undertaken as a precondition for focussing possible governmental interventions on risk groups (Methmann and Oels forthcoming). Scientists identified 'climate hotspots' in which climate impacts might cause major disruptions and trigger uncontrolled mass migration and conflict (German Advisory Council on Global Change 2007). These projections have led to calls for a legal refugee status for populations displaced by the impacts of climate change (Conisbee and Simms 2003; Docherty and Giannini 2009; Biermann and Boas 2010). Here, the law would be used as a means of enabling (and possibly responsibilizing) foreign governments to go out and 'save' populations not of their own nationality.

With the signing of the Kyoto Protocol in 1997, advanced liberal elements joined the biopolitical management of the global climate. The Kyoto Protocol used the form of a contract between its member states to agree upon targets and timetables for emission reductions. Self-reporting on target fulfilment at regular intervals was employed to monitor performance. These are clearly technologies of agency and technologies of performance, which create member states as 'free' agents in the realm of climate policy while at the same time disciplining their freedom along the path of emission reductions (Oels 2005). However, these emission reductions were to be achieved in the most cost-effective way by allowing project-based emissions trading. From 2005 onwards, when the Kyoto Protocol entered into force, certified emission reductions could be traded between industrialized countries (joint implementation) and also between industrialized and developing countries (clean development mechanism). Here, the creation of a market to trade certified emission reductions was employed as a technology of governing, disciplining market participants into a sophisticated bureaucracy of verification and certification prior to realizing market sales (Bäckstrand and Lövbrand 2006).

Climate Apocalypse

From 2003 onwards, sources other than the IPCC began to sound the alarm about runaway climate change. Climate change was considered capable of marking the end of civilization (Schwartz and Randall 2003), triggering millions of climate refugees (Christian Aid 2007) and possibly climate wars (Welzer 2012). The idea of linking climate change and variability to violent conflicts goes back to an earlier debate about environmental security in the 1990s, often using identical methodologies and arguments, now updated with regard to climate change (El-Hinnawi 1985; Homer-Dixon 1994, 1999; Myers and Kent 1995). The 2007 IPCC Report was criticized for presenting overly conservative scenarios which do not reflect the more recent published research (Mabey et al. 2011: 19). In particular, the IPCC was charged with not paying enough attention to tipping elements in the global climate system, such as the instability of the West Antarctic ice sheet (Lenton et al. 2008). Moreover, the inadequacy of the nonbinding mitigation pledges made by countries in Copenhagen 2009 was a cause of concern. According to scientific calculations, even if all these reduction targets were met, this would still result in a greater than 50 per cent chance that warming would exceed three to four degrees Celsius (Rogelj et al. 2010). The new scenario is then a world where a rise in global average temperature by three to four degrees

Celsius is likely and an increase of five to seven degrees Celsius is possible (Mabey et al. 2011). The proponents of the climate security discourse argue that '[w]ith such warming, there is little uncertainty over whether extreme impacts will occur, only when they will happen, and to what extent they will affect specific locales' (Mabey et al. 2011: 43).

Not only active climate securitizers, but also concerned scientists have questioned how realistic it is to define 'safe' levels of climate change because of the remaining uncertainties in modelling carbon cycle dynamics and in quantifying climate sensitivity (Boykoff, Frame and Randalls 2010: 53). As a result, the climate is no longer understood as a system that can be kept in a stable equilibrium. Instead, it is now described more along the lines of a complex adaptive system with multiple equilibria. This system is acknowledged to be nonlinear, as well as inherently unpredictable and radically uncertain. Climate change is part of 'an environment that, operating through uncertainty and surprise, has itself become terroristic' (Duffield 2011: 763). The UN secretary-general reacted with his report *Climate Change and Its Possible Security Implications* in 2009, while the UN Security Council discussed climate change in open sessions in 2007 and 2011.

Here, the problem of climate change becomes one of securing global (capital) circulation from disruptions caused by the unpredictable, yet apparently unavoidable impacts of climate change (Oels 2013). Instead of avoiding the possible consequences of climate change, the new emphasis is on preparing for these contingencies, surfing them, surviving them and making sure that they are dealt with appropriately to prevent them from turning into large-scale disasters. The UN secretary-general's report emphasizes the importance of sustainable development as a key strategy for 'building resilience to physical and economic shocks' (UN GA 2009: 4). The report frames the issue of climate change as a threat to human security without explicitly using the term – it instead uses 'human vulnerability' (UN GA 2009: 2). The report recommends conflict prevention, disaster preparedness and capacity building for disaster risk reduction as ways of preparing for the increasing number of extreme weather events (UN GA 2009: 27). These are demonstrably elements of a politics of preparedness (Collier and Lakoff 2008; Aradau and Van Munster 2011). The new buzzword to be found in all recent reports on climate change is *resilience* (Methmann and Rothe 2012; Methmann and Oels forthcoming).

The concept of resilience originates from ecology and has been defined by Holling as 'a measure of the ability of these systems to absorb changes of state variables, driving variables, and parameters, and still persist' (1973: 17). Applied to climate change, resilience must be built into important social systems called 'critical infrastructure' (Lundborg and Vaughan-Williams 2011). At the individual level, resilience implies capacity building for self-reliance. Resilience can be fostered by diversifying livelihoods and by enhancing the capacity for adaptive emergence (Dillon 2008). The emphasis on empowerment and self-reliance implicit in resilience means that

Western interventions in severely affected regions are likely to take the form of helping the poor to help themselves (Chandler 2012). According to Duffield, there has been a remarkable shift in development policy. While in the past Western interventions sought to raise absolute wealth levels in developing countries, this goal has now been dropped. Instead, the main target of development policy is to make the poor fit for survival in an ever-changing world of 'unavoidable' disruptions (Duffield 2008). The subject is responsibilized to prepare for the next crisis (Methmann and Rothe 2012; Methmann and Oels forthcoming). In line with advanced liberalism, it is addressed as a subject capable of responding and coping on its own. However, the subjects created by this mode of securing are not as empowered as one might think. The resilient subject is one 'which accepts the disastrousness of the world it lives in as a condition for partaking of that world and which accepts the necessity of the injunction to change itself in correspondence with the threats and dangers now presupposed as endemic' (Reid 2012: 75). In this context, migration induced by climate change is reconceptualized as an appropriate strategy of adaptation which 'in many cases will be an extremely effective way to build long-term resilience' (Foresight 2011: 7). However, when interviewed, ambassadors of small Pacific island nations strongly resist the idea of relocation: 'ambassadors envision a future as self-determining nation-states, and thus, strongly resist media/policy discourses that legitimise their possible future displacement en masse' (McNamara and Gibson 2009: 481). The inherent depoliticization of the causes of the need for dislocation disempowers political action and the political voice of affected populations (Farbotko 2005; Farbotko and Lazrus 2011).

However, not all subjects can be governed in a liberal way. Some may resist or simply fail to adapt in ways foreseen by government and instead do so in more radical spontaneous ways, sometimes drawing on nonliberal strategies of coping (Duffield 2007 offers the example of the global drug trade). This form of 'adaptive self-reliance' (Duffield 2010) 'constantly exceeds and resists neoliberal governmental techniques of preparedness' (Grove forthcoming: 17). These resistant subjects are then targeted – for example by sustainable development and resilience programming – 'to produce a docile population that will not threaten the vital circulations of liberal order' (Grove forthcoming: 17). A disciplinary system monitors when the vulnerable are on the brink of becoming dangerous to global circulation, for example by engaging in uncontrolled mass migration which can be spotted in satellite pictures. Acts of 'adaptive self-reliance' may be countered with military violence if necessary. However, such use of violence is then not framed as overriding the sovereignty of affected states. Instead, the intervening forces are represented as working hand in hand with the government which is conceptualized as in need of support:

> It does mean, however, that even when military intervention takes place, it is discursively framed as an act of facilitating, empowering or capacity-building the vulnerable subjects on the ground. (Chandler 2012: 225)

We can recognize here how disciplinary monitoring of vulnerable populations and sovereign violence to cope with radical adaptation are reframed and reconfigured in the light of risk management through contingency as acts of empowerment. Here, sovereign violence can be employed fully in line with human security, if not actually carried out in its name (Duffield and Waddel 2006).

The Added Value of a Topological Governmentality Analysis Compared to Other Approaches

In the previous sections, I suggested a new and different reading of climate security drawing on a governmentality analysis. In contrast to Betsill and Detraz, I demonstrated that there actually has been a discursive shift in international climate policy. This shift is clearly linked to a reconceptualization of climate change. In the 1990s and early 2000s, climate change was conceptualized as a knowable and manageable phenomenon that could be kept at 'safe' levels. From the early 2000s onwards, however, climate change emerges as a largely uncontrollable phenomenon capable of causing major disruptions and which is therefore more along the lines of 'environmental terror' (Duffield 2011). I have shown that, linked to this reconceptualization of climate change as 'climate apocalypse', a new set of practices of building resilience to survive the climate apocalypse is emerging. Betsill and Detraz's discourse analysis was inherently limited by the fact that they contrasted 'environmental conflict' (based on national security) discourse with 'environmental security' discourse (based on human security). Betsill and Detraz are right that the notion of human security is – at least implicitly – dominant in official documents on climate security, while national security is placed in the background (Oels 2013). This is even true of documents published after 2008, when their analysis ended. However, the authors fail to recognize what is new: the apocalyptic conception of climate change and the emerging response in the form of resilience. Resilience is fully in line with – if not implied by – the concept of human security. However, as shown earlier, neither resilience nor human security are in themselves necessarily conducive towards effective action to tackle the causes of climate change. In fact, a governmentality approach allows an informed critique of human security as just another configuration of power which, in the case of resilience, may even facilitate military interventions – in the name of empowerment and self-help. Human security as such is neither benign nor malign – it is the way that such concepts are actually redeployed in a larger configuration of technologies of power that matters for the effects they facilitate.

The governmentality approach has also enabled the development of a more nuanced argument along the lines of Swyngedouw, who declared that 'climate apocalypse' served to ensure that nothing really changes. I have shown that there has been a shift: a shift towards even *less* action than before. Of course, at a fundamental level Swyngedouw is right: the capitalist system and its fossil-fuel-based economic growth are left completely unquestioned in climate security discourses. In fact, as I have argued elsewhere (Oels 2005), what is secured is still capitalist circulation and capital reproduction – in this case from climatic disruptions. However, on the level of practices of government, there have been changes in *how* capitalism is secured. The practices of securing in the face of a changing climate have shifted and the new focus is clearly on preparedness for unpredictable disruptive climate change impacts (Methmann and Rothe 2012). Preventing climate contingencies by engaging in mitigation action has taken a back seat. A configuration of mostly advanced liberal elements is calling on affected populations to become resilient to climate-change-induced disasters, for example by diversifying their livelihoods or by migrating. Interestingly, the more dramatic conceptualization of 'climate apocalypse' marks a retreat of Western interventionism. There is evidence for a post-interventionist paradigm (Chandler 2012) in dealing with potentially catastrophic climate change impacts. I have offered the example of displacement induced by climate change. While the 'tolerable levels of climate change' configuration of power is concerned with creating a legal status for those displaced by climate-change-induced disasters and organizing resettlement in advance, 'climate apocalypse' responsibilizes the affected populations to do the right thing to survive – by either migrating in time or building shelters. It is this shift towards a 'do it yourself' response to 'climate apocalypse' which is remarkable.

Returning to the Copenhagen School, the real scandal is actually *too little* intervention in the face of 'climate apocalypse', not too much. The focus of the Copenhagen School is too limited in the case of climate security: it is only interested in finding out if a political state of exception might be enabled by security discourses, and whether an issue is thereby removed from 'normal' politics into a depoliticized sphere of high politics. In this case, the Copenhagen School can congratulate 'climate apocalypse' for operating in the realm of 'normal' democratic politics. However, as the governmentality approach has shown, 'climate apocalypse' is also a highly depoliticizing affair. The configuration of power termed 'climate apocalypse' tends to naturalize climate change as an unfortunate but unavoidable problem and focuses attention on preparedness for the impacts of climate change (Methmann and Oels forthcoming). A vision of reality is enacted as future perfect (Bigo 2008), in which small island states are submerged and weather-related extreme events like floods, droughts and heat waves have become endemic (McNamara and Gibson 2009). It is the acceptance of this vision of the future which is implied by 'climate apocalypse'.

Conclusion

In this chapter, I have argued that climate security actually signifies a shift in the technologies of power that render climate change governable. Contrary to existing discourse analytical work, I have demonstrated that the discourses of 'climate apocalypse' enable a new set of practices for fostering resilience. These practices are mostly in line with advanced liberal techniques of government. However, they also draw on disciplinary monitoring of the environment and of vulnerable populations. Moreover, they could draw on sovereign violence when liberal government fails to prevent disruptions to global (capital) circulation. Following Chandler (2012), I have argued that sovereign violence can indeed be exercised in the name of human security.

On a theoretical level, I have demonstrated that we need an analytical framework of security that is sensitive to the shifting modes of power enabled in its name. It is not enough to distinguish between essentialized notions of security like 'national security' and 'human security' and then conclude that one is good and the other bad. Instead, security is best studied as a configuration of different technologies of power and their characteristic techniques. It is the correlation of these various technologies that must be studied – a correlation that varies across time and place. Moreover, techniques can be redeployed and serve different purposes under different technologies of power. This requires a constant reassessment of the actual policy implications of the specific configurations of power exercised in the name of security.

Even if climate change is not 'securitized' in the form of a state of exception, 'climate apocalypse' has facilitated scandalous levels of depoliticization of the structural causes of climate change. If used as an approach for studying competing forms of 'thinking' (Collier 2009), a governmentality analysis also offers a great framework for highlighting what is hidden by the current dominant 'conditions of possibility' for thinking and acting. The many acts of resistance and the many existing attempts to transgress dominant 'conditions of possibility' for thinking and acting have often been missing from the picture. Elsewhere, we have discussed the naturalization of these 'conditions of possibility' for thinking and acting as depoliticization (Methmann and Oels forthcoming). For climate security, I have presented the case of small Pacific island populations which strongly object to proposals that they be resettled, while dominant forms of government seek to render them governable as 'climate refugees' and to 'save' them. A governmentality analysis enables us to demonstrate that the very partition of the sensible (Rancière 2004) is the primary locus of resistance. In their research on why ambassadors from small Pacific island states reject the 'climate refugee' category, McNamara and Gibson found that: 'Pacific ambassadors argued that before their people are inundated out of their homes … industrialised countries must act to contain and reduce greenhouse gases' (2009: 482). However, this demand to tackle the causes of climate change is easily brushed aside as long as climate change remains constructed as an uncontrollable phenomenon, where even ambitious emission reductions will not necessarily translate directly into lower sea levels and a lower number of extreme events. Here, the discursive struggle for a new partition of the sensible has to begin with the very conceptualization of climate change itself.

Acknowledgements

I would like to thank editors Harriet Bulkeley and Johannes Stripple for their very helpful comments on an earlier draft and for their encouragement and support throughout this project. Moreover, Chris Methmann deserves special thanks for commenting on various drafts of this chapter. Finally, the language of this chapter has been improved significantly by Logan Penniket's careful editing.

References

Adey, P. and B. Anderson (2012). Anticipating Emergencies: Technologies of Preparedness and the Matter of Security. *Security Dialogue* **43**(2), 99–117.

Aradau, C. and R. V. Munster (2011). *Politics of Catastrophe: Genealogies of the Unknown*. London, New York: Routledge.

Beck, S. (2009). *Das Klimaexperiment und der IPCC: Schnittstellen zwischen Wissenschaft und Politik in den internationalen Beziehungen*. Marburg: Metropolis-Verlag.

Biermann, F. and I. Boas (2010). Preparing for a Warmer World. Towards a Global Governance System to Protect Climate Refugees. *Global Environmental Politics* **10**(1), 60–88.

Bigo, D. (2008). Globalized (In)Security: The Field and the Ban-Opticon. In *Terror, Insecurity and Liberty: Illiberal Practices of Liberal Regimes after 9/11*, ed. D. Bigo and A. Tsoukala. London, New York: Routledge, 10–48.

Boykoff, M. T., D. Frame and S. Randalls (2010). Discursive Stability Meets Climate Instability: A Critical Exploration of the Concept of 'Climate Stabilization' in Contemporary Climate Policy. *Global Environmental Change* **20**(1), 53–64.

Brauch, H. G. (2009). Securitizing Global Environmental Change. In *Facing Global Environmental Change: Environmental, Human, Energy, Food, Health and Water Security Concepts*. Hexagon Series on Human and Environmental Security and Peace 4, ed. H. G. Brauch, U. O. Spring, J. Grin, C. Mesjasz, P. Kameri-Mbote, N. Chadha Behera, B. Chourou and H. Krummenacher. Berlin: Springer, 65–104.

Buzan, B., O. Waever and J. de Wilde (1998). *Security: A New Framework for Analysis*. Boulder, CO: Lynne Rienner.

Bäckstrand, K. and E. Lövbrand (2006). Planting Trees to Mitigate Climate Change: Contested Discourses of Ecological Modernization, Green Governmentality and Civic Environmentalism. *Global Environmental Politics* **6**(1), 50–75.

Chandler, D. (2012). Resilience and Human Security: The Post-Interventionist Paradigm. *Security Dialogue* **43**(3), 213–29.

Christian Aid (2007). *Human Tide. The Real Migration Crisis*. London et al.: Christian Aid.

Collier, S. J. (2009). Topologies of Power. *Theory, Culture & Society*, **26**(6), 78–108.

Collier, S. J. and A. Lakoff (2008). Distributed Preparedness: The Spatial Logic of Domestic Security in the United States. *Environment and Planning D* **26**(1), 7–28.

Conisbee, M. and A. Simms (2003). *Environmental Refugees: The Case for Recognition*. London: New Economics Foundation.

Dean, M. (2003). *Governmentality. Power and Rule in Modern Society*, first edition. London, Thousand Oaks, CA, New Delhi: SAGE.

Detraz, N. and M. M. Betsill (2009). Climate Change and Environmental Security: For Whom the Discourse Shifts. *International Studies Perspectives*, **10**(3), 303–20.

Dillon, M. (2007). Governing through Contingency: The Security of Biopolitical Governance. *Political Geography* **26**(1), 41–7.

Dillon, M. (2008). Underwriting Security. *Security Dialogue* **39**(2–3), 309–32.

Docherty, B. and T. Giannini (2009). Confronting a Rising Tide: A Proposal for a Convention on Climate Change Refugees. *Harvard Environmental Law Review* **33**(2), 349–403.

Duffield, M. (2007). *Development, Security and Unending War: Governing the World of Peoples*. Cambridge: Polity Press.

Duffield, M. (2008). Global Civil War: The Non-Insured, International Containment and Post-Interventionary Society. *Journal of Refugee Studies* **21**(2), 145–65.

Duffield, M. (2010). The Liberal Way of Development and the Development-Security Impasse: Exploring the Global Life-Chance Divide. *Security Dialogue* **41**(1), 53–76.

Duffield, M. (2011). Total War as Environmental Terror: Linking Liberalism, Resilience, and the Bunker. *South Atlantic Quarterly* **110**(3), 757–69.

Duffield, M. and N. Waddell (2006). Securing Humans in a Dangerous World. *International Politics* **43**(1), 1–23.

Edwards, P. N. (2010). *A Vast Machine: Computer Models, Climate Data, and the Politics of Global*. Cambridge, MA: MIT Press.

El-Hinnawi, E. (1985). *Environmental Refugees*. Nairobi: UNEP.

Farbotko, C. (2005). Tuvalu and Climate Change: Constructions of Environmental Displacement in the *Sydney Morning Herald*. *Geografiska Annaler: Series B, Human Geography* **87**(4), 279–93.

Farbotko, C. and H. Lazrus (2011). The First Climate Refugees? Contesting Global Narratives of Climate Change in Tuvalu. *Global Environmental Change* **22**(1), 382–90.

Foresight (2011). *Migration and Global Environmental Change*. London: The Government Office for Science.

Foucault, M. (1978). *The History of Sexuality. An Introduction. Vol. I*. New York: Vintage.

Foucault, M. (2003). *Society Must be Defended. Lectures at the Collège de France 1975–76*. New York: Picador.

Foucault, M. (2007). *Security, Territory, Population. Lectures at the Collège de France 1977–78*. New York: Picador.

German Advisory Council on Global Change (2007). *World in Transition: Climate Change as a Security Risk*. London: Earthscan.

Grove, K. (forthcoming). Biopolitics. In *Critical Environmental Politics*, ed. C. Death. London: Routledge Intervention Series.

Holling, C. S. (1973). Resilience and Stability of Ecological Systems. *Annual Review of Ecology and Systematics* **4**, 1–23.

Homer-Dixon, T. (1994). Environmental Scarcities and Violent Conflict: Evidence from Cases. *International Security* **19**(1), 5–40.

Homer-Dixon, T. (1999). *Environment, Scarcity and Violence*. Princeton, NJ: Princeton University Press.

Lenton, T. M., H. Held, E. Kriegler, J. W. Hall, W. Lucht, S. Rahmstorf and H. J. Schellnhuber (2008). Tipping Elements in the Earth's Climate System. *Proceedings of the National Academy of Sciences (PNAS)* **105**(6), 1786–93.

Lentzos, F. and N. Rose (2009). Governing Insecurity: Contingency Planning, Protection, Resilience. *Economy and Society* **38**(2), 230–54.

Luibhéid, E. (2002). *Entry Denied: Controlling Sexuality at the Border*. Minneapolis; London: University of Minnesota Press.

Luke, T. W. (1999). Environmentality as Green Governmentality. In *Discourses of the Environment*, ed. E. Darier. Oxford, Malden: Blackwell, 121–51.

Lundborg, T. and N. Vaughan-Williams (2011). Resilience, Critical Infrastructure, and Molecular Security: The Excess of 'Life' in Biopolitics. *International Political Sociology*, **5**(4), 367–83.

Mabey, N., J. Gulledge, B. Finel and K. Silverthorne (2011). *Degrees of Risk: Defining a Risk Management Framework for Climate Security*. London: Third Generation Environmentalism Inc. (E3).

McNamara, K. E. and C. Gibson (2009). 'We do not want to leave our land': Pacific Ambassadors at the United Nations Resist the Category of 'Climate Refugees'. *Geoforum* **40**(3), 475–83.

Methmann, C. and D. Rothe (2012). Politics for the Day after Tomorrow: The Logic of Apocalypse in Global Climate Politics. *Security Dialogue* **43**(4), 323–44.

Methmann, C. and A. Oels (forthcoming). Securing the Environment: From Defence to Resilience. In *Dialogues on Security: Theoretical, Methodological, and Empirical Advances and Challenges*, ed. C. Daase, G. Schlag and J. Junk. London: Routledge.

Myers, N. and J. Kent (1995). *Environmental Exodus. An Emergent Crisis in the Global Arena*. Washington: Climate Institute.

Oels, A. (2005). Rendering Climate Change Governable. From Biopower to Advanced Liberal Government. *Journal of Environmental Policy and Planning* **7**(3), 185–207.

Oels, A. (2012). Comparing Three Theoretical Perspectives on Climate Change as a Security Issue: From the 'Securitisation' of Climate Change to the 'Climatisation' of the Security Field. In *Climate Change, Human Security and Violent Conflict: Challenges for Societal Stability*, ed. J. Scheffran, M. Brzoska, H.-G. Brauch, P. M. Link and J. Schilling. Berlin: Springer, 185–206.

Oels, A. (2013). Rendering Climate Change Governable by Risk: From Probability to Contingency. *Geoforum* **45**(March 2013), 17–29.

Rancière, J. (2004). Who is the Subject of the Rights of Man? *The South Atlantic Quarterly* **103**(2–3), 297–310.

Reid, J. (2012). The Disastrous and Politically Debased Subject of Resilience. *Development Dialogue* **58**, 67–80.

Rogelj, J., J. Nabel, C. Chen, W. Hare, K. Markmann, M. Meinshausen, M. Schaeffer, K. Macey and N. Höhne (2010). Copenhagen Accord Pledges Are Paltry. *Nature* **464**, 1126–8.

Rose, N. (2001). The Politics of Life Itself. *Theory, Culture & Society* **18**(6), 1–30.

Rose, N., P. O'Malley and M. Valverde (2006). Governmentality. *Annual Review of Law and Social Sciences* **2**, 83–104.

Schwartz, A. D. and P. S. Randall (2003). *An Abrupt Climate Change Scenario and Its Implications for United States National Security*. Washington, DC: Environmental Media Services.

United Nations Security Council (UNSC) (2007a). Minutes 5663rd meeting, Tuesday, 17 April 2007, New York, S/PV.5663.

United Nations Security Council (UNSC) (2007b). Minutes 5663rd meeting, Tuesday, 17 April 2007, 3 p.m., New York, S/PV.5663 (Resumption 1).

United Nations Security Council (UNSC) (2011a). Minutes 6587th Meeting, Wednesday, 20 July 2011, 10 a.m., New York, S/PV.6587.

United Nations Security Council (UNSC) (2011b). Minutes 6587th Meeting, Wednesday, 20 July 2011, 3 p.m., New York, S/PV.6587 (Resumption 1).

UN GA [United Nations, General Assembly] (2009). *Climate Change and its Possible Security Implications: Report of the Secretary-General*, A/64/350. New York: UN.

Stripple, J. (2002). Climate as a Security Issue. In *Human Security and the Environment: International comparisons*, ed. E. Page and M. Redclift. Cheltenham: Edward Elgar, 105–27.

Swyngedouw, E. (2010). Apocalypse Forever? Post-Political Populism and the Spectre of Climate Change. *Theory, Culture & Society* **27**(2–3), 213–32.

Trombetta, J. (2008). Environmental Security and Climate Change: Analysing the Discourse. *Cambridge Review of International Affairs* **21**(4), 585–602.

United Nations Framework Convention on Climate Change (UNFCCC) (1995). Geneva: United Nations.

United Nations General Assembly (UNGA) (2009). *Climate Change and Its Possible Security Implications. Report of the Secretary-General*. A/64/350. New York: United Nations.

Welzer, H. (2012). *Climate Wars: What People Will be Killed for in the 21st Century*. New York: John Wiley & Sons.

Part III

Future Directions

12

The Rise and Fall of the Global Climate Polity

Olaf Corry

Introduction

Not so long ago the idea that a global climate polity could exist would have seemed bizarre or simply nonsensical. 'The climate' was effectively just patterns of weather over time. Though there is a long history of attempts at affecting weather, these were generally limited to engineering local and temporary effects on rainfall, and historically many schemes ended in failure or even ridicule (Fleming 2012). Few if any people seriously entertained the idea that people, states, corporations and international organizations would mobilize and operate giant monitoring and regulatory systems in concerted attempts to change (or preserve) the chemical composition of the global atmosphere.

This raises not only the question of how the idea of governing something like the climate so rapidly became a matter of course but also how sure we can be that it will remain so in, for example, another thirty years. Other objects today considered immutable or irrelevant may also become objects of governance of central importance while things we today regard as governable may disappear as targets of governance just as other objects of governance such as 'Mesopotamia' straddling modern-day Iraq or 'Danelaw' – a power structure in northern and eastern England in the ninth and tenth centuries – have come and gone. It is too early to write the definitive story of 'the rise and fall' of the global climate polity, but this chapter makes a start by considering the bigger picture of a climate polity, what one is and what its future prospects might be.

In the following, first the question is asked whether examinations of localized governmental power and micro-techniques in the chapters of this book need to be accompanied by a consideration of a bigger picture or structural context. Next, the global climate polity is theorized beyond the models of hierarchy and anarchy that underlie the Westphalian paradigm. A generic theory of what constitutes a polity is elaborated from Foucault's ideas about how changing objects of governance are central to the shape and logic of political entities, and the criteria for a 'global climate polity' are defined. The third part looks back briefly asking when and how a global climate polity thus conceived could be said to have evolved. The final section peers towards the future to consider differing visions of the demise of the global climate polity as we know it: Will it be superseded by a more encompassing 'earth system polity' centred on governing not only the climate but also other interrelated geophysical systems? Or could it splinter as it becomes increasingly clear that the climate is too complex and feral to be an object of governance? In this latter scenario the climate polity will split into smaller and more local polities based on carving out certain favourably located areas (not necessarily the nation-states as we know them) as enclaves of human survival in a new, hotter and less hospitable climate on Earth in a new climatic equilibrium five to eight degrees Celsius above preindustrial levels.

Governmental Techniques and the Bigger Picture

Whereas most of the other chapters in this volume go down to the nitty-gritty where subjects and objects are constructed, where micro-practices constitute truths or domains of governance and where assemblages of technical and human networks are forged, this section aims to take a step up to elucidate a clearer picture or model of what a global climate polity is in terms of its totality: How are the elements of a global polity structured and what makes a polity a polity? Rather than beginning from a grand narrative or structural model of international politics, the chapters of this book tend to foreground practices of governing and how these shape and render subjects and objects real and governable in particular ways. Thus Tom Hargreaves examines smart meters in terms of 'techniques and tactics through which individual subjectivities are constantly made and remade to achieve particular ends' (Chapter 7). For Sally Eden, 'The power to shape carbon behaviour therefore resides neither in the state nor the individual consumer, but is distributed through the complex assemblage of humans and nonhumans' (Chapter 6). Juliet Fall explores 'how calculative practices and scientific discourses on biodiversity frame how mobile species are governed' (Chapter 9) while Anders Blok emphasizes that instead of assuming an integrated global climate polity, 'climate governance is a shifting meshwork of superimposed local-global "problem geometries"' (Chapter 2). Benjamin Stephan, Delf Rothe and Chris Methmann usefully explore the concept of hegemony as 'the missing link between Foucault's everyday and mundane micro-practices of power and broader macro-structures of governmental power' (Chapter 3), but leave these broad macro-structures largely implicit.

This tendency to bracket structural questions and the wider context of governmental techniques in favour of local techniques and tools of governing is common to many Foucault-inspired analyses of international relations intent on escaping the usual assumptions of states and anarchy. For Michael Dillon, for example, with a Foucauldian analytic of power '(t)he grand narratives of geopolitics, sovereignty, self-determination, imperial conquest and ideological conflict would also be bracketed-off, replaced with accounts of the micro-practices by which individual and collective bodies are inscribed with power' (2004: 79). In an early statement on global governmentality, Larner and Walters also saw governmentality as a way of '"bracketing" the underlying forces and causes' (2004: 16), although theirs is also a project which ultimately aims to illuminate 'how these [underlying] institutions, forces and realms come to be known, and to know themselves, in particular ways' (2004: 15).

While bracketing can be a fruitful temporary methodological move to perform what James Rosenau called a 'conceptual jail-break' from statism (2006: 14), a failure to provide alternative overall accounts of the structural context of governmental formations and assemblages bears the risk of subsequently interpreting them through conventional models of world order, only by default. This volume characterizes the global climate polity itself primarily in terms of what it is not: something 'in-the-making' or a 'meshwork' (Chapter 2), 'complex assemblages' and 'imperfect systems' (Chapter 6), the 'heterogeneity and "messiness" of governmental regimes' (Chapter 3). The complexity and messiness of a climate polity is in turn conceived in terms of an underlying simple or neat model of what climate politics is *not*: just the business of states, working in ordered multilateral form within settled constitutional settings. The structural models of the Westphalian Age creep back in to fill the structural gap, in other words. To many, 'global climate polity' might evoke images of an emerging global political system that authoritatively distributes values and burdens connected with regulating the world's climate (however imperfectly). For others, a global climate polity may refer to a situation where independent actors are engaged, via international institutions but with no central authority, in governing the climate through negotiation, competition and compromise. Hierarchy and anarchy are easy to recognize behind these two ways of understanding a 'climate polity'. With these models the usual distinctions (domestic/international), actors (states and nonstate actors) and practices (diplomacy, negotiation, bargaining and institution building) soon retake centre stage (Walker 1993; Bartelson 1995; Corry 2006, 2010).

But bracketing the structural context of governmental techniques also risks marginalizing the governmental approach itself. Jonathan Joseph has argued that Foucault's governmentality approach focuses on the 'how' of governing as opposed to the 'why', leaving aside structural conditions for governmental techniques: 'To complete the picture, we need to move from strategies and rationalities to conditions of governmentality and to underlying social structures and processes' (2012: 14). He interprets governmentality not as constitutive of social relations per se but as a neoliberal set of governing techniques conditioned by an underlying capitalist system of production and a system of states. The state system generates realist 'structural dynamics' including 'concerns of foreign policy and ensuring security in an uncertain world of rival states' (2012: 44), and governmentality becomes one tool in this. In this way the anarchic structure of the international system and an underlying economic structure considered constitutive of governmentality are reintroduced to fill a lack in the governmentality approach which allegedly cannot account for what makes governmentality possible in the first place. This is a particular reading of governmentality in which it is largely stripped of strong claims to constitutive effects and is confined to examining the uneven spread of liberal forms of power in an international system. Choosing this reading of governmentality has its merits (although it does not take on board the central idea of governmentality that the intimate meshing of mentalities, technologies and materialities condition how politics works (Brass 2000)), but also illustrates how the absence of a model of overall structure of its own invites the usual models back into governmental analysis, somewhat defeating the original methodological move.

Here the aim is to examine the global climate polity as a structure in a fashion that does not bracket wider structural questions and does not render governmentality epiphenomenal leaving the theoretical stage empty for a reentry of standard models of international order.

What is a Global Climate Polity?

So how is a global climate polity to be understood if not in the usual way as a system of major actors and international institutions? Like others in this book, I draw on Foucault's governmentality lectures to develop an answer, but my aim is to use a model focussing on the role of objects in the constitution or 'anchors' of political structure. What if the object of governance is what organizes (or constitutes) the structure itself – that is what if the governance-objects are what constitute the polity?

Foucault suggested that regimes of knowledge about governing such as the new science called 'political economy' '[that] arises out of the perception of new networks of continuous and multiple relations between population, territory, and wealth' (2002: 217) helped to advance the art of government. What emerged with the art of government was not just a new ideology, but a whole new range of 'objects', spheres and technologies of government. These depended on and in turn fed a new mentality of governing, displacing amongst other things mercantilism 'trapped within the inordinately vast, abstract, rigid framework of the problem and institution of sovereignty' (2002: 214). This allowed new technologies of governing to be developed such as political economy and statistics. It is not just the emergence of social objects of 'the economy' and 'the population' but ultimately the emergence of a domain of 'society' as such, that is so characteristic of the modern liberal art of government that preoccupied Foucault's later work.

> What was discovered at that time … was the idea of society. That is to say, that government not only has to deal with a territory, with a domain, and with it subjects, but that it also has to deal with a complex and independent reality that has its own laws and mechanism of distribution. This new reality is society. From the moment that one is to manipulate a society,

one cannot consider it completely penetrable by police. One must take into account what it is. (2002: 352)

What Foucault describes is effectively how the governance-object of 'territory' was gradually replaced by 'society'. Sovereign power over territory merged with disciplinary and police power over a new set of governance-objects such as economies and populations.

Moving an analytical step up, Foucault's narrative can be taken to imply that the existence of governance-objects can be constitutive of polities (Corry 2006, 2010; Methmann 2011). In these terms, 'climate polity' can be defined as a situation in which the climate is constituted and then treated as an object that can and should be manipulated purposively. In this way governmental techniques not only enact power over subjects but also help constitute the objects of governance that in turn structure political relations. Thus, if the global climate is emerging as a governance-object through apparatuses, discourses and technologies rendering the climate governable, including knowledge and disciplining regimes that frame the world as essentially one place ('global'), then a 'global climate polity' can also be said to be congealing. The very existence of the climate as a governance-object is constitutive for political relations. What is common to the infrastructure and actors (or 'actants' in ANT terminology) of a climate polity is that they define themselves or are oriented towards governing 'the climate'. They may disagree about how, and they may be located anywhere, and they may be state or nonstate. But minimally they share an orientation towards the global climate (see also Corry 2013).

This approach suggests an understanding of the global climate polity that differs in several ways from the two standard models of political structure that saturate accounts of 'the international'. Firstly, conceived in this way, the idea of a 'global' polity should not immediately trigger a sterile debate about whether state sovereignty is being eroded or pooled, but about *whether 'global' objects – those understood within a frame of the world as one place – are beginning to become the focus of political struggles*. States will likely be major players in governing global objects, so it is useful to be able to debate a global polity without having to imply that state power has eroded or disappeared.

Secondly, with this understanding of polity we can debate the existence of a global climate polity *prior* to or separately from any debates about whether a global hierarchy of authority is emerging and whether states are still the 'main actors'. Issues of power differentials (and forms of power) will affect how the climate is governed of course, but they are not the 'way in' to questions concerning whether a global climate polity exists, what it consists of and where to look for one. A governance-object may be governed by an all-powerful monarch or in a giant muddle of unstable power relations between multiple actors of different kinds. Hierarchy is not the precondition for a polity.

Thirdly, this Foucault-inspired polity approach shifts focus to include the 'things' being governed and the 'how' of climate governance – to the mechanisms of governance – and *via that* to the question of 'who'. In effect, polity is recast as a structure in which an object becomes a target of power practices that organizes a field of politics. Anarchy and hierarchy are of course also viewed as structures, but of a slightly different kind. Kenneth Waltz's theory of international politics started from the question: 'in defining structure, the first question to answer is this: What is the principle by which the parts are arranged?' (1979: 81), but he limited his idea of 'parts' to actors (states). Realism is often labelled 'state-centric' while pluralist global governance approaches draw in other nonstate actors with both remaining *actor*-centric. Even most constructivist and poststructuralist approaches focus on (subject) identities (Lapid and Kratochwil 1996; Wendt 1999). This means that political structure is normally reduced to the constellation of actors, and thereby fails to register fundamental changes such as the transition from governing territories to governing populations or economies – or the governing of earth systems such as the climate. Such changes are not necessarily ignored by subject-centric approaches to structure but are effectively excluded from their notions of structure. In contrast, a polity is conceived here as a structure, though not one defined solely by relations between actors, between certain governance-subjects and governance-objects. This chimes with Anders Blok's suggestion of an 'issue-centric' perspective focussed on 'new hybrid associations of humans and non-humans' where he also prefers to remain agnostic concerning the overall structure of 'nascent and as yet unpredictable chains of climatic authority and control' (Chapter 2).

Finally, the global climate polity is not a territorial structure of power, as a state or an empire is, but neither is it unbounded. Its limits are drawn instead by discourses, practices and apparatuses of climate governance that potentially criss-cross spatial (including national) borders as well as disciplinary and institutional boundaries. The global climate polity 'stretches' (to use a physical metaphor) from the negotiations between President Obama and the Chinese representative in Copenhagen in 2009 to the carbon-conscious individual who exercises climate discipline by taking the train rather than the airplane. They may be subject to different discourses about global warming, tipping points and carbon emissions, but they both orient themselves towards governing 'the climate' in some way. Yet the climate polity is not 'everywhere' either. In terms of subjects, the key distinction is not between state and nonstate actors but between those who do and those who do not orient themselves to governing the global climate. But a global climate polity is also constituted by physical and socio-technical structures: satellites that orbit the earth supplying data to models and research communities; measuring equipment in the air (at various altitudes), at sea and on land that monitors the climate; the carbon calculators we find online when planning minute details of our lives. Together in a giant assemblage of material and social (Latour 2007), the climate is being governed, albeit seemingly without the risks of catastrophic climatic change diminishing (the increased governance of the climate does not necessarily translate into a growing 'solution' to climate change). Along with the direct measurement of the climate as an object through the growth of apparatus and mentalities of governing it, 'carbon control' has become 'a primary objective of political and economic governance at urban, regional and international scales' (Bridge 2010: 821; see Chapter 3) and carbon is now the defining unit of account for a cartography of carbon resources (oil, gas, methane etc.), an ethnography of carbon management (systems of governing carbon, see Chapter 6) and a biopolitics of carbon that provides 'conditions of possibility for the living of life' (Bridge 2010: 821; Chapter 7).

How Did a Global Climate Polity Emerge?

How did a 'global climate polity' thus conceived emerge? This question has begun to be explored in the literature in two main ways corresponding broadly to the two elements in Foucault's compound term 'governmentality' which was designed to signal the intimate mix of technologies and mentalities of governing: one focuses on the emergence of technical climate governance infrastructure, the other on the role of discourse positing the global climate as something that can and should be governed. Climate polity subjects inhabit a space defined by these two elements. The following presents illustrative examples of studies focussed on each.

The most comprehensive account of how the climate became a global governance object can be found in Paul N. Edwards's history of the ever-evolving 'vast machine' of infrastructure of climate governance. The global climate monitoring and modelling complex is an example (arguably the leading example) of what he also calls *infrastructural globalism*, that is 'projects for permanent, unified, world-scale institutional-technological complexes that generate globalist information not merely by accident, as a by-product of other goals, but by design' (2010: 25). The climate polity has, in these terms, taken shape gradually through the emergence of a global climate governance-object constituted by the vast machine intertwined with 'the very idea of a planetary climate as something that can be observed, understood, affected by human wastes, debated in political processes, cared about by the general public and conceivably managed by deliberate interventions such as reforestation or gigantic Earth-orbiting sunshades' (2010: 8).

This is a relatively new phenomenon, especially if we take the governance of the *global* climate rather than local weather as the governance-object of a climate polity. Attempts at affecting the weather in localities have a longer pedigree (Fleming 2012) closely connected to agricultural and military projects. Diverse theories have arisen throughout history about the local effects on weather of clearing forest, swamp drainage and other domestic practices. Weather monitoring infrastructure arose from time to time, such as the seventeenth century pan-European weather network of (only ten) stations with comparable measuring equipment set up by Grand Duke of Tuscany Ferdinand II (Edwards 2010: 33). Early climatological maps of the entire climate system include Edmond Halley's map of the trade winds published in 1686 in the article 'An historical Account of the Trade Winds, and Monsoons, Observable in the Seas between and near the Tropicks, with an Attempt to Assign the Physical Cause of the Said Wind' (Halley 1686).

Such grand schemas of global climate were rare while attempts at local weather forecasting useful for shipping and military endeavours began to accumulate. Over decades and centuries such practices spread and gradually and through a tortuous (and often flawed) process of standardization and data harmonization became integrated. This formed bodies of data of ever-growing geographical reach and temporal scope. This eventually led to renewed concern with the global system as data covered more of the Earth and longer-term trends led to climate data (as opposed to weather data). This in turn reinvigorated efforts directed at theoretical and graphic mapping of circulation patterns, this time based on empirical observations as well as models of the systems. Only by the 1930s had theoretical meteorology 'set its sights firmly on the planetary scale as the most fundamental level of explanation' (Edwards 2010: 140). Today Global Circulation Models (GCMs) integrate a truly vast array of data and theories to model the entire global climate, drawing in more and more systems such as those describing ocean currents, changes in vegetation and ice cover.

Although this mapping and measuring gradually constituted the global climate as an object of study, this was not initially done with a view to governing it – indeed the rationale for considering whether or how to govern the climate came only really with concern over human influence of the global climate. The history of anthropogenic global warming has been written several times elsewhere (Weart 2003) usually beginning with Joseph Fourier's 'greenhouse effect' and John Tyndall and Svante Arrhenius's work on greenhouse gases from around the turn of the twentieth century. Arrhenius's 1896 publication of a paper that earned him the Nobel Prize is often cited as the starting shot of the anthropogenic global warming debate, but early efforts building on this were usually more concerned with global cooling and the fear of an approaching ice age. For our purposes, the idea that we can and should govern the makeup of the atmosphere with a view to regulating the climate was a later arrival and can arguably be dated to the post–World War II era. With the means and the idea of the climate as an object that can and should be governed widely established, however, we may say that the climate polity had been formed.

This process of polity formation has been driven primarily (though mostly inadvertently) by scientists, but states became intermittently interested in weather and climatic knowledge because of the lure of military or economic gain. Scientific bodies pragmatically piggybacked on these whims when they could in order to expand and consolidate their studies and informational infrastructure. They did this first to understand, then to model, predict and finally put forward the idea of regulating the global climate. This idea is reaching what is so far its purest form in the idea of 'geoengineering' – the deliberate manipulation of the global climate using solar radiation management (planetary sun shields and changes in the Albedo effect) or carbon trapping methods (Shepherd 2009). Eventually, the global climate became something in principle governable and has begun to structure politics to an ever greater degree – it has become a polity.

Turning briefly from technologies to mentalities (or from physical to informational infrastructure as Edwards calls it) framing the global climate as a governance-object, the idea of the globe as an operational and unified entity has its own discursive history. This was closely linked to first religion, then science, environmentalism and in more recent years particularly to space exploration and now climate change. According to Edwards, 'meteorology participated in the larger scientific project of envisioning "the world" as a whole – a single, dynamic, coherent physical system' (2010: 40) and space exploration was a significant milestone in envisioning the world as a whole (Jasanoff 2001). Lövbrand, Stripple and Wiman show more widely how scientific practices 'constituted the global climate as an object of study' and emphasize in particular the role played by 'sophisticated mathematical representations – i.e. models – of ecological processes' in establishing the earth system as a governable object (Lövbrand et al. 2009: 9).

In popular discourse a recent account of the transformation of the 'world picture' makes the often heard case that a globalist perspective of the world as one place gained dramatically in strength and diffused beyond narrow scientific forums when pictures of the Earth taken from space circulated in the 1960s and 1970s. In a fascinating article, Benjamin Lazier points to the 'rise of an "Earthly vision," or a pictorial imagination characterized by views of the Earth as a whole' (2011: 605) flowing principally from two photographs taken by the Apollo missions: the 'Earthrise' picture taken of the Earth rising over the horizon of the moon and the 1972 'Blue Marble' picture of the Earth, showing 'the disk of our terraqueous planet suspended in the void' (2011: 605–6). 'As a stand-in for the idea of the Whole Earth itself', Lazier concludes, 'it has acquired an iconic power that helps organize a myriad of political, moral, scientific, and commercial imaginations as well' (2011: 606). For Lazier, this is a part of a longer history of the world picture, but with the Earthrise era, the first part of the slogan 'Think globally, act locally' became more a general condition than something we have to be exhorted to do. Climate change has arguably slotted into – but also accelerated – the compulsion to 'think globally', although this has a long and complex history (see also Litfin 1997; Litfin 1998; Jasanoff 2001; Cosgrove 2003; Helmreich 2011).

The central tension of that Earthrise era is between understandings of the globe as natural on one hand and as man-made – or at least as manipulable by human activity – on the other. The recent idea of the Anthropocene – the geological age of humans – suggested by Nobel Laureate Paul Crutzen (Crutzen and Stoermer 2000) sums up the idea that human influence on the Earth and its systems – not least through climate change – has made humans the main drivers of geological time, taking us out of the climatically relatively benign and stable Holocene era. Not just in specialist scientific circles but in philosophy and eventually politics and popular culture, then, the climate is now often seen as something governed significantly by human activity, yet also as something potentially spinning out of human control.

Thus, via technological and symbolic structures constituting objects and subjects in certain ways, the climate polity now comprises the globalist infrastructure of climatology, international organizations such as the IPCC and the UNFCCC but also the International Metrological Organization (IMO), as well as a melange of other systems of knowledge, epistemic communities and their artefacts focussed on ocean currents, biological processes, states, conjoining scientific disciplines, regions, cities, nongovernmental organizations (NGOs), companies, individuals and micro-practices. Identities generated by these elements are all predicated on the existence of the global climate as a governable entity fit for steering operations. A political mass mobilization around that idea, that perhaps reached its most recent peak in 2009 at the UN Conference on Climate Change in Copenhagen, represents just the tip of the iceberg of an emerging but consolidating global climate governmentality that renders the climate governable and – perhaps – increasingly central to political identities.

Beyond the Global Climate Polity?

This sketch of the formation of a global climate governance-object is of course incomplete, and in terms of the future, nothing is inevitable. That previously unthinkable entities such as a global climate polity can become taken for granted as natural (and perhaps one day will be thought of as timeless) testifies to the idea that other such entities will inevitably arise. If we instead cast our eyes briefly forwards speculatively, more limited or even more encompassing governance-objects can be seen competing for political prioritization with the climate polity, indicating that there are other possible polities in the making. One could potentially subsume the global climate polity in a greater 'earth system polity' in which climate becomes just one element in a greater governance-object – a planetary boundary or tipping point interlinked with other systems in an aggregated human-natural earth aystem governance-object (Lövbrand et al. 2009). Alternatively, the climate polity may fragment into more local polities such as national or regional ones if global mitigation efforts currently in vogue lose credibility and more local 'lifeboat' politics and adaptive strategies gain discursive and institutional strength. In the following, Mark Lynas's *The God Species* (2011) and James Lovelock's *The Vanishing Face of Gaia: A Final Warning* (2009) are considered harbingers of each of these potential postclimate polity scenarios, respectively.

In the first scenario, the climate polity is subsumed within what we might term an earth systems polity. The governance-object that actors orient themselves towards would be transformed such that 'the climate' ceases to be regarded and treated as a meaningful object on its own. Only as a part of a greater 'earth system' is it governable. Having identified and posited an object – 'the climate' – through the emergence of global climate infrastructure and information, scientists have increasingly begun pointing to linkages and connections between the climate and other physical systems. That climate is regulated by complex interactions among components of the earth system is the second of seven *Essential Principles of Climate Sciences* suggested by the United States Global Change Program (U.S. Global Change Program 2009). According to one characterization of the earth system approach, '(t)he atmosphere, hydrosphere, biosphere and geosphere form the simplest collection, though some would add the cryosphere as a special element dealing with polar regions and processes, and others would add the anthroposphere emphasizing human dimensions and impact on the planet' (Ruzek 2010). A conference including the four international global change research programmes – the International Geosphere-Biosphere Programme (IGBP), the International Human Dimensions Programme on Global Environmental Change (IHDP), the World Climate Research Programme (WCRP) and the international biodiversity programme DIVERSITAS – issued the so-called Amsterdam Declaration, declaring that:

> the Earth System behaves as a single, self-regulating system, comprised of physical, chemical, biological and human components. The interactions and feedbacks between the component parts are complex and exhibit multi-scale temporal and spatial variability. (Earth System Science Partnership 2001)

A scientific journal has been renamed *Journal of Earth System Science* and the earth systems approach has been described as 'the concern of the Geological Society of London (with which the Gaia Society recently merged)' and as 'a major subject of inquiry and research' that 'no longer has to justify itself' (Tickell 2010).

Whether an equivalent political and discursive infrastructure constituting the earth system as a governance-object will be established remains to be seen, but a related initiative indicates that this may be happening. The idea of planetary boundaries that 'define the safe operating space for humanity with respect to the earth system and are associated with the planet's biophysical subsystems or processes' (Rockström et al. 2009: 472) has emerged with some fanfare. The earth system is conceived as having nine critical boundaries (biodiversity, climate change, nitrogen boundary, land use, freshwater, toxics, aerosols, ocean acidification and the ozone layer) that cannot be crossed without potentially disastrous consequences, and each one potentially interacts with the other eight (hence the link to earth systems science) to form limits to legitimate human activity. Mark Lynas has elaborated from this framework a rough blueprint of 'managing the planet' (2011: 229) but points out that two of the nine boundaries have yet to be identified and quantified (chemical pollution boundary and atmospheric aerosol loading). In time, however, globalist infrastructure related to the nitrogen cycle, the ozone layer, the climate, global water use and the other systems with 'planetary boundaries' (Figure 12.1) may well become better developed, linked up and framed by an 'Earth Systems govern-mentality' (Lövbrand et al. 2009) that further establishes the earth systems polity where actors are oriented principally towards governing the earth system as one entity. The climate as a governance-object would in this case be effectively swallowed up, just as the governance object of 'territory' was overlaid by governmental power over economies, populations and societies with the advent of Foucault's 'art of government'.

Alternatively, this envisaged expansion of governance capacity may be a hubristic illusion that may be crumbling. Sooner than we perhaps think, the climate governance-object may unravel, because, though the global climate may be a meaningful entity, it may be deemed an effectively ungovernable one. Could the climate be undone as a governance-object?

Founder of Gaia theory James Lovelock comes close to this position in his recent books, *The Revenge of Gaia* (2006) and *The Vanishing Face of Gaia: A Final Warning*

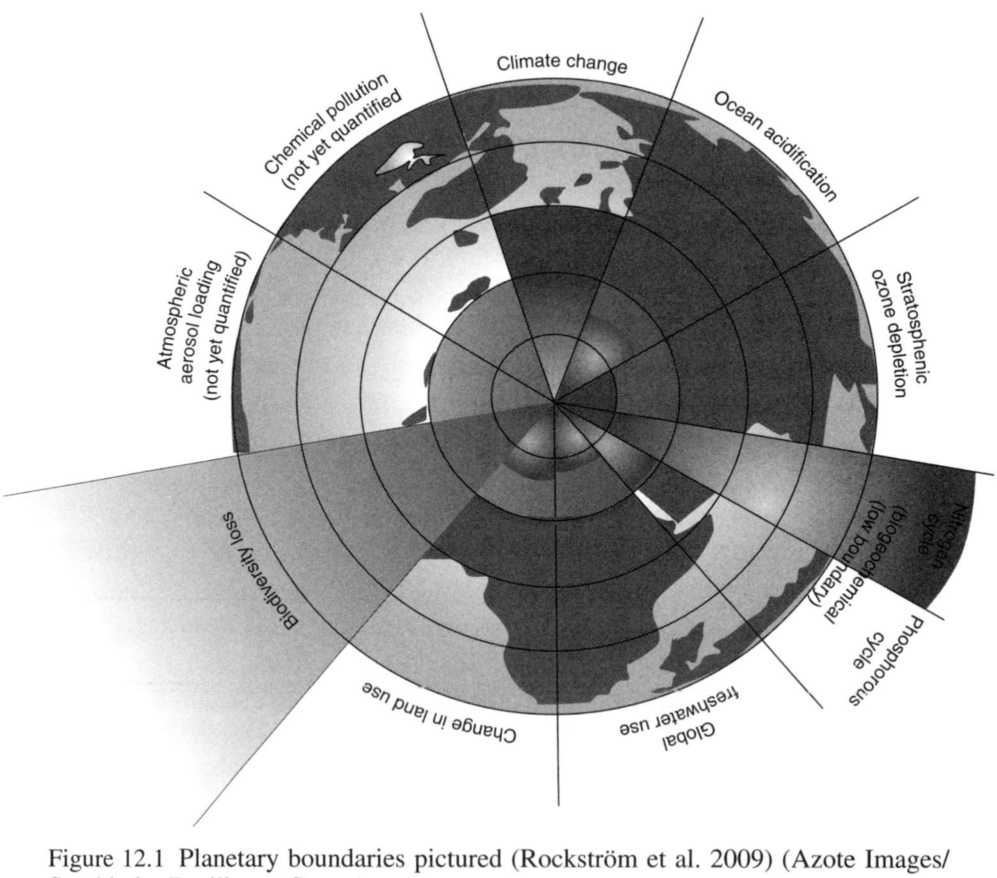

Figure 12.1 Planetary boundaries pictured (Rockström et al. 2009) (Azote Images/ Stockholm Resilience Centre).

(2009). In these, Lovelock pours scorn on the idea that mitigation is feasible, on the climate models that allegedly predict future climate scenarios, as well as on claims that they provide meaningful guidance to the future (he thinks they are far too optimistic). The International Panel on Climate Change (IPCC) has 'been able to make intelligent guesses about future climates' but 'these guesses have been unable to match the observed changes in climate closely enough for us to be confident about IPCC forecasts decades into the future' (2009: 3). Efforts at mitigation have so far been hopelessly inadequate and the idea that the entire energy system of the world's first global civilization might be redesigned in the short time span now available is a project of social engineering of such eye-watering proportions that only the delusional or politically motivated can still maintain that it is possible. Much more likely is the conclusion that the Earth is heading for a much hotter equilibrium much less conducive to human life and that only pockets of human life will therefore survive. A lifeboat strategy of adaptation in certain, mostly northern countries (including the British Isles, Scandinavia and northern Canada), may be possible and governments should concentrate on planning for the transition to this decimated version of human civilization in manageable units. But 'the human world of the lifeboat islands and continental oases will be constrained by limited food, energy and living space' (Lovelock 2009: 12). Lovelock's account is not a lone voice as narratives of abrupt or 'chaotic' climate change are now on the ascendancy replacing earlier linear accounts previously used to frame climate politics (Methmann 2011; Mayer 2012). Essentially, he advocates a form of 'risk management through contingency' where sovereign power is used to pursue human security (Table 11.1).

In this scenario, the governance-object is no longer the global climate, which is already out of kilter and way too complex to be meaningfully predictable and governable. Instead the governance-objects implied are more local and possibly still national entities such as the United Kingdom or regional ('northern Canada') as well as other 'continental oases' (which of course like everywhere else would look somewhat different geographically as sea levels rose and populations migrated).

The fall of the global climate polity follows logically from Lovelock's Gaia theory that holds that the Earth is effectively one giant organism of almost unfathomable dynamic complexity (but with some self-regulating capacity). Gaia is not described by Lovelock in terms of a governance-object, but rather as the framework within which we as humans carve out other more limited objects of control (and hence polities). Lovelock does not rule out that humans may effectively one day become 'Gaia's brain' such that if we survive 'there is a chance that our descendants will one day serve Gaia and assist her in the fine-tuned self-regulation of the climate and composition of our planet' (2009: 21). But currently 'our ignorance of the Earth system is overwhelming and intensified by the tendency to favour model simulations over experiments, observation and measurement' (2009: 103). Of course the climate polity may develop further and become more central as more governance techniques and identities become more centred on the climate as a governance-object – or alternative polities may also develop in as yet wholly unforeseeable ways.

Rather than view governmentality as one particular type of liberal kind of power or as something only relevant at the micro level of power techniques, this chapter has aimed to glean a model of structure from it: governing not only regulates actors and their behaviour in multiple possible ways, it constitutes objects of governance that structure politics into polities. If we consider the local governmental power techniques described in other chapters within a wider structure of polities constituted through the production of governable objects, then we will not be as bound by force of habit or the conventions of, for example 'the logic of anarchy' in our understandings of the past and present of climate politics, nor in our predictions of the future. Anarchy and hierarchy as models concentrate solely on the relations of super- and subordination between subjects and fail to draw attention to the changing objects of governance. What is emerging is neither a global climate hierarchy nor just an anarchy modified by international cooperation. In a period in which materiality appears to be coming back into academic fashion (Connolly 2011), and human civilization is just beginning to face up to the task of 'managing the Earth', a structural model of politics open to changing constructions of objects of governance could come in useful as various increasingly wobbly earth systems appear in need of deliberate steering efforts.

References

Bartelson, J. (1995). *A Genealogy of Sovereignty*. Cambridge: Cambridge University Press.

Brass, P. R. (2000). Foucault Steals Political Science. *Annu. Rev. Polit. Sci.* **3**, 305–30.

Bridge, G. (2010). Resource Geographies 1: Making Carbon Economies, Old and New. *Progress in Human Geography* **35**(6), 820–34.

Connolly, W. E. (2011). *A World of Becoming*. Durham, NC: Duke University Press.

Corry, O. (2006). *Constructing a Global Polity*. Copenhagen: University of Copenhagen.

Corry, O. (2010). What is a (Global) Polity? *Review of International Studies* **36**(S1), 157–80.

Corry, O. (2013). *Constructing a Global Polity. Theory, Discourse and Governance*. Basingstoke: Palgrave.

Cosgrove, D. (2003). *Apollo's Eye: A Cartographic Genealogy of the Earth in the Western Imagination*. Baltimore, MD: Johns Hopkins University Press.

Crutzen, P. J. and E. F. Stoermer (2000). The Anthropocene. *Global Change Newsletter* **41**(1), 17–18.

Dillon, M. G. (2004). The Security of Governance. In *Global Governmentality: Governing International Spaces*, ed. W. Larner and W. Walters. London: Routledge, 76–94.

Earth System Science Partnership (2001). *The Amsterdam Declaration*. [Online] Available at: [http://www.essp.org/index.php?id=41]. Accessed 02 January 2013.

Edwards, P. N. (2010). *A Vast Machine: Computer Models, Climate Data, and the Politics of Global Warming*. Cambridge MA: MIT Press.

Fleming, J. R. (2012). *Fixing the Sky: The Checkered History of Weather and Climate Control*, Columbia Studies in International and Global History. New York: Columbia University Press.

Foucault, M. (2002). *Power: The Essential Works of Michel Foucault 1954–1984 v. 3*. Essential Works of Foucault 3. London: Penguin.

Halley, E. (1686). 'An historical Account of the Trade Winds, and Monsoons, Observable in the Seas between and near the Tropicks, with an Attempt to Assign the Physical Cause of the Said Wind', *Philosophical Transactions* **16**, 153–68. Available online (accessed 05 March 2013) at http://www.clidyn.ethz.ch/ese101/Papers/halley1686.pdf.

Helmreich S. (2011). From Spaceship Earth to Google Ocean: Planetary Icons, Indexes, and Infrastructures. *Social Research*, **4**, 1211–42.

Jasanoff, S. (2001). Image and Imagination: The Formation of Global Environmental Consciousness. In *Changing the Atmosphere Expert Knowledge and Environmental Governance*, ed. C. A. Miller and P. N. Edwards. Cambridge MA: MIT Press, 309–37.

Joseph, J. (2012). *The Social in the Global: Social Theory, Governmentality and Global Politics*. Cambridge Studies in International Relations. Cambridge: Cambridge University Press.

Lapid, Y. and F. Kratochwil (1996). *The Return of Culture and Identity in IR Theory*. Critical Perspectives on World Politics. Boulder, CO: Lynne Rienner Publishers, Inc.

Larner, W. and W. Walters (2004). *Global Governmentality: Governing International Spaces*. Routledge Advances in International Relations and Global Politics. London: Routledge.

Latour, B. (2007). *Reassembling the Social: An Introduction to Actor-Network-Theory*, Clarendon Lectures in Management Studies. Oxford: Oxford University Press.

Lazier, B. (2011). Earthrise; or, The Globalization of the World Picture. *The American Historical Review* **116**(3), 602–30.

Litfin, K. T. (1997). Sovereignty in World Ecopolitics. *Mershon International Studies Review* **41**(2), 167–204.

Litfin, K. T. (1998). *The Greening of Sovereignty in World Politics*. Cambridge, MA: MIT Press.

Lovelock, J. (2006). *The Revenge of Gaia: Why the Earth is Fighting Back and How We Can Still Save Humanity*. London: Allen Lane.

Lovelock, J. (2009). *The Vanishing Face of Gaia: A Final Warning*. London: Allen Lane.

Lynas, M. (2011). *The God Species: How the Planet Can Survive the Age of Humans*. London: Fourth Estate.

Lövbrand, E., J. Stripple and B. Wiman (2009). Earth System Governmentality. *Global Environmental Change* **19**(1), 7–13.

Mayer, M. (2012). Chaotic Climate Change and Security. *International Political Sociology* **6**(2), 165–85.

Methmann, C. P. (2011). The Sky is the Limit: Global Warming as Global Governmentality. *European Journal of International Relations*. Forthcoming.

Rockström, J. (2009). Planetary Boundaries. *New Perspectives Quarterly* **27**(1), 72–4.

Rockström, J., W. Steffen, K. Noone, A. Persson, F. S. Chapin, E. F. Lambin, T. M. Lenton, M. Scheffer, C. Folke, H. J. Schellnhuber, B. Nykvist, C. A. De Wit, T. Hughes, S. Van Der Leeuw, H. Rodhe, S. Sörlin, P. K. Snyder, R. Costanza, U. Svedin, M. Falkenmark, L. Karlberg, R. W. Corell, V. J. Fabry, J. Hansen, B. Walker, D. Liverman, K. Richardson, P-Crutzen and J. A. Foley (2009). A Safe Operating Space for Humanity. *Nature* **461**(7263), 472–5. Available at: [http://dx.doi.org/10.1038/461472a]. Accessed 18 January 2013.

Rosenau, J. N. (2006). *The Study of Word Politics: Volume 1: Theoretical and Methodological Challenges*. Abingdon: Routledge.

Ruzek, M. (2010). *Earth System Science in a Nutshell*. [Online] Available at: [http://serc.carleton.edu/introgeo/earthsystem/nutshell/index.html]. Accessed 4 January 2013.

Shepherd, J. G. (2009). Geoengineering the Climate: Science, Governance and Uncertainty. *Design* **1**, 1–98.

Tickell C. (2010). *Earth Systems Science: Are We Pushing Gaia Too Hard?* [Online] Available at: [http://www.le.ac.uk/ebulletin-archive/ebulletin/features/2000–2009/2006/11/nparticle.2006–11–20.html]. Accessed 2 January 2013.

U.S. Global Change Program (2009). *Climate Literacy. The Essential Principles of Climate Sciences*. Washington, DC: U.S. Global Change Program. [Online] Available at: [http://downloads.climatescience.gov/Literacy/Climate Literacy Booklet Low-Res.pdf.]. Accessed 18 January 2013.

Walker, R. B. J. (1993). *Inside/Outside: International Relations as Political Theory*. Cambridge: Cambridge University Press.

Waltz, K. N. (1979). *Theory of International Politics*. New York: McGraw-Hill, Inc.

Weart, S. R. (2003). *The Discovery of Global Warming*. Cambridge, MA: Harvard University Press.

Wendt, A. (1999). *Social Theory of International Politics (Cambridge Studies in International Relations)*. Cambridge: Cambridge University Press.

13

Climate Change Multiple

Samuel Randalls

Introduction

Concerns about the climate are not new to the twentieth and twenty-first century. Historians have highlighted the ways populations have been governed to live with, work in and manage different kinds of climates (e.g. see Fleming and Jankovic 2011). Yet the concept of 'global climate change' has become undoubtedly powerful in the latter half of the twentieth century and has been adopted in diverse calls for societal intervention in the name of defending sound science, a stable climate and enhancing sustainability. The contributions in this book illustrate the diverse mechanisms, practices, discourses and rationalities of policy interventions to govern for and through the global climate. While a substantive policy literature explores these topics, the contributions here provide innovative and conceptually convincing demonstrations of the value of Foucauldian governmentality approaches (on their own or in coordination with other conceptual frames) to understanding climate governance.

Authors in this volume explore climate governance in diverse ways, for example as processes of rationalization and calculation (Chapter 10), global visions (Chapter 4 and Chapter 12), post-political neoliberal rationality (Chapter 3) and new experiments, whether with urban form (Chapter 2) or the everyday use of energy (Chapter 7). They raise important questions such as how climates, forests and energy meters are conceptualized, modelled and categorized for the purposes of climate governance. They also reflect the ways subjects are constructed through these processes and the various effects of these new modes of governance and self-governance. Climate governance is thus about governing climate, governing with and through climate, governing sets of practices tied to climate change through temporally and spatially specific associations of actors and governing populations as climate citizens or subjects. There are thus many forms of climate governance being discussed, which raises an important question. Do these represent different facets of *a* political form or different kinds of interventions and experimentation in practice?

This question is taken as the challenge to be explored in my contribution. Blok (Chapter 2) suggests that there are 'multiple climatic problems' and presents a series of problematizations, interdependent areas in which a problem becomes delineated and made potentially resolvable, but which do not encompass nor accrue to a form of totalizing climate change governance. Given this, however, the question is how climate change becomes tied to these sets of practices and whether what emerges is thus climate change multiple. Here I draw from Mol's (2002) work on the body in medical practice, which suggests that the variety of practices regarding atherosclerosis remakes the body and disease not as a singular entity or as pluralist bodies. Mol (2002) rather suggests 'the body multiple' – more than one, less than many. She delinks the question of what to do from the question of what is real, and argues that this politics of what must 'assume that the end points of trials, the goals sought for, are political in character' (2002: 175). Politics opens up questions rather than solving them through facts or argument. The question is where is the political here?

While governmentality perspectives may be aligned methodologically with actor-network theory, as most clearly shown in Whitehead (2009), there is still a question about the theorization of the political therein. As Blok (2011) points out, some of the performativity approaches within science and technology studies have failed to articulate a suitable theory of the political that is actually amenable with actor-network concepts. Is there a role for governmentality here or would this, as Blok suggests, undermine precisely what is distinctive about actor-network approaches? Remaining agnostic on this question for the present time, here I want to explore this question through the different ways climate change is enacted in the governance practices identified in previous chapters. In particular, three primary forms of climate change are enrolled in these practices, namely ones that focus on security, individualization and economics. These overlap in some ways and diverge in other ways, and the ways these practices rub up against each other are particularly informative. To illustrate that these practices reorder climates and populations, the following sections provide necessarily short examples of these three diverse enactments of climate change before moving on to a broader reflection on the question of climate governance in the conclusion.

Climate Change Security

Climate change has been enacted as a security risk through assemblages of security officials, risk models and the legitimating power of rhetoric of threat that has been reinvigorated since 9/11. For all this seeming authority, however, as Oels (Chapter 11) suggests, there has been rather less in the way of dramatic interventions than might be expected from climate change as exceptional state, and rather more adaptive, market-oriented governance that enables a depoliticized resilience mantra to become established. It is nonetheless clear that security practices shape the object of climate change in a particular way that stresses potential nonlinearity and a requirement for continual vigilance and monitoring of unsettled peoples and ecosystems. Monitoring movement, as Fall (Chapter 9) suggests, becomes a key organizing principle for these emerging forms of climate security governance.

Governments are not only anticipating potential worst-case scenarios but also actively imagining a phantasmagoria that becomes real (de Goede and Randalls 2009). Examples include the collections of pictures of future scenarios of climate change (London under floods, or in tropical conditions, or completely iced over), the classic disaster film imaginary of immediate climatic cataclysm (most well known in *The Day After Tomorrow*) and the prospect of geoengineering to deliberately alter the global climate. In the case of biodiversity changes, the lists of (invasive) species works both to confirm particular species as threats as well as to legitimate managerial interventions to confine or restrict the spread of these invasive species in the name of enhancing biosecurity. As de Goede (2012) illustrates in the case of terrorist finance, these lists then become performative. Being added to a list literally turns that person or organization into a risk and defines and restricts the future circulations of finance. In climate change governance, particular places and populations become risky: the North African and Middle Eastern water-deprived areas that might foster Islamic radicalism or the Bangladeshi migrants fleeing rising waters into neighbouring India. Climate threats become defined through scientific modelling – indeed it is interesting that in climate health research, while south-east Asian countries are predicted to have deaths from malnutrition the European Union is not, primarily because of an assumption that the European Union will supply food to all its member states whereas the south-east Asian countries will be less co-operative with each other (Randalls 2011).

Climate change is not a neutral scientific script that lends authority to policy makers to establish governance interventions vis-à-vis a global climate polity. Rather climate change is always already a coproduced science-political hybrid and as such is enacted in particular ways by different assemblages of practices. The good outcome is a world secured from climate change (impacts), but that does not necessarily mean interventions to prevent climate change. Subjects that are compensated through insurance mechanisms for climate change impacts are an equally important outcome of securitization (Stripple 2012). Thus climate change security is both multiple and constitutes multiple subjects (Stripple 2012).

Individualization

For other groups, climate change is primarily an issue of responsibility and consumption, with citizens needing to take the lead to encourage carbon-friendly resourcing of required goods and services. This is implemented through a variety of engagements with consumers from carbon calculators (Chapter 6), to energy-smart meters (Chapter 7) to interventions to 'nudge' people to make 'correct' decisions (Chapter 5). This approach is characterized by an individualization of responsibility, an attitude of self-governance and a concerned subject that never knows if he or she has done enough to stave off climate disaster.

Hargreaves (Chapter 7) provides a particularly in-depth exploration of the use of smart meters in households. Here one learns that rather than technologies simply delivering emissions reductions, they play more complicated and nuanced roles in lifestyles. As Mol writes: 'technologies do not subject themselves to what we wish them to do, but interfere with who we are' (2008: 58). Smart meters enact a particular kind of climate-concerned subject, which enacts climate change as a phenomenon that is to be managed through reductions in carbon usage and guilt. Climate change here is less the dramatic security personae and more a day-to-day practice, an attention to the ways consumers in richer countries (at least) manage their lives in the expectation that changes can and will be made to deliver a less carbon-intensive lifestyle (Paterson and Stripple 2010). If marketized in the form of personal carbon trading, where each individual would have a carbon allowance and then buy/sell credits as required, then this equally rewards those that can competently economically manage carbon. The marketplace resolves individual choices by putting a market price on carbon and letting consumers decide how they wish to spend their portion of the atmospheric pie.

Again, this is not a neutral threat of climate change and one of many options to resolve the problem. In this case, climate change is modelled and counted by establishing 'carbon equivalency' as the commodity that is to be managed to save the planet. Climate change is enacted as a problem of carbon management, legitimating interventions into better governing the use of carbon in people's lives. The good outcome here is the rational carbon consumer, who will enable the reduction of carbon dioxide emissions to prevent climate change.

Economic Agenda

Similar to the individualization focus, the practice of using economic instruments, especially markets, as the primary governance mechanism for climate change narrows the scope of action to 'carbon dioxide equivalent', valuing and pricing this new commodity. Here, though, rather than individuals being entrusted to govern themselves, the focus is much more on the producers of emissions rather than the consumers. Of necessity, the introduction of carbon markets requires strong state involvement. This is no laissez-faire approach, but rather one characterized by scientific authorities and legislators working together to measure, monitor and map carbon production and absorption for the purposes of establishing mechanisms of trade that, according to their stated objectives, enable the overall reduction in emissions. Good examples of this are the REDD+ schemes for rewarding the prevention of destruction to forests as well as the creation of new forests (Chapter 10 and Chapter 3) and feed-in tariffs to enable renewable energy development (Chapter 8).

The case of forestry governance through carbon markets illustrates the considerable expansion of a new economic sector, carbon, which relies on carbon market professionals as sources of authority and expertise and establishes financial circuits as ideal modes of governance. As Lovell (Chapter 10) notes, this connects a whole set of scientific expertise in remote sensing and carbon science with the informational needs of a market. Climate change is here enacted as a financial problematic, defining a new commodity and set of trading relationships to stabilize the object, climate change, as something amenable and solvable through market relations. Rather than presenting an irruptive threat to the market, climate change comes to be governed through the market. One can associate such practices with discourses such as cost-effectiveness, externalities, optimal policies and a belief in the market as primary information processor.

For Methmann, Rothe and Stephan (Chapter 3), these micro-practices of carbon governance through commoditization are allied to broader power relations that reinscribe forms of neoliberal governmentality that, like the security network described earlier, depoliticizes debates about climate change. At the same time, however, this market-based approach is but one enactment of climate change circulating at present, albeit a particularly dominating one if the focus is on international policy. This economic agenda has at its core a stated goal of achieving emissions reductions at the lowest possible cost, but its goals seem to be as much about achieving some kind of economic rationality that incorporates and resolves external critique as it is about managing climate change. While even the most dedicated carbon market enthusiasts agree that changes are needed if markets are to achieve emissions reductions, there still remains a faith that given the right conditions markets will eventually deliver on this promise. Paterson and Stripple (2012) deploy the term 'virtuous carbon' to emphasize the ways virtuality and morality are entwined in carbon markets such that the moral goodness of carbon trading to 'save the planet' outweighs specific critiques of market failures. If climate change represented such an imminent threat (as the security network might suggest), then carbon markets hardly appear an ideal solution in terms of time scale.

Conclusions

The usefulness of the concepts from governmentality analyses for understanding contemporary climate governance has been established in previous chapters, but it is clear that different authors are interpreting and reading these interventions in different ways. Do these micro-practices of governance totalize into a neoliberal governmentality or are there just too many differences to allow for hegemony? Blok (Chapter 2) suggests that rather than a global polity there is a meshwork of assemblies in the making. I would concur with this conclusion, but it is nonetheless important to stress that these assemblies always remake things and the way they are remade is not neutral.

Death (Chapter 4), like Blok, would critique a governmentality approach that adopts a global 'programmer's view', but nevertheless retains the value of charting particular analytical frames that are observed in climate governance. Some interventions share very strong commonalities with themes and discourses that appear rather similar to those developed in the neoliberal thought collective as outlined by Mirowski (2009), for example the redefinition rather than the removal of the state, the market as the natural and most advanced information processor and solution to any perceived market problems, and the recoding of freedom as autonomous self-governance. At the same time, there is no singular neoliberal narrative on climate change: promoting scepticism and carbon markets is not exactly a consistent position. So to call something neoliberal should not necessarily be to appeal to a structural determinant that defines the analysis, but rather to call attention to specific practices that bear remarkable similarities to stated neoliberal thinkers' philosophies.

Each of the three sets of practices or assemblages identified earlier in this chapter can and do share some neoliberal tropes, most evidently in the carbon commodity interventions and perhaps less so in the security network, notwithstanding the power this has for geoengineering. Fundamentally, however, these practices do not seem to be different solutions to the same problem – climate change – but rather function as different ways of thinking, writing, doing, practising and enacting climate change. In other words, climate change is not a singular entity that is then made governable in different ways; rather climate change is inherently multiply constituted through different assemblages of people, organizations, ecologies and much more. To demonstrate this more empirically, the good outcome for the market-oriented strategies of a cost-effective climate change is not necessarily consistent with the need to secure populations and ecologies from climatic harm. Likewise geoengineering interventions to secure the planet seem to demobilize the need and legitimacy of interventions in everyday lives and practices to reduce emissions. These goals and objectives are not simply commensurable with each other. Yet they are held together in tension, or with frictions, by the ways they conflict materially and discursively with each other. As such, solving climate change cannot just be about reducing carbon emissions.

The good outcomes of climate policies therefore are different and may be conflictual, even within the same policy intervention. In forests around the world, for example, the goal of reducing carbon emissions through REDD+ schemes or mitigating pollution through carbon offsets creates significant new economic incentives to plant trees (Chapter 10 and Chapter 3). Inasmuch as this might be a rational solution to reducing carbon equivalent concentrations in the atmosphere; and it might be ecologically good if one considers that more trees may enhance biodiversity; and it might be socially good if local populations are provided with necessary means to manage the trees; there are serious criticisms. Monocultures may not enhance biodiversity. Contracts with local populations may privatize land in ways that function to exclude previous users of lands from damaging controlled carbon sinks (the national park scenario in conservation). Carbon offsets may simply function to enable people to continue with their lifestyles, paying a small charge to repair the damage, which might not ever be realized in measurable emission reductions. Each climate assemblage, therefore, portrays certain goods that are sought for, while obscuring others that are considered negative, while in different assemblages the goods and bads might be reversed. What do we make of this in terms of governmentality (some questions) and policy (some responses)?

First in terms of academic literatures, it asks researchers to consider how emerging issues are enacted as part of new forms of governance mechanisms and prompts us to question why some seem to become dominant and others much less visible. So an important question is: How are goods enrolled within particular networks as the central concern while other goods are marginalized? It is hard to critique particular governmental interventions as enforcing post-political frames when it is so difficult to construct a viable alternative, especially when any alternative (socialism maybe?) can be caricatured as having its own set of sought-for goods that are not democratically agreed upon. On the other hand, actor-network theorists appear more amenable to contemporary governance interests in showing how things work, for example, in financial markets. What, then, are the politics of our critiques? A second important question for governmentality and actor-network scholars is how to deal with similarities and uniqueness across different fields. The literature on climate governmentality has emerged alongside environmental literatures on forestry and biodiversity, also exploring similar themes of calculation, self-governance, exclusion/inclusion and marketization. As such it seems incongruous to think that climate interventions are unique. How is it that assemblies in the making (Chapter 2) formulate and derive ideas and legitimation from other assemblies in the making? Why do certain forms of governance persist from issue to issue, or good sought to good sought, while others are challenged and may fail?

Second, in terms of policy, this multiplicity opens up new questions and challenges in decision-making. There can be no simple way of 'communicating' climate change to engender particular forms of climate actions, as there is no singular climate change that crosses diverse sets of climate associations in practice. This focus forces us to consider climate policies as multiple interventions that have as their goal different kinds of goods to be achieved that are not mere choices for us to decide between: shall we take the economic good or the security good? It is not that simple. As Mol puts it: 'Presenting the *body multiple* [climate change multiple] as the reality we live with is not a solution to a problem, but a way of changing a host of intellectual reflexes' (2002: 184). The classical conception of solving climate change through emissions reduction is opened up to a broader political problematic that considers the good outcome to be achieved. Even within this frame, the coalescing of political focus on preventing a global mean average temperature increase of 2°C above pre-industrial temperatures has struggled at times to legitimate this goal as well as prove how it is even deliverable in emission changes. It is a fixation on a statistical outcome rather than considering a multiplicity of goods: quality of life; quality of ecosystems; types of interventions; and political questions about what kinds of freedoms we want. The good life to be lived needs to be more directly questioned in these interventions appreciating the reasons 'we disagree about climate change' (Hulme 2009). Climate change multiple requires openness, continual intervention, continual questioning and potential resistance. This is a qualitatively different debate to figuring out a global political solution to best solve climate change singular.

References

Blok, A. (2011). Clash of the Eco-Sciences: Carbon Marketization, Environmental NGOs and Performativity as Politics. *Economy and Society* **40**(3), 451–76.

De Goede, M. (2012). *Speculative Security: The Politics of Pursuing Terrorist Monies*. Minneapolis: University of Minnesota Press.

De Goede, M. and S. Randalls (2009). Precaution, Pre-Emption: Arts and Technologies of the Actionable Future. *Environment and Planning: D* **27**(5), 859–78.

Fleming, J. R. and V. Jankovic (ed.) (2011). *Klima*. Osiris, 26. Chicago, IL: University of Chicago Press.

Hulme, M. (2009). *Why We Disagree about Climate Change*. Cambridge: Cambridge University Press.

Mirowski, P. (2009). Postface: Defining Neoliberalism. In *The Road from Mont Pelerin: The Making of the Neoliberal Thought Collective*, ed. P. Mirowski and D. Plehwe. Cambridge, MA: Harvard University Press, 417–55.

Mol, A. (2002). *The Body Multiple: Ontology in Medical Practice*. Durham, NC: Duke University Press.

Mol, A. (2008). *The Logic of Care: Health and the Problem of Patient Choice*. Abingdon: Routledge.

Paterson, M. and J. Stripple (2010). My Space: Governing Individuals' Carbon Emissions. *Environment and Planning: D* **28**(2), 341–62.

Paterson, M. and J. Stripple (2012). Virtuous Carbon. *Environmental Politics* **21**(4), 563–82.

Randalls, S. (2011). Climate Change Pathology. *Environment and Planning: A* **43**(6), 1242–7.

Stripple, J. (2012). The Subject of Security in a Warming World. *Brown Journal of World Affairs*, **18**(2), 181–94.

Whitehead, M. (2009). *State, Science and the Skies: Governmentalities of the British Atmosphere*. Chichester: Wiley-Blackwell.

Conclusion

Towards a Critical Social Science of Climate Change?

Harriet Bulkeley and Johannes Stripple

Introduction

In his foreword to this book, Ronnie Lipschutz calls attention to the question of order in relation to climate change, how social orders are created and maintained and how they might fall apart. That question – what makes social order possible – is a foundational one for the social sciences and the humanities. This book includes many investigations into different kinds of social orderings around climate change and carbon, from households' energy-saving activities to different agencies' preparedness practices in the face of climate impacts; from neurologically based ideas about behaviour change to international agreements on how to capture the carbon content of tropical forests. Governing the climate, as we insist in this book, is about interrogating the many diverse sites in which a carbon constrained and/or rapidly warming world is problematized, represented, categorized and ordered. Our aim is to offer a new perspective on 'climate governance', one concerned with the governing of ourselves and others in relation to climate change. Sites such as households, markets, forests, migratory species, renewable energy and displaced persons are all examples of locations where subjectivities around climate change and carbon are formed, forged and contested.

Such new perspectives are, we suggest, vital if the social sciences are to critically engage with climate change. The climate issue has come a long way since the early 1990s when it became a topic for intergovernmental negotiations or when it showed up in surveys of individual attitudes. But social science research has been lagging behind. In the early 2000s, the concept of climate governance broadened the field of study, aiming to understand the ways climate change was governed 'beyond the state'. The focus turned to different kinds of emerging institutionalized arenas (not just the state or the UNFCCC), taking into account governance mechanisms with public, hybrid and private sources of authority. While scholarship in this area has pointed to the important roles of such arenas and institutions in governing climate change, particularly in relation to emerging carbon markets, we have yet to develop our understanding of how governing is achieved in these arenas in the absence of the formal authority of the state. Climate governance, it seems to us, is increasingly an empty concept – empty because it is often approached in the general. With a Foucauldian approach to governance, engaged with the practices, techniques and rationalities employed to shape conduct in particular places, comes the possibility of arriving at the 'specificity of governance' (Walters 2012: 56). As Lövbrand and Stripple argue in Chapter 1, a governmentality perspective challenges the fixedness and rigidity of what climate governance is all about. From this starting point, this book begins to raise questions about the ways rationality, power and politics are constituted and practised in the climate arena, and the implications for how we understand the potential and limits for addressing the climate problem.

There is, however, no one orthodox climate governmentality presented in this volume. As Death points out in Chapter 4, there are many different interpretations of governmentality, within and beyond Foucault's own work. Governmentality has sometimes been used almost synonymously with neoliberalism, as a particular political economic era, while alternatively it has been used more as a general framework for analysis of different power relations. In this book, it is the latter interpretation that most authors adopt, albeit to different degrees. Furthermore, governmentality itself does not seem to offer a sufficient account of climate politics. Instead, we can see that governmentality is fruitfully engaged in conjunction with other approaches, ANT, behavioural science, science studies and poststructural approaches to discourse and security. When climate governance is encountered from the vantage point of governmentality, it is clear that this is a plural field – there is not one 'climate governmentality'. This is in line with Walters's appeal to how we should encounter contemporary governmentalities; 'Beware of applicationism. Beware of panopticism. Be suspicious of "Foucault's triangle"' (2012: 144). There appears to be much then to learn from the engagement between governmentality and climate. Yet, as Lövbrand and Stripple (Chapter 1, this volume) point out, while climate scholarship has increasingly been informed by general debates in Foucauldian governmentality studies, climate governmentality scholars have, in general, not been speaking back to that community. In this chapter we explore – by considering first the question of political order and then engaging the concepts of rationality, power and politics – what they would say if they were.

Ordering the Social (and Natural) World

As reflected in the foreword to this volume, the turn to concepts of governmentality and other critical political and social theories within climate change scholarship has been driven by rather more long-standing concerns within the social sciences about how political order comes to be constituted and sustained. In April 2010, Harvard University held a symposium where it asked scholars to agree on the 'top ten questions' facing the social sciences in order to inform research agendas and funding agencies. Interestingly, the question of order looms large in many of the challenges identified at the symposium, for example, 'how do societies create effective institutions?' or 'why do social processes, such as civil violence, either persist over time or suddenly change?' (Giles 2011). Thomas Hobbes famously posed the question of order in his 1651 book *Leviathan.* For the sake of his argument, he imagined people living in a situation prior to the establishment of society – the state of nature – where every person feared losing his life to another. It was a fundamental situation of distrust, which made most forms of collaboration impossible. In this situation, argued Hobbes, it was a rational decision for men to give up their natural rights in exchange for common security, that is life in a civil society. Badersten shows how Hobbes's question has a much longer lineage (2002: 37–47). It is a theme in the Greek tragedies and it was a question for Antique philosophy overall. To Aristotle, we are political animals – *zoon politicon* – implying that we are social beings that look after each other. Hence, in contrast to Hobbes's idea that social orders break down when left to individuals, for Aristotle people will do 'the right thing' because they know what is right and what is good. Machiavelli, on the other hand, argued that 'it must needs be taken for granted that all men are wicked' (Machiavelli, cited in Skinner 1978: 186) and that this must be taken into account when laws and constitutions are formed.

Many theories and concepts in social science relate, in various ways, to these questions of how and why social order comes into being, for example the different approaches to the 'logic of collective action' (Olson 1965), formalized as in game theory (Kuhn and Tucker 1950), articulated as the tragedy of the commons (Hardin 1968) or as a cooperative management of common-pool resources (Ostrom 1990). Contrary to such intentional interpretations of social order we find a more spontaneous account in Adam Smith and the invisible hand of the market, which rendered governmental intervention to establish order unnecessary. There are also those approaches to (political) order where order is seen as natural – as emanating from within the social system. Ethnicity, race and ties of kinship have frequently been put into use as ordering principles (Badersten 2002: 51). Finally, Lipschutz, in his foreword to this book, warns against thinking about order and social power as somehow exogenous to social relations. Religious apocalypse as well as apocalyptic narratives of climate change understand order as an exogenous given, that is they locate the question of order beyond the social sphere and therefore as unrelated to practices and identities of individuals.

From these historical roots and emerging principles the social sciences have frequently approached the question of order through master narratives, or overarching concepts, that can give an account of the organizing principles of the social realm. The concept of anarchy, which has been central to the discipline of International Relations, for example, regards the international sphere as a particular political order that plays out among sovereign authorities. Other prominent examples are the accumulation of capital in capitalism; the Marxian focus on the social division of labour; the Weberian bureaucratization of society; or rational choice theory. In contrast, Foucault and scholars of governmentality have taken a rather different approach to social order. In their retrospective book, *Governing the Present*, Miller and Rose start with the large questions of 'governing economic and social life' and 'how, and to what ends, did so many socially legitimated authorities seek to interfere in the lives of individuals in sites as diverse as the school, the home, the workplace' (2008: 1). Central to such an approach was not only a new understanding of the way social and political order was constituted, but a different means of approaching such questions through the 'precise, meticulous and scholarly tracing of the small and dispersed events that brought something new into existence' (2008: 4). For Miller and Rose, and other scholars inspired by Foucault's work, the insistence to start analysis in unfamiliar places, with minor figures, with documents such as manuals or with technical devices, is crucial. Political order, or governance if you will, was therefore not approached as an abstract entity but as a practical question – how is it that some things appear as problems that need management and regulation? – and as a specific technology – by what means, mechanisms, procedures, instruments and vocabularies is authority constituted and rule accomplished?

The chapters in this book share this ethos both in terms of how questions of political order are approached and the means through which they are investigated. Looking back at thirty years of work, Rose, O'Malley and M. Valverde (2006) note that the language and analytics of governmentality has dispersed, become mixed with other approaches and gone in many different directions. What remains salient and challenging, however, is the insistence on moving away from grand theories (e.g. globalization, modernization, risk society) and instead investigating the practices through which we govern and are governed in our everyday lives. Against the political science habit of starting with the state and the political apparatus, or those of sociologists and philosophers who turn to the great names of their canons, this book shows how the analysis of our carbon-constrained present can be approached from elsewhere, starting with mundane practices, things and inconsequential moments. The chapters provide many such examples – carbon footprints, smart energy meters, forest carbon modelling, disaster preparedness plans, neurological research – through which objects and subjects of climate governance are constituted, stabilized and contested.

Rationality

When Miller and Rose picked up Foucault's concept of governmentality they teased out two aspects of governing; on one hand, the representation and knowing of a phenomenon, and, on the other hand, the acting upon the same phenomenon so as to transform it. Miller and Rose called the former aspect 'rationalities of government' and the latter 'technologies of government' (2008: 15). They wanted to underline that problems are not just out there, they have 'to be rendered thinkable in such a way as to be practicable or operable' (2008: 16). The basic premise is that government is to be seen as a rational and thoughtful activity, where we pay particular attention to the forms of thought, knowledge, expertise, strategies or rationality employed in the practices of governing. Within the climate change domain, there is a clear sense that such rationalities have been shifting. Long gone are the times when climate change was just a question for international diplomacy, a sector covered by national regulatory authorities or an issue that mobilized a small set of highly motivated nongovernmental organizations (NGOs) and businesses. For Hulme the idea of climate change is now:

> circulating with mobilising force in the fields of business, of law and of international trade. It is an idea circulating with potency in the fields of knowledge and invention, of development and welfare, of religion and ethics, and of public celebrity. And it is an idea circulating creatively in the fields of art, of cinema, of literature, of music and of sport. (2010: 171)

Hulme makes a compelling case for why we need to understand that the climate issue is everywhere, but yet it is never alone. Climate change is articulated in very different contexts alongside many different concerns and problematics. These include enduring questions of politics and development, from African civil wars and Arctic shipping routes to 'carbon fasting' where churches have connected personalized emission reductions to well-established religious practices and subjectivities. Climate change also appears as an opportunity to reinvigorate campaigns for active lifestyles, especially via transport modes, reframed as 'low carbon', or in response to obesity, either through constructing overweight subjects as a problem for climate change or by problematizing low-density urban environments as origins of problems of climate change and obesity (Roberts and Edwards 2010; Paterson and Stripple 2013). Whitehead, Jones and Pykett (Chapter 5) highlight how a broader history of liberal government and biopolitical desire provide the context for the 'Behaviour Change Agenda' – the fusion of psychology and public policy – that is shaping climate policy throughout the world. Of importance is the creation of the 'real human' as the subject of governmental intervention, which needs to be governed in an advanced liberal manner. In this way, particular rationalities lead to the creation of subjects, and in so doing summon the world.

While the climate issue is therefore becoming interwoven in the fabric of modern societies and the way we live and organize our lives (eating, moving around, warming or cooling our houses), we have yet to grasp this condition and what it means for climate science and policy. To be sure, different policy or scholarly communities will continue to insist that governing the climate is about designing and implementing one particular programme of government, for example Greenhouse Development Rights (Baer 2013), an architecture for a global agreement (Aldy and Stavins

2007) or a set of behavioural change policies (Chapter 5). Yet the chapters in this book suggest that conceiving of the climate issue as one problem that can readily be governed from the top down is to miss the crucial multiplicity of the rationalities that have come to structure climate change as an issue. From this starting point, Randalls (Chapter 13) asks if the emerging governance of climate change indeed represents different facets of *a* political form or is actually more appropriately conceived as different kinds of interventions and experimentation in practice. Hence, he calls for considering 'climate change multiple' and, similarly, Blok (Chapter 2) calls for the appreciation of 'multiple climatic problems' and Stephan, Rothe and Methmann (Chapter 3) highlight the importance of retaining heterogeneity and messiness of government.

Approaches based on governmentality, in enabling an alternative approach to questions of political order, can begin to open up such forms of multiplicity climate issue. However, such approaches also contain a totalizing tendency. As Death argues in Chapter 4, the danger is that if scholars are starting with analysing the strategies and programmes of government agencies they risk adopting the 'programmers view', in which singular rationalities are rolled out to govern at a distance, smoothing space and negating difference and contestation. In particular there has, in some Foucauldian climate change research (Higgins 2001; Oels 2005; Letell, Sundqvist and Elam 2011), been a tendency to interpret many matters of climate policy as expressions of a neoliberal rationality. Blok (Chapter 2) points to the ever-present danger of *all* totalizing narratives underneath governmentality, which deny plurality and the 'always in the making' nature of assemblages. Instead of reading ever more phenomena as evidence of advanced liberal government, actor-network theory (ANT), with its focus on the shifting meshwork of multiple climate problematizations, paves the way for research on how those problematizations are rendered governable, for example through calculation. Discourse approaches (Chapter 3) are, in this respect, also helpful because they seek to interrogate the social and material constitution of these problematizations. In contemporary climate governance, there are important heterogeneities in the ways governing is conducted in terms of spatiotemporal dynamics, material devices and forms of political rationality (Chapter 2). These chapters suggest that what is important is not only the question of *what* rationalities are adopted in particular instances through which to order climate change, but *how* this is accomplished.

In this vein, many of the chapters urge us to move away from a sense that rationality is free floating or operates at a distance, and instead consider the objects and issues through which it is made to cohere. The book includes many examples of governmental rationalities 'at work', for example, the measurement, reporting and verification (MRV) of forest carbon (Chapter 10), online carbon calculators to measure carbon footprints (Chapter 6); practices for fostering resilience in the face of 'climate apocalypse' (Chapter 11); the behaviour change agenda (Chapter 5). Rather than adopting a 'programmers view', this work pays attention instead to the heterogeneities of governing and how rationalities are negotiated, contested and resisted in practice. Oels (Chapter 11) conceptualizes this process as a discursive struggle around the 'partition of the sensible', in her case how small Pacific island populations strongly object to how dominant forms of government seek to render them governable as climate refugees and thereby to 'save' them through resettlement proposals. Hargreaves (Chapter 7) explores how people interact with feedback on their domestic energy consumption from smart energy monitors. While the monitors enable householders to 'see' their energy use (and its associated carbon emissions) and thus take steps to reduce it, they are also 'domesticated into the physical domain, social relations and cultural practices of each household', generating new patterns of both cooperation and contestation. Even though energy savings became a normal aspect of using energy in everyday life, the monitors also created fatalism, despondency and anxiety in the households because the savings seemed futile compared to what would be required to mitigate climate change. Overall, this work suggests that carbon rationalities around real-time energy displays do not appear very powerful in relation to other forms of rationality and practice based on economic concerns and daily forms of comfort (Chapter 7; Hargreaves, Nye and Burgess 2010).

Through these chapters we find that the domain of climate governance is not only characterized by multiple rationalities, and hence climate problems that need to be resolved, but that such rationalities can also conflict. One such tension is evident in the ways climate change has come to be represented as a matter of individual responsibility, of behaviour change and the creation of new forms of energy use in the home, yet at the same time regarded as a matter to be controlled through particular forms of technical, autonomous interventions within the domestic environment (Chapters 5 and 7). A similar tension between climate rationalities that seek to create self-governing subjects and the desire to control and regulate individuals and populations is apparent in the case of climate change migration (Chapter 11) and in relation to particular species (Chapter 9). Furthermore, the rationalities of governing climate change are not uniform in themselves, and may mutate or, chameleon-like, adopt new guises as they move between different contexts. In the case of the governing of biofuels in Finland (Chapter 8), logics concerning the need to attend to the remnant tree stumps left in the ground from the emerging renewable energy industry in terms of reducing overall carbon emissions were continually translated through logics of the development of regional economies and wider economic circulations. The contestation and mutation of rationalities appears to be central to the governing of climate change, and indeed the process of translation and circulation central to the ways governing takes place in practice. Such findings add further weight to the argument that the 'programmers view' is an inadequate vantage point from which to engage with the process of climate governance, and to the contention that climate change is indeed a 'multiple' problem.

Power

At the heart of our concern with the rationalities and practices of governing is the appreciation that perspectives from critical political and social theories offer an alternative approach to the nature of power, which highlights the assembling of political order through the continual rearticulation of the social and material world. As studies of climate governance have sought to move 'beyond' the state, questions of power – of who has it, how it is manifest, where it lies and how it might be shared – have loomed large. In the absence of the usual trappings of power associated with the state – sanction and the rule of law – how might other nonstate actors come to be authoritative? Even where such authority proves apparent, will its largely voluntary nature and fragile institutional basis necessarily give way to the will of more concrete forms of power held by the state? A governmentality perspective allows us to take a different perspective on such questions. Rather than starting from the question of how power might be shared between actors, or from an understanding that some actors, sites and arenas are more powerful than others (Chapter 1), such a perspective, also adopted within ANT and discourse theory, asks that we examine how power itself is constituted and resisted. In this sense, power is not a capacity that inheres in (particular parts of) the social world, but rather is produced through social relations. As Corry (Chapter 12) argues, practices of power do not operate as if the object to be governed – in this case, climate change – is already fully formed, but rather come to constitute the object and its political field simultaneously. In this view, the 'polity is recast as a structure in which an object becomes a target of power practices that organizes a field of politics'. Whether or not the climate object constitutes one field of power relations or is multiple, such insights suggest that we need to direct our attention to the ways power is practised in the constitution of the political field of climate change.

In so doing, the chapters in this book point to the important ways power is constituted not as a capacity or resource, but through the relational practices of circulation, translation and assemblage (Chapter 2, Chapter 8). For Kortelainen and Albrecht (Chapter 8), for example, power works topologically, through a series of 'translation loops' such that distinct forms of power come to operate and be felt in different arenas and through different regimes of practice. At the same time, the authors in this volume, while focussed on forms of governmental power, recognize that power may operate in different modes and registers (Allen 2003). Oels (Chapter 11) argues that 'Foucault's work enables us to distinguish between at least four modes of securing, each linked to a technology of power: sovereign power, discipline, liberal biopower and advanced liberal government'. Drawing attention to the multiple modes of power evident in Foucault's work, Oels goes on to identify the ways such different forms of power are exercised in relation to climate change migration, and the different effects that such processes have both in terms of constituting the domain and subject to be governed. Adopting such conceptual positions, where power is seen as productive and as working through multiple modalities, enables social science scholarship on climate change to move beyond the rather stale debates about the transfer of authority between different actors and different sites and instead to start to engage with the ways power is actually produced through the intention to govern climate change (Bulkeley 2012). Lovell (Chapter 10) applies Foucault's ideas about biopower to forests, thus viewing forests as a kind of 'carbon population'. Biopolitics is about the construction of aggregate objects – populations – through specific forms of knowledge (e.g. statistics, demography) which are then governed on the basis of those aggregated measures. Carbon populations, as other populations, are no natural entities. Forest carbon populations are established through, for example, modern advances in remote sensing and modelling. The combined work of organizations such as UNFCCC, UN-REDD and IPCC is thus a technology of power that aims to measure, understand and therefore control forest ecosystems. Furthermore, by opening up the question of how power comes to be constituted, such approaches also require that we engage with the ways it is realized and enacted. Here, we find the chapters in this volume make three important contributions. First, many of the chapters point to the ways governing is practised through particular techniques and objects that appear either mundane or innocuous. Forms of codification and the development of standards, whether this is in the form of carbon footprints (Chapter 6), REDD (Chapter 3), the carbon content of tree stumps (Chapter 8) or codebooks for calculating forest carbon (Chapter 10), provide the means through which power is produced and circulated. Equally, accomplishing climate governance requires both the insertion of a whole host of new objects into the social realm – from carbon calculators (Chapter 8) to smart meters (Chapter 7) – and the redefinition of everyday material practices as globally significant. Blok (Chapter 2) illustrates this with the translation of climate change into tangible domains of local transport (biking in Copenhagen) and housing (vernacular wooden eco-houses in Japanese Kyoto). Such techniques and materials become from this perspective not merely the carriers of predetermined rationalities or discourses, but the means through which power is configured and contested. Second, several of the chapters point to the critical process of constituting the subjects of climate governance as essential to the working of power. Here, the critical conclusion is that subjectivities do not exist in the social realm ready to be aligned to particular rationalities, but are in themselves produced through the operation of governance and the practice of assemblage that accompany the constitution of particular climate problematics. Importantly, the chapters suggest that this is not achieved only through the appeal to the rational 'self-government' of individual citizens, but works through the constitution of particular communities or populations (invasive species, the forest, climate migrants). Furthermore, in Chapter 5, Whitehead, Jones and Pykett argue that the Behaviour Change Agenda actually reflects the extension of governmentality into the more-than-rational registers of human behaviour such that power comes to be constituted 'neurologically'.

Third, the chapters in this book raise questions about where the limits of what can be governed might lie. Conventionally, governmentality scholarship has tended to analyse the governing of the social world, albeit that the importance of particular material artefacts, objects and devices is increasingly recognized. This in turn has meant limited engagement with the properties and dynamics of such materials, and in particular little consideration as to how the emergent characteristics of the 'natural' world might shape the possibilities for conducting conduct. For several of the authors in this volume (Chapter 2, 5 and 10), it is precisely this dynamic quality that requires a move away from governmentality per se to an engagement with those approaches that have put the 'more than human' at the centre of their analysis, in particular ANT. For Lovell (Chapter 10), 'actor-network theory is used to assess the implications of forests being living, rather unstable things, not easily enrolled into networks centred on the policy goal of accurate, long-term measurement'. Fall (Chapter 9) shows how the emerging field of biosecurity studies, which examines mobile threats to human, animal and plant life, combines governmentality and biopolitics with accounting paradigms and a critical geography of spatial processes of categorization and boundary making. This enables Fall to interrogate categories such as 'native species' and 'invasive species', and overall to show how nature is rendered manageable as nationally listed biodiversity – thus inevitably and inescapably planned and governed through and by states. Bringing governmentality together with ANT seems to offer a deeper understanding than is usually brought to bear in the study of climate governance of the capacity of nonhuman things such as forests to 'bite back'. At the same time, such perspectives raise questions about the limits of a governmental notion of power – if the subjects to which governmentality pertains are 'more than human', does the conception of such entities as capable of some form of 'self-governing' of their conduct in relation to particular rationalities still hold? Or does this point to the need to regard forms of governmentality as more emergent, unstable and contingent than is conventionally argued?

Central to any such account is an understanding that the production and practice of power always operates within a field that is traversed by existing forms of power and rationality, such that specific programmes are not 'invented ab initio … [but rather] draws upon and is situated within a heterogeneous assemblage or *dispositif*' (Murray Li 2007: 6). As Hargreaves (Chapter 7) argues, for Foucault new formations of power are not separate from those already in operation, but rather the creation of new constellations of problems, rationalities, subjects and techniques represents a means through which social relations are reconfigured and renewed. Further, the operation of power through particular subjects is always open to contestation and resistance. For Foucault, rather than being a matter of control or discipline, the operation of governmental power depends on the condition of freedom, so that 'individual or collective subjects … are faced with a field of possibilities in which several kinds of conduct, several ways of reacting and modes of behaviour are available' (2000: 342). Such a perspective reinforces the need to examine the workings of power as it operates not only through particular logics or the formation of subjectivities, but in terms of the practice of power in the messy actualities of governing climate change in particular sites and spaces.

Politics

Acknowledging the workings of power means that we must also attend to the ways governing the climate entails engaging with questions of politics. For the most part, the approaches taken to such questions within social sciences are structured through the particular ontologies, that is ideas about the kind of stuff the world is made out of and what kind of stuff matters in the world, which they adopt. Is it states? Perhaps qualified as welfare states, failed states, rough states, advanced liberal states or virtual states? Or is it rather those actors that are lumped together because they are 'not states', the NGOs – for example multinational firms, international organizations, lobby groups, networks of individuals, communities, regions, scientific advisory groups or terrorists? Or is it nations and classes? Or is the world rather shaped through the structures of capitalism, colonialism, empires, civilizations, centre and peripheries or the patriarchy? Discovering and negotiating the basic features of the world is part of an ongoing conversation within the social sciences. Usually, such ontological assumptions are often posed (implicitly or explicitly) before any enquiry starts. It is about what constitutes the knowledgeable reality, and hence the matter of what constitutes the political realm is structured by what it is that we can come to know.

Walters (2012: 56) characterizes modern social theorizing as dominated by the assumption of rational actors (individuals, firms, states etc.) *or* by approaches that posit a political world that is fundamentally characterized by socio-economic structures of class, economy, gender, race and so on. While the former base their analysis on the choices and interactions these actors make as they pursue their respective interests, the latter consider that political behaviour can often be modelled as rational action, but always inside a world profoundly shaped by structural asymmetries of power. Against these dominant modes of theorizing politics, Walters (citing Veyne 1997) argues that studies of governmentality instead approach 'the analysis of politics in a way that makes at best only the most minimal assumptions about the ontology of that which they are investigating' (Walters 2012: 57). Ontology is thus not a theoretical foundation, rather the focus is on the 'regimes of truth, the practices and strategies that ontologize the world in the first place' (Walters 2012: 57). Lövbrand and Stripple (Chapter 1) follow such an approach, suggesting the need to eschew a priori assumptions about the distribution of power in world politics, a certain location of government (the state) or a certain understanding of the locus of the political (local, national, international) and instead focus on how different locales are constituted as powerful, how agents are assembled and how particular domains are constituted as governable and administrable.

Where then do we find an appropriate and usable starting point for approaching 'climate governance'? The answer given by studies in governmentality and ANT is to focus on 'problematization'. According to Rose and Miller, government is a problematizing activity (1992: 181). By this they meant how the articulation of government is always tied to the identification of difficulties and failures of ways of governing. Hence, problems are not pregiven, but have to be constructed and made visible. Miller and Rose term this process 'problematization' – 'issues and concerns have to be made to appear problematic, often in different ways, in different sites, and by different agents' (2008: 14). It is thus possible to think about the history of 'climate government' as a history of specific problematizations, which emerge at different sites mobilizing different sorts of knowledges, actors and practices – from domestic energy consumption, invasive species and climate refugees. Blok (Chapter 2) regards problematization as being about the creation of new political spaces as well as the reworking of the old. Through problematization, new things come to be subject to politics, like eating meat, driving SUVs or carbon markets. But there is nothing 'natural' in any specific problematization. Stephan, Rothe and Methmann (Chapter 3) read problematization as a hegemonic struggle, highlighting processes of politicization and depoliticization, thereby accounting for contingency and resistance. Some forms of problematization are, in this view, more political than others – those that involve 'dislocation', uncontrollable moments that disrupt the mundane and thereby questions hegemonic knowledge (such as Hurricane Katrina). Dislocations can arise from disruptions to the sociomaterial world as well as from acts of protest or contestation, and each brings with it the possibilities of opening up new forms of political space and engagement.

In contrast to such moment of disruption and dislocation, critical social and political theorists have argued that the advent of governmental power has bought with it a retreat of the space of the political and in its place that we have witnessed the advent of 'police' (Swyngedouw 2010); a term that could be taken as reflecting the means through which problematics come to be 'rendered technical' (Murray Li 2007: 7), through processes of calculation and other governmental techniques, and through which new subjects are formed and charged with taking responsibility through governing their own conduct. This reworking of questions of politics as questions of technique through the formation of new 'climate' subjects is a reason one of the central issues for climate politics has become that of subjectivity – what types of people (what sorts of desires, daily routines and so on) we are and need to become to address climate change. Whitehead, Jones and Pykett (Chapter 5) draw attention to the 'more-than-rational' context for human decision-making and identify a key shift to how the subjects of climate policy are conceived, from an idealized homo economicus to a neuroliberal subject responsive to emotional pressures and subconscious prompts to action. Oels (Chapter 11) elaborates on how the poor are now made fit for survival in an ever-changing world of climatic disruptions. In line with advanced liberal government, the subject is responsibilized to prepare for the next crises – a subject capable of responding and coping on its own. This is a process of depoliticization where subjects do not any longer question why there are 'climatic disruptions' and accept the injunction to adapt, migrate, buy insurance or whatever needed to be resilient in the face of endemic climate dangers. Through such a reading, the very act of governing climate change itself could be regarded as part of the 'post-political' condition itself (Swyngedouw 2010). The 'rendering technical' of climate politics is not something that goes uncontested. In Finland, Kortelainen and Albrecht (Chapter 8) demonstrate how contestation over the technical dimension of bioenergy is orchestrated through 'two camps of scientists which have presented quite opposite arguments and calculations of the climate change benefits of the removal and burning of stumps and roots'. In the case of smart meters in the United Kingdom, Hargreaves finds that participants ignore carbon calculations in favour of monetary metrics, and that the presence of the monitors in households led 'to new forms of both cooperation and conflict between household members'. Governmental projects, and by extension attempts to make climate change 'nonpolitical' are subject to contestation, counter rationalities and alternative conducts (Murray Li 2007).

Furthermore, other authors suggest that we need to move beyond a conception of 'the political' in which it is only regarded as disruptive, charismatic and explicit. Rather, as several of the chapters in this book illustrate, politics, as the contestation of existing power relations, as the formation of new forms of solidarity and social engagement and as claims that things could be otherwise, can be found located in practices that are not usually thought of in such terms – the mundane, the technical, the ecological or the neurological. Eden's (Chapter 6) investigations into voluntary carbon calculators as attempts to govern consumption outside public regulation find that those devices of 'carbon surveillance' only work if they merge the social and the technical, making the subjects of government also active users and co-creators of environmental calculation. Even nonhuman actors have the capability to disrupt carbon governmentalities through these quieter, more mundane, registers. Lovell (Chapter 10) shows how forest carbon becomes governable through a particular governmentality that hinges on the durability and stability of forests and the possibility to collect data, standardizing and quantifying them. But this governmentality of surveillance and control might 'be resisted' and fall apart because of the dynamic, diverse and fragile features of many forest ecosystems.

What then might a proper political position on climate policy consist of? While the objective of study is critical, the governmentality concept does not produce critique in the conventional sense of the term. Much of social theory assumes a particular ground from where critique can be leveraged. This assumed (and hence privileged) ground could be a certain subject of history (e.g. classes, civilizations), a certain ethics (e.g. Habermasian discourse ethics, Rawlsian veil of ignorance) or a certain ontological condition (e.g. utilitarian rationality, eternal anarchy). Foucault's suspiciousness of all ethical foundations has attracted criticisms and frustration from progressives and those at the left of politics. But eschewing firm ethical foundations is not the same as being normatively neutral. According to Dean, Foucault's critique denotes 'the exposure and contestation of assumptions rather than to express a general oppositional stance to the putatively pathological character of a social or cultural totality' (1994: 119). Death (Chapter 4) finds three ways Foucault's work on ethics could be used for reflecting on the ethics and politics of climate change. First, there is scope for embracing critique as an 'ethics of resistance' that discomforts and unsettles instead of seeking resolution and reconciliation. Second, Foucault's appreciation for a plurality of forms of power and politics, from the structural to the more mundane, underscore the need to pursue power relations with as little domination as possible. Finally, Foucault's later writings on the possibility of the care of the self as an ethical project is an underexplored dimension of studies in climate governmentality. Will we see an ethic of care emerge around climate change? Skoglund's account of 'Homo Clima' (2011), which suggests the emergence of a new ideal citizen, one that invests in its own vulnerability, adaptability and changeability to amend as well as protect a climate-authorized, aestheticized life, is an interesting new direction in this respect.

A New Agenda?

When we first encountered the climate issue in the mid-1990s, governing the climate was seen as occurring primarily in relation to two principal sites – the intergovernmental negotiations under UNFCCC and the design and implementation of 'climate policies' in affluent democracies. In the early 2000s, the introduction of the concept of 'climate governance' showed how other sites needed to be taken into account, for example the activities of environmental NGOs (e.g. carbon disclosure, standard setting) or the work of international organizations (e.g. the World Bank) and business lobby groups (e.g. International Emission Trading Association) on the function and development of carbon markets. In brief, governing the climate was seen as either a question of what the state did on its own, what states did together or what was done 'beyond the state'. Even though the concept of climate governance liberated and energized many empirical efforts, it seems increasingly unable to account for the manifold ways the climate is encountered and governed. As vividly illustrated throughout this book, the sites at which a carbon-constrained and/or rapidly warming world are represented, categorized and ordered are just about everywhere, from the obvious and theatrical (intergovernmental negotiations) to the mundane and routine (eating and travelling).

In bringing this volume together, our intention has been to explore what an engagement with critical social and political theory could bring to the study of climate change. We suggest that these approaches, drawing on governmentality but including a wider range of theoretical positions, have offered a new set of insights into the ways climate change is creating new forms of social order, and the ways these are structured through the workings of rationality, power and politics. Governing the climate, as we insist in this book, is about making sense of very diverse set of practices, techniques and strategies of governance. We need to discover and interrogate them as particular problematizations that call for and enable specific techniques of power, forms of rationality and subjectivity. The sites that need to be encountered range from our brains, our bodies and our homes to urban landscapes (streets, shopping centres, airports) and spaces of nature (e.g. tropical rainforests, the Arctic, extreme weather events) to the mobility/circulation of people, species and capital. Climate governance, when approached in the general, tends to become an empty concept unable to grapple with the way the climate issue becomes interwoven in our lives and how we organize our societies. We need instead to populate this terrain and explore the specificities through which climate governance takes place.

Towards this end, our ambition with pursuing this book has been to offer a new agenda for the study of 'climate governance', one concerned with the manifolds of sites and practices mentioned earlier – that is the governing of ourselves and others in relation to climate change. Governing the climate is a moving target under constant variation and mutation. As scholars, we need to arrive at the specificity of governance and identify how certain ways of governing the climate recur across sites. We need to name such commonalities and inject an element of difference between them, to open space for new thinking around these sites and practices of governing (Walters 2012: 144). We have no doubt that some will resist this kind of agenda, arguing that it is not properly political, rational, sociological, behavioural, ethical and so forth. But updating our analytical grids around questions of rationality, power and politics is crucial if social science shall remain a vital enterprise.

Such an engagement with climate change will also offer insights into issues at the limit of current thinking – of the ways we need to engage not only with the 'programmers view' but to explore the 'messy actualities' of governmentality, of the multiplicity of (climate) governmentality and the ways it accords with, resists and exceeds forms of 'advanced liberal government' and of the ways the materiality of the objects of climate governance – the 'more than human' components of the world to be governed – provide both agency and structure to the field of intervention within which climate politics operates. Climate governance is not only a field in which we need to invent and apply new concepts, but it is also an important social, political and material phenomenon in its own right, which may challenge existing theoretical conventions, approaches and methods. There is much for climate change to offer to mainstream social science, and vice versa.

References

Aldy, J. E. and R. N. Stavins (2007). *Architectures for Agreement: Addressing Global Climate Change in the Post-Kyoto World.* Cambridge; New York: Cambridge University Press.

Allen, J. (2003). *Lost Geographies of Power.* RGS-IBG book series. Malden, MA: Blackwell Publishing.

Badersten, B. (2002). *Medborgardygd. Den europeiska staden och det offentliga rummets etos.* Stockholm: Natur och Kultur 2002.

Baer, P. (2013). The Greenhouse Development Rights Framework for Global Burden Sharing: Reflection on Principles and Prospects. *Wiley Interdisciplinary Reviews: Climate Change* **4**(1), 61–71. doi: 10.1002/wcc.201.

Bulkeley, H. (2012). Governance and the Geography of Authority: Modalities of Authorisation and the Transnational Governing of Climate Change, *Environment and Planning A* **44**(10), 2428–44.

Dean, M. (1994). *Critical and Effective Histories: Foucault's Methods and Historical Sociology.* London; New York: Routledge.

Foucault, M. (2000) The Subject and Power. In *Power: Essential Works of Foucault 1954–1984 Volume 3*, ed. J. D. Faubion. New York: New Press, 326–48.

Giles, J. (2011). Social Science Lines up Its Biggest Challenges, *Nature* **470**, 18–19. doi:10.1038/470018a.

Hardin, G. (1968). Tragedy of the Commons. *Science* (**162**) 3859, 1243–8.

Hargreaves, T., M. Nye and J. Burgess (2010). Making Energy Visible: A Qualitative Field Study of How Householders Interact with Feedback from Smart Energy Monitors. *Energy Policy* **38**, 6111–19.

Higgins, V. (2001). Calculating Climate: 'Advanced Liberalism' and the Governing of Risk in Australian Drought Policy. *Journal of Sociology* **37**(3), 299–316. doi: 10.1177/144078301128756355.

Hobbes, T. (1981). *Leviathan.* London: Penguin Books.

Hulme, M. (2010). The Idea of Climate Change: Exploring Complexity, Plurality and Opportunity. *Gaia* 1**9** (3), 171–4.

Kuhn, H. W. and A. W. Tucker. (1950). *Contributions to the Theory of Games.* 4 vols. Annals of Mathematics Studies. Princeton, NJ: Princeton University Press.

Letell, M., G. Sundqvist and M. Elam (2011). Steering through the Neighbourhood: Towards an Advanced Liberal Risk Society? *Environment and Planning A* **43**(1), 106–25.

Miller, P. and N. Rose (2008). *Governing the Present.* Cambridge: Policy Press.

Murray Li T. (2007). *The Will to Improve: Governmentality, Development, and the Practice of Politics.* Durham, NC: Duke University Press.

Oels, A. (2005). Rendering Climate Change Governable. From Biopower to Advanced Liberal Government. *Journal of Environmental Policy and Planning* **7**(3), 185–207.

Olson, M. (1965). *The Logic of Collective Action; Public Goods and the Theory of Groups, Harvard Economic Studies.* Cambridge, MA: Harvard University Press.

Ostrom, E. (1990). *Governing the Commons: The Evolution of Institutions for Collective Action.* Cambridge: Cambridge University Press.

Paterson, M. and J. Stripple (2013). Governing Subjectivities in a Carbon Constrained World. In *Culture, Politics and Climate Change*, ed. M. Boycoff and D. Crow. London: Routledge Earthscan, in press.

Roberts, I. and P. Edwards. (2010). *The Energy Glut: Climate Change and the Politics of Fatness.* London; New York: Zed Books.

Rose, N. and P. Miller (1992). Political Power beyond the State: Problematics of Government. *The British Journal of Sociology* **43**(2), 173–205.

Rose, N., P. O'Malley and M. Valverde. (2006). "Governmentality." *Annual Review of Law and Social Science* **2**(1), 83–104. doi:10.1146/annurev.lawsocsci.2.081805.105900.
Skinner, Q. (1978). *The Foundations of Modern Political Thought. 2 vols*. Cambridge New York: Cambridge University Press.
Skoglund, A. (2011) *Homo Clima: Climate Man and Productive Power – Government through Climate Change as Bioaesthetic Frame*. PhD thesis, KTH Royal Institute of Technology, Stockholm. http://urn.kb.se/resolve?urn=urn:nbn:se:kth:diva-34285.
Swyngedouw, E. (2010). Apocalypse Forever?: Post-Political Populism and the Spectre of Climate Change. *Theory Culture Society* **27**(2–3), 213–32. doi: 10.1177/0263276409358728.
Veyne, P. (1997). Foucault Revolutionizes History. In *Foucault and His Interlocutors*, ed. A. Davidsson. Chicago, IL: University of Chicago Press, 147–82.
Walters, W. (2012). *Governmentality: Critical Encounters, Critical Issues in Global Politics*. New York: Routledge.

Index